SOCIETY FOR EXPERIMENTAL BIOLOGY
SEMINAR SERIES: 38

HERBICIDES AND PLANT METABOLISM

SOCIETY FOR EXPERIMENTAL BIOLOGY SEMINAR SERIES

A series of multi-author volumes developed from seminars held by the Society for Experimental Biology. Each volume serves not only as an introductory review of a specific topic, but also introduces the reader to experimental evidence to support the theories and principles discussed, and points the way to new research.

1. Effects of air pollution on plants. *Edited by T.A. Mansfield*
2. Effects of pollutants on aquatic organisms. *Edited by A.P.M. Lockwood*
3. Analytical and quantitative methods. *Edited by J.A. Meek and H.Y. Elder*
4. Isolation of plant growth substances. *Edited by J.R. Hillman*
5. Aspects of animal movement. *Edited by H.Y. Elder and E.R. Trueman*
6. Neurones without impulses: their significance for vertebrate and invertebrate systems. *Edited by A. Roberts and B.M.H. Bush*
7. Development and specialisation of skeletal muscle. *Edited by D.F. Goldspink*
8. Stomatal physiology. *Edited by P.G. Jarvis and T.A. Mansfield*
9. Brain mechanisms of behaviour in lower vertebrates. *Edited by P.R. Laming*
10. The cell cycle. *Edited by P.C.L. John*
11. Effects of disease on the physiology of the growing plant. *Edited by P.G. Ayres*
12. Biology of the chemotactic response. *Edited by J.M. Lackie and P.C. Williamson*
13. Animal migration. *Edited by D.J. Aidley*
14. Biological timekeeping. *Edited by J. Brady*
15. The nucleolus. *Edited by E.G. Jordan and C.A. Cullis*
16. Gills. *Edited by D.F. Houlihan, J.C. Rankin and T.J. Shuttleworth*
17. Cellular acclimatisation to environmental change. *Edited by A.R.Cossins and P. Sheterline*
18. Plant biotechnology. *Edited by S.H. Mantell and H. Smith*
19. Storage carbohydrates in vascular plants. Edited by *D.H. Lewis*
20. The physiology and biochemistry of plant respiration. *Edited by J.M. Palmer*
21. Chloroplast biogenesis. *Edited by R.J. Ellis*
22. Instrumentation for environmental physiology. *Edited by B. Marshall and F.I. Woodward*
23. The biosynthesis and metabolism of plant hormones. *Edited by A. Crozier and J.R.Hillman*
24. Coordination of motor behaviour. *Edited by B.M.H. Bush and F. Clarac*
25. Cell ageing and cell death. *Edited by I. Davies and D.C. Sigee*
26. The cell division cycle in plants. Edited by *J.A. Bryant and D. Francis*
27. Control of leaf growth. *Edited by N.R. Baker, W.J. Davies and C. Ong*
28. Biochemistry of plant cell walls. *Edited by C.T. Brett and J.R. Hillman*
29. Immunology in plant science. *Edited by T.L. Wang*
30. Root development and function. *Edited by P.J. Gregory, J.V. Lake and D.A. Rose*
31. Plant canopies: their growth, form and function. *Edited by G. Russell, B. Marshall and P.G. Jarvis*
32. Developmental mutants in higher plants. *Edited by H. Thomas and D. Grierson*
33. Neurohormones in invertebrates. *Edited by M. Thorndyke and G. Goldsworthy.*
34. Acid toxicity and aquatic animals. *Edited by R. Morris, E.W. Taylor, D.J.A. Brown and J.A. Brown.*
35. Division and segregation of organelles. *Edited by S.A. Boffey and D. Lloyd.*
36. Biomechanics in evolution. *Edited by J.M.V. Rayner.*
37. Techniques in comparative evolution: An experimental approach. *Edited by C.R. Bridges and P.J. Butler.*
38. Herbicides and plant metabolism. *Edited by A.D. Dodge.*
39. Plants under stress. *Edited by H.G. Jones, T.J. Flowers and M.B. Jones.*

HERBICIDES AND PLANT METABOLISM

Edited by

A.D. Dodge,

School of Biological Sciences, University of Bath, UK.

CAMBRIDGE UNIVERSITY PRESS

Cambridge

New York   Port Chester   Melbourne   Sydney

CAMBRIDGE UNIVERSITY PRESS
Cambridge, New York, Melbourne, Madrid, Cape Town, Singapore, São Paulo

Cambridge University Press
The Edinburgh Building, Cambridge CB2 8RU, UK

Published in the United States of America by Cambridge University Press, New York

www.cambridge.org
Information on this title: www.cambridge.org/9780521344227

First published 1989
This digitally printed version 2008

A catalogue record for this publication is available from the British Library

Library of Congress Cataloguing in Publication data

Herbicides and plant metabolism.
(Seminar series / Society for Experimental Biology ; 38)
Includes index.
1. Herbicides. 2. Plants, Effect of herbicides on. 3. Herbicide resistance. 4. Plants–Metabolism. I. Dodge, A.D. II. Series: Seminar series (Society for Experimental Biology (Great Britain)) ; 38.
SB951.4.H454 1989 632'.954 88–25813

ISBN 978-0-521-34422-7 hardback
ISBN 978-0-521-05034-0 paperback

CONTENTS

CONTRIBUTORS

Barry, P.
Department of Biochemistry, University of Liverpool, PO Box 147, Liverpool L69 3BX, UK.

Britton, G.
Department of Biochemistry, University of Liverpool, PO Box 147, Liverpool L69 3BX, UK.

Coggins, J.R.
Department of Biochemistry, University of Glasgow, Glasgow G12 8QQ, UK.

Collin, H.A.
Department of Botany, University of Liverpool, PO Box 147, Liverpool L69 3BX, UK.

Dodge, A.D.
School of Biological Sciences, University of Bath, Bath BA2 7AY, UK.

Gressel, J.
Department of Plant Genetics, The Weizmann Institute of Science, PO Box 26, 76100 Rehovot, Israel.

Harwood, J.L.
Department of Biochemistry, University College, Cardiff CF1 1XL, Wales, UK.

Hawkes, T.R.
ICI Agrochemicals, Jealott's Hill Research Station, Bracknell, Berkshire RG12 5EY, UK.

Howard, J.L.
ICI Agrochemicals, Jealott's Hill Research Station, Bracknell, Berkshire RG12 5EY, UK.

Kerr, M.W.
Shell Research Ltd., Sittingbourne Research Centre, Sittingbourne, Kent ME9 8AG, UK.

Lea, P.J.
Department of Biological Sciences, University of Lancaster, Lancaster LA1 4YQ, UK.

Owen, W.J.
Biochemistry Department, Royal Holloway and Bedford New College, Egham Hill, Egham, Surrey TW20 0EX, UK.

Parry, K.P.
ICI Agrochemicals, Jealott's Hill Research Station, Bracknell, Berkshire RG12 5EY, UK.

Pontin, S.E.
ICI Agrochemicals, Jealott's Hill Research Station, Bracknell, Berkshire RG12 5EY, UK.

Putwain, P.D.
Department of Botany, University of Liverpool, PO Box 147, Liverpool L69 3BX, UK.

Ridley, S.M.
ICI Agrochemicals, Jealott's Hill Research Station, Bracknell, Berkshire RG12 5EY, UK.

van Rensen, J.J.S.
Laboratory of Plant Physiological Research, Agricultural University Wageningen, Gen. Foulkesweg 72, 6703 BW Wageningen, The Netherlands.

Walker, K.A.
Department of Biochemistry, University College, Cardiff CF1 1XL, Wales, UK.

Young, A.J.
Department of Biochemistry, University of Liverpool, PO Box 147, Liverpool L69 3BX, UK.

PREFACE

This volume represents the proceedings of a Symposium of the Plant Metabolism Group of the Society for Experimental Biology, held at the University of York in April 1987. I am most grateful to the chairman of this Group, Dr Curt Givan and the Chairman of the Publications Committee of the SEB, Professor Ken Bowler, for their encouragement to publish these proceedings. The generous financial support of the SEB is also gratefully acknowledged. Particular thanks are also due to all contributors who not only provided excellent verbal presentations but also produced manuscripts in good time.

The allocation of a Symposium to this area reflects the current interest in the physiology and biochemistry of herbicide action. Workers in the field of photosynthesis have used herbicides such as monuron and diuron as experimental tools for nearly 40 years, and as a consequence have extended our detailed knowledge of the mechanism of action of these compounds. In recent years the discovery of inhibitors of the shikimic acid pathway, of branched chain amino acid biosynthesis and of fatty acid biosynthesis, for example, has focussed much more attention on these areas of plant metabolism than there might otherwise have been.

In spite of the inevitable time lag between the presentation of papers and the publication of this volume, it is hoped that many undergraduates, research students and workers in academia and industry will find this volume of use and interest.

Alan Dodge
Bath

K.P. PARRY

Herbicide use and invention

This chapter will deal with herbicide use and invention. The first part will cover the background of the agrochemical industry, i.e. the 'use' side, and in the second part an outline of herbicide discovery will be given.

Background to agrochemical industry

'Why use herbicides?' There are three prime reasons for using herbicides: to increase yield of a crop by reducing competition for light and water from other species (weeds); to improve crop management – particularly harvesting, raising the quality of the final harvested crop/seeds, fruits; to reduce the risk of cross-infection from fungi and insects, from weeds to the crop.

Some idea of the effects of specific weed infestations of maize on yield can be seen in Figure 1 (Aldrich, Scott & Long, 1975). The data show that failure to control weeds, particularly early on in the growing season, can lead to losses of up to 20%.

Fig. 1. Yield. (Bu = Bushels)

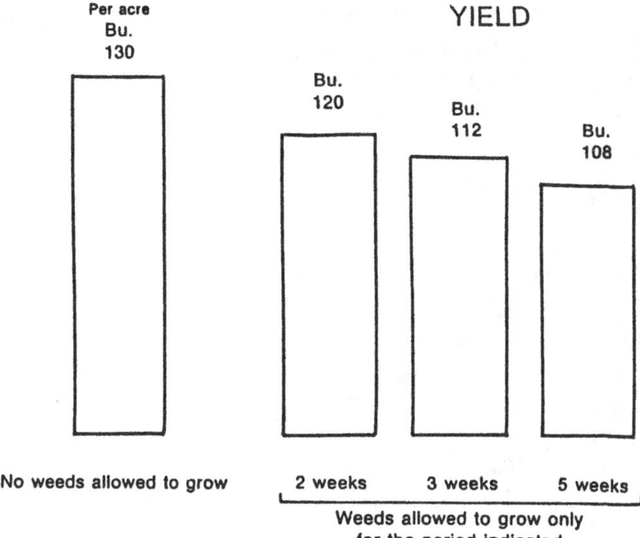

Other data from the US, in soya, indicate similar effects (Figure 2) (Walker *et al.*, 1984; Shurtleff & Coble, 1985). In both cases, yields were deleteriously affected by increasing infestation either in total or over time.

In the UK, infestation in cereals occurs with both grass weeds, e.g. *Alopecurus myosuroides*, and broad-leaved weeds like charlock. Again the effects on yield of a treatment particularly early in the season can have dramatic effects, particularly so in the case of the grass (Figure 3) (Wilson, Thornton & Lutman, 1985).

There are different advantages in controlling broad-leaf weeds or grasses. If control of *Alopecurus myosuroides* was delayed, yields declined more rapidly than by delaying the control of broad-leaved weeds (see Figure 3). In the early stages of competition there is a tendency for the broad-leaved weeds (*Galium aparine*) to increase the total dry weight of the biomass whilst the grass *Alopecurus myosuroides* replaces the crop (Wilson, Thornton & Lutman, 1985). At harvest, weights of material other than grain were similar, irrespective of treatment, weeds, etc but large differences in grain yield arose. The cost efficiency of combine harvesting is therefore much impaired (Figure 4) (Elliott, 1981).

Overall effects on harvesting are difficult to quantify as they depend to a considerable extent on the nature of the combine harvester used and its method of

Fig. 2. Effect of season-long weeds on yield.

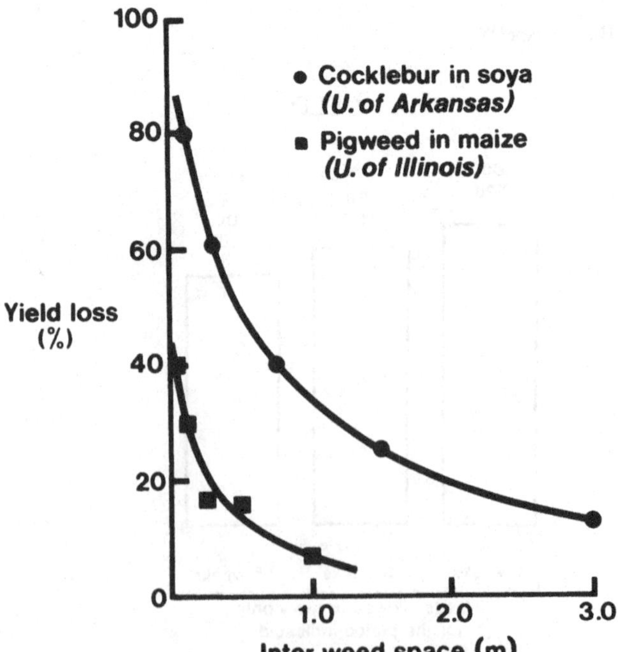

Fig. 3. Effect of delaying control on mean % yield response (excluding Newington).

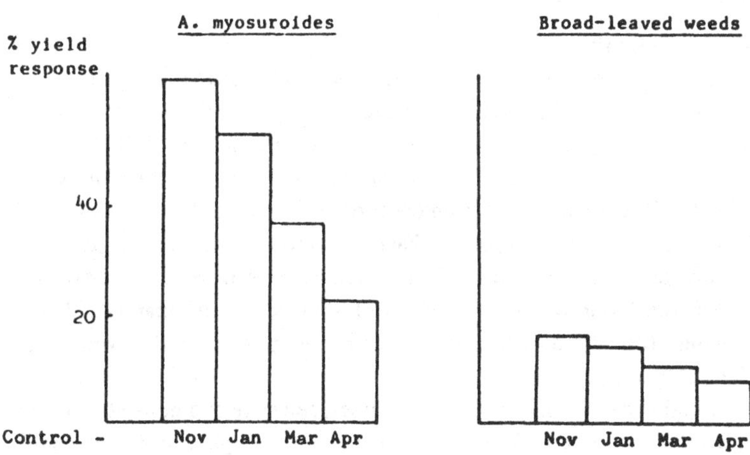

Fig. 4. Stem and dry weight assessments in early May at Merriscourt (*A. m* = *A. myosuroides*).

Timing of Control A.m.	Dicots	Barley Stems/plant	Dry wt g/m² Barley	A. myosuroides	Dicots	Total
Nov.	Nov.	4.7	1145	0	0	1145
Nov.	Jan.	4.7	1020	0	0	1020
Nov.	Mar.	4.5	938	0	14	952
Nov.	Apr.	4.7	1083	0	50	1133
Nov.	None	4.9	1013	0	64	1077
Jan.	Nov.	4.5	892	7	0	899
Mar.	Nov.	3.7	785	135	0	920
Apr.	Nov.	3.8	685	330	0	1015
None	Nov.	3.8	687	354	0	1041
None	None	3.5	547	491	18	1056
S.E. ±		0.25	64.3			60.3

B. J. Wilson et al.

operation. Elliott (Elliott, 1981) has attempted to quantify the effects and some of his results have been used in Figure 5 (MOG = material other than grain).

World market sizes

Figure 6 shows the growth of the agrochemical market over the last few years. There has been a particularly large expansion from 1970–80 with a slight reduction in growth in recent years (Wood, Mackenzie & Co, 1981–6). It is interesting to note the slight fall in 1983 – attributed to the PIK (Payment in Kind) programme in the USA where the government legislated to effect a reduction in land usage. The agrochemical market appears to have recovered so limitations on land use about to be introduced in the EEC and USA may not have a serious long-term effect. The herbicide market has grown in a very similar fashion to the total market and now constitutes about 40% of the total. Insecticides make up another 40% whilst the balance is fungicides.

Markets for herbicides fall into two major classes: selective and non-selective. In the former, weed control in a major world crop, e.g. maize, soya, cereals, rice is the prime objective whilst, in the latter, the paraquat and glyphosate market, total vegetative control is desired. Some idea of individual markets is given in Figure 7 showing the value of particular herbicides used in crops. Thus the total market for herbicides used in maize for a wide variety of uses is in excess of $1000m. A different

Fig. 5. Cost of processing MOG put at £10/t. Grain selling price roughly £80–100/t.

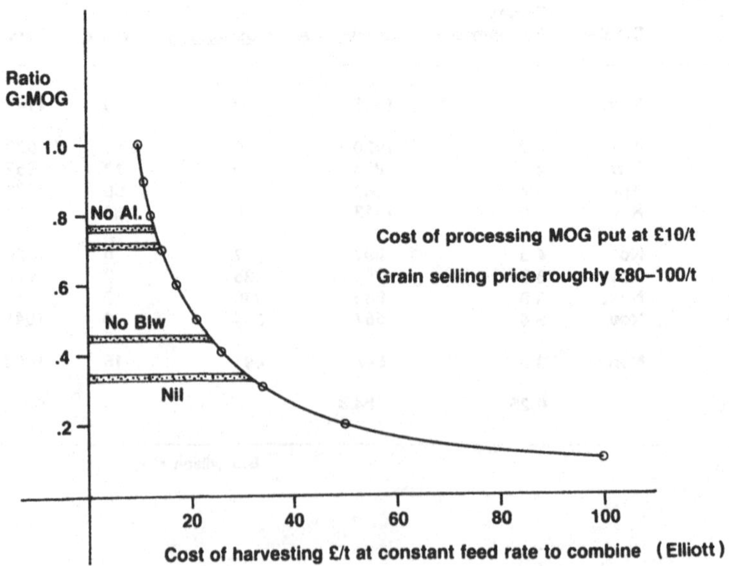

segment of the selective herbicide market is a herbicide which kills a particular range of weeds, e.g. a grass killer or graminicide approximately $2000m. The market for total vegetative control, the paraquat and glyphosate market, is believed to be somewhere about $800m.

Market structure of products

A considerable range and number of products are now available. Figure 8 shows the number of products available in the UK over time. A significant increase in

Fig. 6. Agrochemical market size with time.

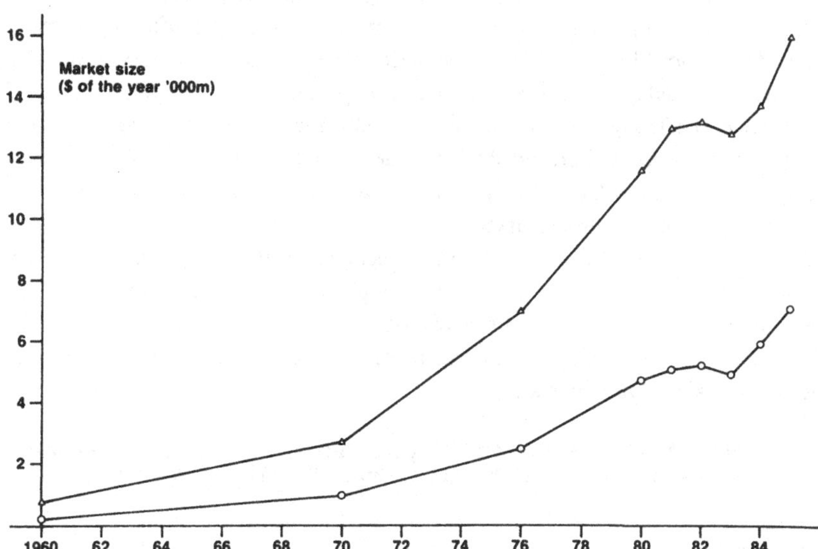

Fig. 7. Herbicide usage per crop sector (end user).

	1979	1985
Maize herbicides	**1050 M$**	**1575 M$**
Soya herbicides	**760 M$**	**1475 M$**
Cotton herbicides	**350 M$**	**340 M$**
Graminicides	**1930 M$ (1982)**	
Total Vegetative Control	**>800 M$ (1982)**	

Wood. Mackenzie

the late 60s and early 70s is followed by a plateau (Corbett, 1979; Hill, 1982), with a total of 561 products now available (Makepeace, 1986).

If current chemicals are classified in terms of a chemical toxophore, some interesting observations are apparent. First there are a smallish number of toxophores which have produced a large number of products. For instance, Figure 9 illustrates some very common herbicide structural types and Figure 10 the number of products which have come from each of these groups (Wood, Mackenzie & Co, 1981–6). Some idea of the size of production can be gained from Figure 11.

Figure 12 shows that some 50% of herbicides are single compound toxophores and a remaining 30% arise from those groups already illustrated (Figures 9 and 10).

Another interesting feature is to look at the growth rates of these different groups of herbicides, Figure 13 (Wood, Mackenzie & Co). One very general observation is that the older herbicidal groups are least profitable. A product matures until it is displaced by a more cost-effective material in the technically advanced parts of the world. The developing countries will presumably continue to use the older commodity products for cost reasons although eventually the newer replacements will become commodity products and be used in these markets.

An examination of Figure 13 leads to the speculation that the hormones of 1945 are probably about to be displaced by Du Pont's sulphonylureas. We could speculate that the next set of compounds to disappear from the agrochemical scene would be triazines, carbamates and ureas. Replacements will need to be active at much lower rates and be safer environmentally.

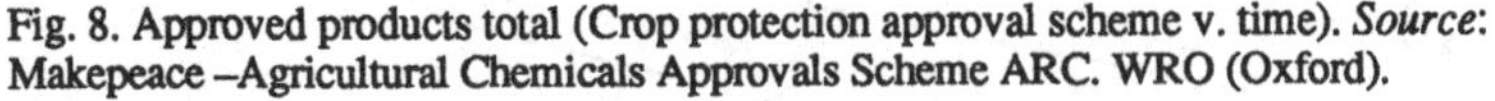
Fig. 8. Approved products total (Crop protection approval scheme v. time). *Source*: Makepeace –Agricultural Chemicals Approvals Scheme ARC. WRO (Oxford).

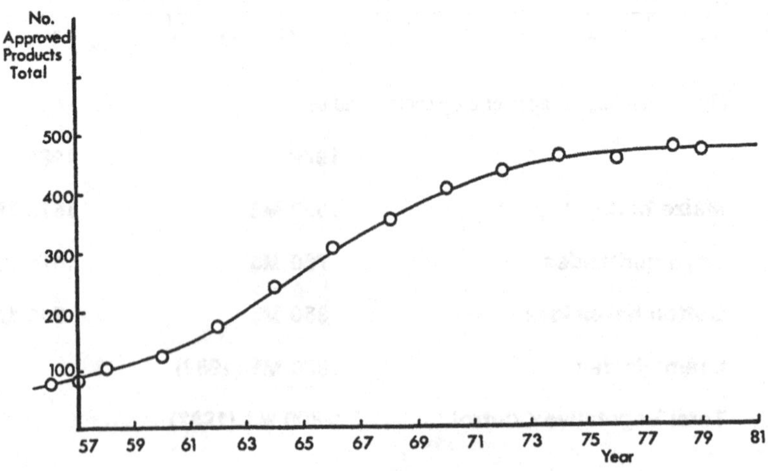

General trends within the industry seem to indicate that application rates are falling (Figure 14) (Graham-Bryce, 1981).

What is the lowest rate that could be used? A value of 2.5 x 10^{-6} g/ha has been estimated assuming one molecule of herbicide is required per cell of living tissue to effect death (Corbett, Wright & Baillie, 1984). Some of the assumptions are considered to be debatable by these authors and the figure possibly inaccurate by a factor of 10^3, but application rates of the order of mg/ha should be possible.

Another view is to examine the toxicological values for a number of well-known poisons, Figure 15 (Registry of Toxic Effects of Chemical Substances, 1976).

Fig. 9. Major herbicide types.

Generic Group	Est. Market Size ($m)	Typical member	
Triazines	1425	Atrazine 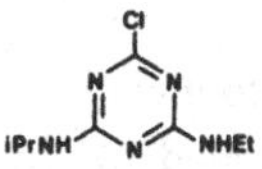	Cl; N; N; iPrNH; N; NHEt
Amides (haloacetanilides)	940	Alachlor 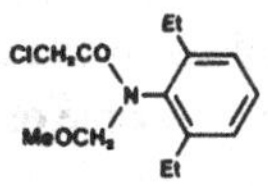	$ClCH_2CO$; N; $MeOCH_2$; Et; Et
Carbamates	765	EPTC 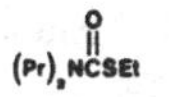	$(Pr)_2NC(O)SEt$
Toluidines (dinitroanilines)	660	Treflan 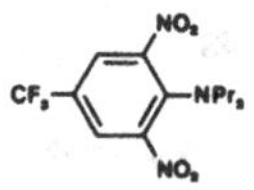	CF_3; NO_2; NPr_2; NO_2
Ureas	670	Linuron 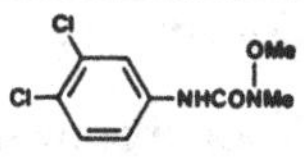	Cl; Cl; NHCONMe; OMe
Hormones	435	2.4.D 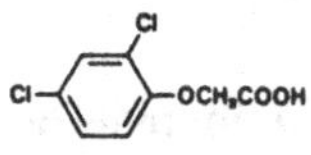	Cl; Cl; OCH_2COOH
Diphenyl Ethers	345	Flex 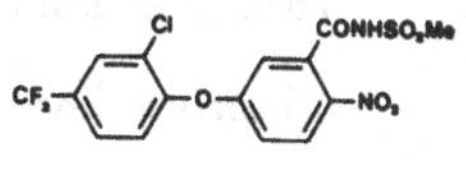	CF_3; Cl; O; $CONHSO_2Me$; NO_2
Others	1835	includes paraquat, glyphosate, diazines etc.	

Assuming a biomass at the time of spray of roughly 1–5 tonnes/ha, and a tox of 20–50 μg/kg, we could get application rates of about 20–250 mg/ha. Irrespective of method, it would appear that there is still scope for increasing activity levels.

Pesticide development

A general outline of the processes involved in developing a new pesticide is outlined in Figure 16.

Initially, exploratory testing in the glasshouse works synergistically with chemical synthesis. When active analogues have been selected, these are then tested on the field and given limited toxicological studies (simple studies of behaviour in soil and rats). Patenting is also initiated, and if the performance of a compound is good, further field testing, more toxicological studies and research for a manufacturing route are initiated. If the compound continues to show good activity and is cost effective, development will lead to marketing.

An idea of the timescale involved in development is shown in Figure 17 (Braunholtz, 1981).

Exploratory work at the early stages is the least expensive whilst later development and associated toxicological work makes major cash demands. The cash flow, illustrated in Figure 18 (Green, 1977) shows that there is a large financial outlay between years 4 and 6 and the product may not start to pay back until some ten years after discovery. A couple of points to note are that , although research is a fairly cheap

Fig. 10. Products and development compounds, arising from a particular toxophore.

Ureas	26	37
Alkoxyureas	11	
Carbamates	16	34
Thiocarbamates	19	
Triazines		29
Diphenylethers		29
Haloacetanilides		24
Dinitroanilines		22
Phenoxyacetics		17
Aryloxyphenoxys		17

Fig. 11. Estimated US production 1984.

	thousand tons
Atrazine	**39**
Cyanazine	**13**
Diuron	**2–3**
2.4.D	**30**
Alachlor	**50**
Glyphosate	**12**

Fig. 12. Distribution of products/toxophore.

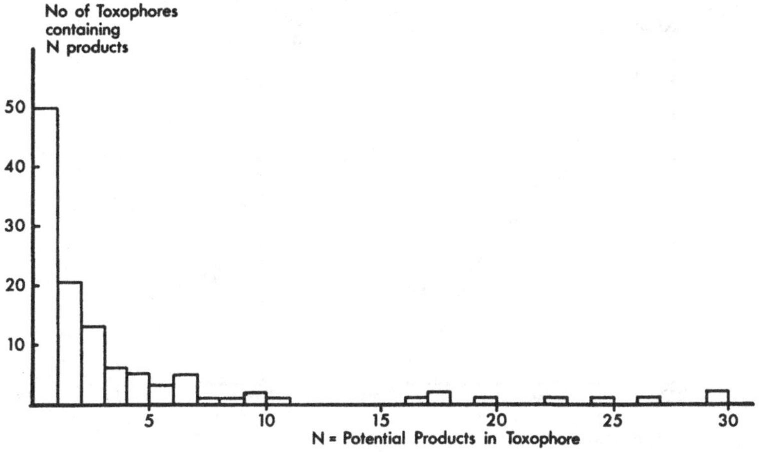

Fig. 13. % growth rate – herbicides.

	Year of introduction	1972–85 (real)	1985–90 (est)
Hormones	1945	0.2	−5.0
Triazines	1956	4.0	−3.1
Carbamates	1954–8	4.9	0
Ureas (inc. S.U.)	1960	1.7	5.2
Toluidines	1965	6.4	−2.0
Amides	1966	8.1	2.2
Diphenylethers	1970	19.2	9.3
Diazines	1974	19.1	5.4
Market		4.6	4.2

Wood. Mackenzie

Fig. 14. Application rates for representative crop protection chemicals v. time.

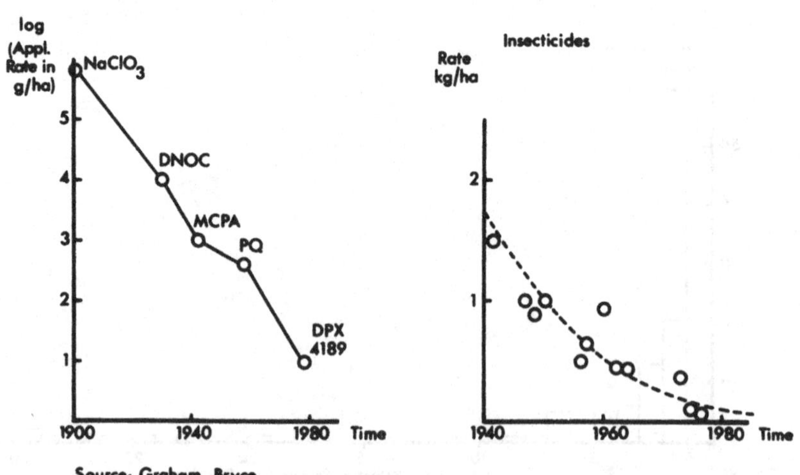

Fig. 15. Toxins.

	Min lethal dose µg/kg	LD50 µg/kg	Method
Botulinus. Toxin A	**0.16**		**Ip mouse**
Ricin	**~20**		**Ip mouse**
Tetrodotoxin		**10**	**oral rat**
Aflatoxin		**40**	**oral rat**
Amanatin α		**100**	**Ip mouse**
Sarin		**450**	**Ip mouse**
Sodium cyanide		**5000**	**Ip mouse**

Fig. 16. Discovery and development of a new pesticide.

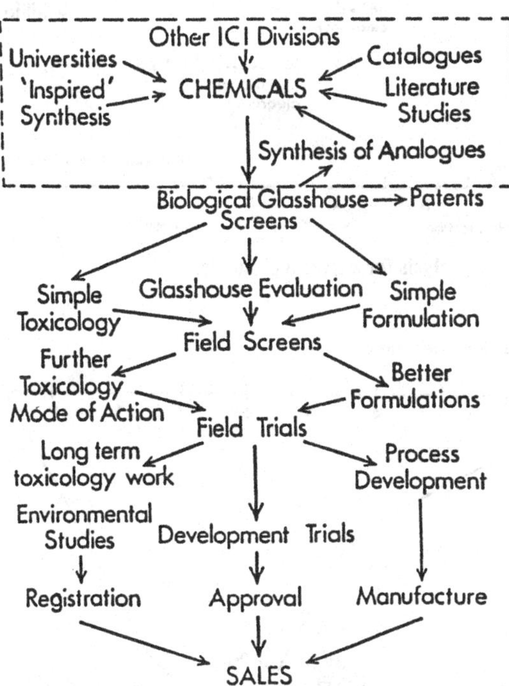

process, it occurs early on in the cash flow so it can have a disproportionate effect on the end value. Secondly, the major cash demands occur at about year 4 and any hold up thereafter would have a very dramatic effect on the project's profitability (Gilbert, 1978; Green, 1977).

With increasing requirements for both toxicological work and for standards for biological efficacy, the cost of developing a pesticide is rising and is now of the order of $20 m or more (Figure 19). Likewise, expenditure by the industry on Research & Development (R & D) is also rising rapidly, Figure 20 (Braunholtz, 1981; Gilbert, 1978; NACA, 1984; Agrow, 1985). One source of potential worry for industry is that R & D costs are rising faster than the rate of market expansion. These increases,

Fig. 17. Registration of new chemicals in the UK.

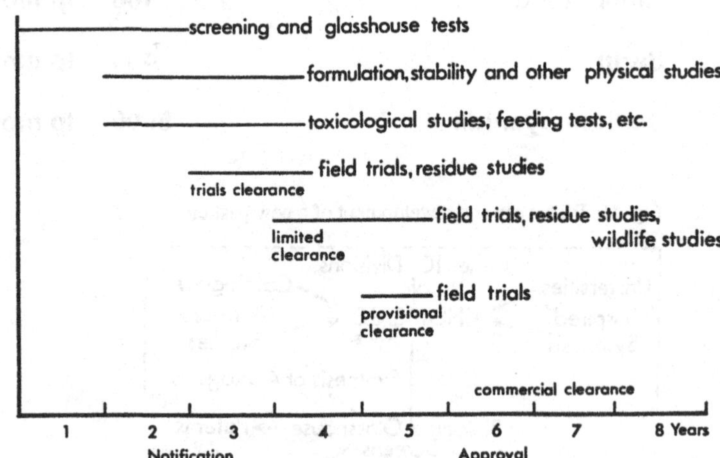

Fig. 18. Cash flow analysis for a typical pesticide.

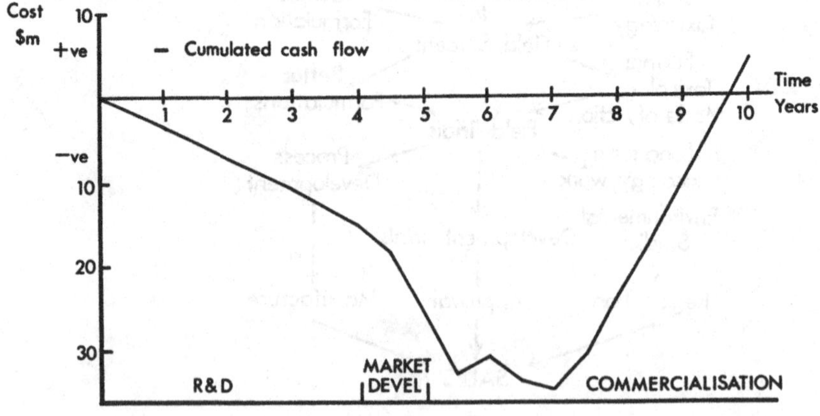

together with increasing costs of registration, will inevitably lead to some business rationalisation where companies expand by merging. A number of mergers have occurred in the last ten years and others are expected to follow until there are only about 20 major companies.

Pesticide discovery

There are four major approaches possible for discovering new pesticides – random screening; imitative chemistry; natural products; biorational development.

Random screening

In random screening, a wide range of chemicals, acquired from a large number of sources, is tested against a biological screen which represents market opportunities. Occasionally a compound of good activity appears and this can then act as a lead for the chemist. As time has progressed it has become harder to find such leads, probably because of the rising standards of biological criteria required. At present, an average of more than 20,000 chemicals are tested to produce one product, Figure 21 (Büchel, 1979; Storck, 1984; Hill, 1982; Menn & Henrick, 1981; Menn, 1980; Chapman, 1978; Hill, 1983). There is no evidence to support the contention that numbers of potential chemicals are falling. Figure 22 shows the number of new chemicals (organic and inorganic) which have been indexed into Chemical Abstracts

Fig. 19. Cost of pesticide development with time per product.

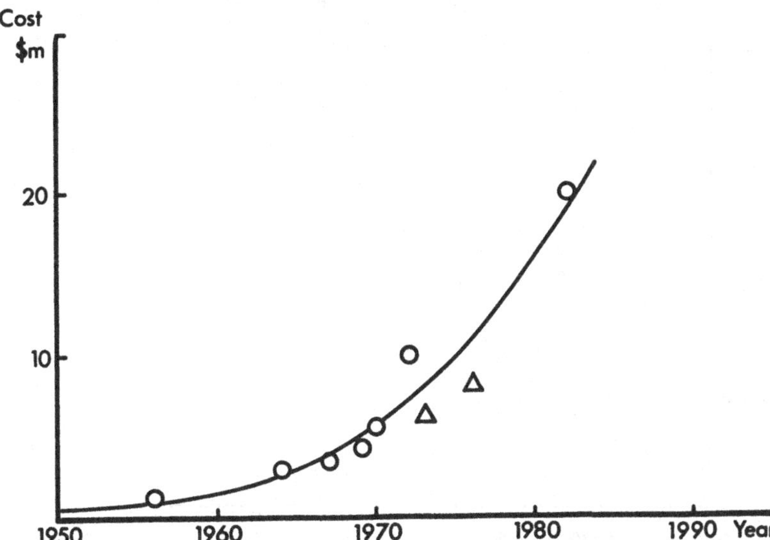

Registry. Over the last ten years this has averaged just under 400,000 chemicals per annum.

In random screening, chemicals that show activity on one screen, either pesticidal, PGR or pharmaceutical, can often act on another. Three good examples are the imidazolidinones, the sulphonylureas and glyphosphate, all of which were initially found to show some PGR activity (Sauers & Levitt, 1984; Los, 1984; Grossbard & Atkinson, 1985).

Although random screening has been shown to be an effective method for producing about 50% of current herbicides, it does have a number of disadvantages. A steadily increasing number of chemicals are required to discover a lead and the time taken to discover a new molecule is progressively longer. By testing 8000 chemicals

Fig. 20. US industry wide R & D expenditure.

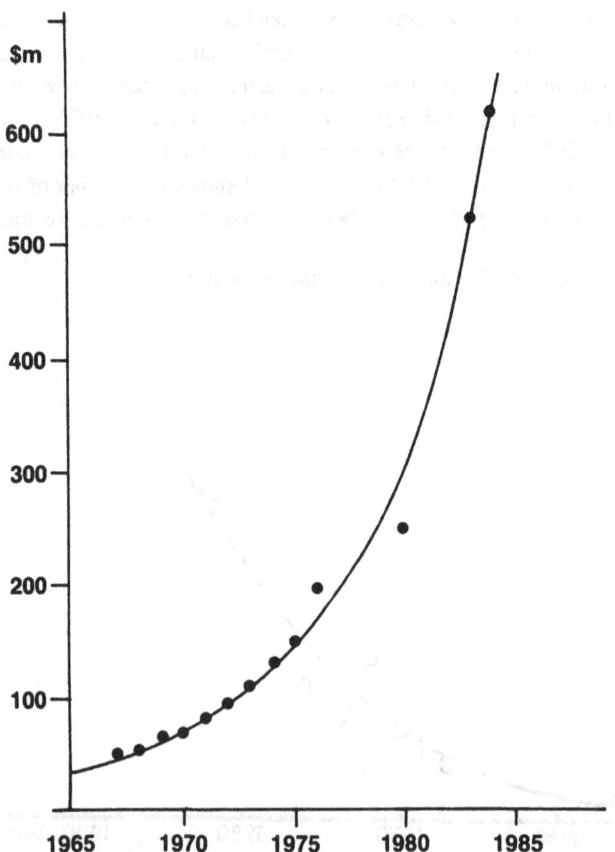

Source: Gilbert, Braunholtz, NACA – 1984

Fig. 21. Numbers of chemicals screened for one commercial product v. time.

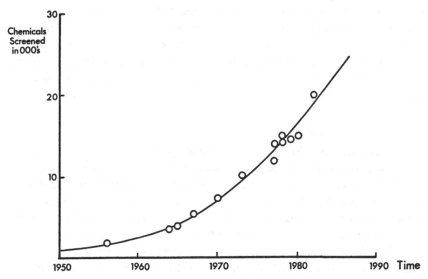

Fig. 22. Number of chemicals given registry numbers in CA.

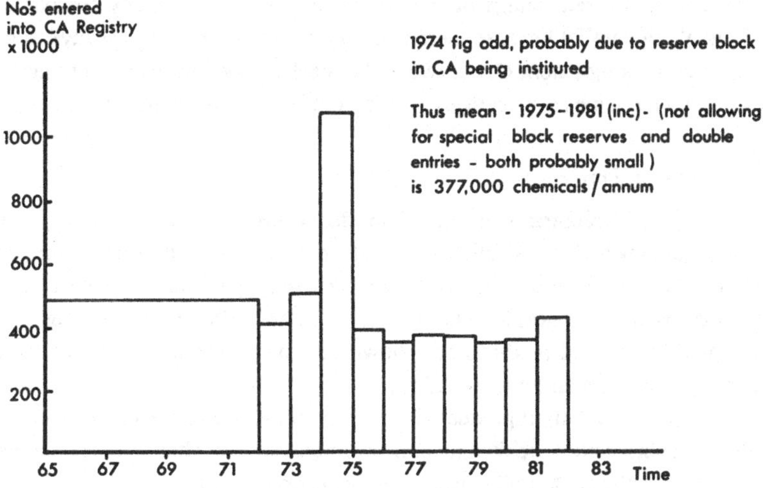

per annum, it might take 2–3 years to find one new product. The principal disadvantage is the production of a large number of false positives which require chemical effort to clarify often with little chance of success. Currently, synthesis of 30–50 chemicals in an area may be needed to be reasonably certain about the nature of the biological activity. A balanced judgement of this activity clearly takes into account the level and novelty of biological activity, the chemical novelty and the scope for chemical exploitation.

Imitative chemistry

Imitative chemistry is well illustrated by Figure 23. This shows that a number of diphenyl ethers produced by a wide range of companies are structurally very similar. In this sort of work, companies examine competitors' patents and look for 'chemical holes' in their claims which will give scope to generate potential products. The area of aryloxyphenoxy propionates has been similar, e.g. compare fluazifop-butyl, diclofop-methyl quinofop-ethyl and fenthiaprop-ethyl. Extension with the sulphonylureas is, no doubt, proceeding in a number of companies.

The principal advantages of this approach which has produced around 50% of herbicidal products is that there is a very high chance of discovering biological activity, though not necessarily novelty. Furthermore, the chemist can get some limited structure–activity information from the literature to guide him to a new structure.

The principal disadvantage of imitative chemistry is that many companies are working along parallel lines, so there is a very high risk of losing the invention as a result of competing patent claims. So a chemical compound may be produced that does exactly the same as one already existing but is some 2–3 years too late.

Natural products

The development of natural products has not yet been widely used in discovery of new herbicides. With other pesticides such as the pyrethroid insecticides, their activity was modelled on the naturally occurring pyrethrum. The most important herbicide example is phosphinothricin (see Chapter 7), Figure 24. Although a wide range of other natural products has shown activity, it is usually of an indifferent specificity or at too high a concentration.

A variant on the natural product approach is to use a mycoherbicide (Templeton, 1985). Mycoherbicides are formulations of fungal spores which kill specific weeds. They can be applied in a similar way as current chemical products.

At present, their principal advantage is that registration for use is somewhat easier than conventional pesticides. The principal disadvantage is that they are usually restricted to one weed. Thus Collego controls only northern jointvetch whilst Devine controls stranglevine.

The principal advantages of the natural products approach are that it has been shown to work in a restricted number of cases and there is a high chance of chemical novelty.

The principal disadvantages are, however, that the isolation and identification of the actives is rarely straightforward and the toxicity of the natural product may be highly specific, possibly limited not only to one plant species but to one variety.

Biorational approach

This is the most recent approach and has probably achieved increasing prominence in recent years as a result of developments in biochemistry.

The biorational approach is to identify a particular mechanism within a plant which will lead to death and then, by a study of this mechanism, to identify and synthesise potential inhibitors which should be capable of being herbicides. Selection criteria for pathways to inhibit can be illustrated using photosynthesis and amino acid biosynthesis. In both cases inhibition is known to cause death. For instance, ureas,

Fig. 23. Imitative chemistry, e.g. diphenyl ethers.

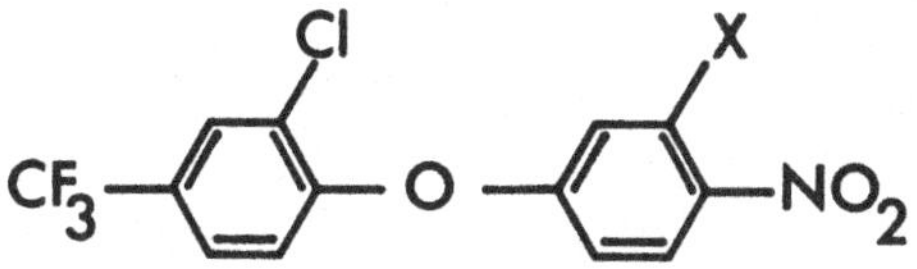

X =	
H	Rohm & Haas
OEt	------ " ------
COOH	------ " ------
COOEt	------ " ------
COOMe	Mobil
$CONHSO_2Me$	I C I
$OCHCH_2OCH_2OCH_2$	U B E
$OCHCH_2CH_2OCH_2$	Mitsui
$OCH(Me)COOCH_2CH_2OMe$	Ciba Geigy
$COOCH(CH_3)COOEt$	PPG

triazines and uracils all inhibit photosynthesis and are known families of herbicides. Likewise, glyphosate and the sulphonylureas inhibit amino acid production and again cause death. These areas illustrate two other useful features, first these vital areas of metabolism are common to all plants and secondly are plant specific, so toxicity to mammalian species could be low. In these areas, the chemistry of the biochemical process has been the source of much mechanistic research and should therefore provide the chemist with enough information to design enzyme inhibitors.

Other areas could well include sugar metabolism, lipid metabolism, nucleoside synthesis and possibly inhibition of use of essential metal ions.

The principal advantages are that there is a good chance of being first in a particular area and hence able to establish a position – and ideally the chemistry and possibly the biology is likely to be novel.

However, in spite of a quite considerable effort (Corbett & Wright, 1971; Jangaard, 1974; Baillie *et al.*, 1980; Baillie, Corbett, Dowsett & McCloskey, 1972; Wright *et al.*, 1973; Geissbuhler, 1983) no commercial pesticide has been discovered by this approach. Furthermore, it may produce a very active compound biochemically but the effects could be nullified by uptake, translocation and metabolism. The whole approach is therefore one of high risk.

Fig. 24. Natural products.

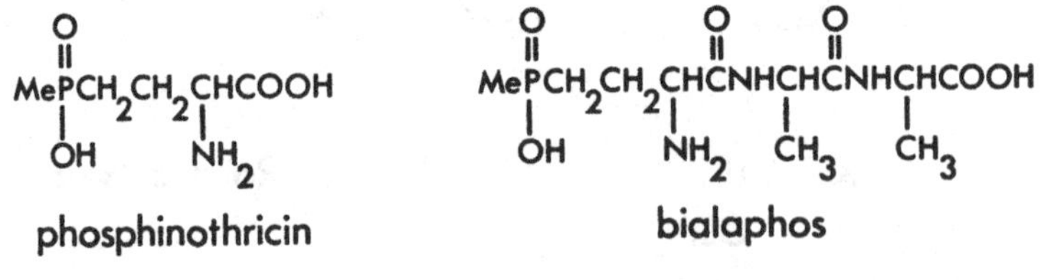

```
   O                          O          O      O
   ||                         ||         ||     ||
MePCH2CH2CHCOOH            MePCH2CH2CHCNHCHCNHCHCOOH
   |     |                    |     |      |      |
   OH    NH2                  OH    NH2    CH3    CH3
```

phosphinothricin

bialaphos

References

Agrow –*The World Agrochemical Market*, George St Publications Ltd, 1985.

Aldrich, S.R., Scott, W.O. & Long, E.R. (1975). *Modern Corn Production*, 2nd Edition, A and L Publishers, Champaign, USA.

Baillie, A.C., Corbett, J.R. & McCloskey, P. (1972). Inhibitors of shikimate dehydrogenase as potential herbicides. *Pestic. Sci.* **3,** 113–20.

Braunholtz, J.T. (1981). Crop protection – the role of the chemical industry in an uncertain future. *Phil. Trans. Roy. Soc. London.* **B295,** 19–34.

Büchel, K.H. (1979). Synthese-konzeptionen zu neuen biologischen Wirkstoffen. *Naturwissenschaften* **66,** 173–81.

Chapman, T. (1978). Pesticides – a period of progress. *Span*, **21**(2), 62–3.

Corbett, J.R. & Wright, B.J. (1971) Inhibition of glycollate oxidase as a rational way of designing a herbicide. *Phytochemistry* **10,** 2015–24.

Corbett, J.R. (1979). Technical considerations affecting the discovery of new pesticides. *Chem and Ind.*, 17 Nov, 772–82.

Corbett, J.R., Wright, K. & Baillie, A.C. (1984). *The Biochemical Mode of Action of Pesticides*. Academic Press. 2nd Edition London.

Elliott, J.G. (1981). The economic significance of weeds in the harvesting of grain. *Proc. 1980 British Crop Protection Conference – Weeds Abstracts*, pp. 787–97.

Geissbuhler, H. (1983). CHEMRAWN, 643–56. *Chemistry and World Food Supplies: The New Frontiers*. Ed. L.W. Schemilt. Int. Conf. Manila, 6–10 Dec 1982, Pergamon.

Gilbert, G.H. (1978). The increasing riskiness of the pesticide business. *Farm Chemicals* (April), 20–7.

Graham-Bryce, I.J. (1981). Current status and future potential of chemical approach to crop protection. *Phil. Trans. Roy. Soc. London.* **B295**, 6.

Green, M.B. (1977). *Chemicals for Crop Protection and Pest Control*, Chapter 2, 2nd ed. Oxford: Pergamon Press.

Grossbard, E. & Atkinson, D. (1985). *The Herbicide Glyphosate*. Butterworths. London. Chapter 1. Discovery, Development and Chemistry of Glyphosate. J.E. Franz, 3–17.

Hill, G.D. (1982). Herbicide technology for integrated weed management systems. *Weed Sci.* **30** (Suppt.), 35–9.

Hill, G.D. (1983). Thirty years of progress in weed science – what's next? *Proc. 36th Southern Weed Science Meeting*, 8–17, Beloxi, Mississippi.

Jangaard, N.D. (1974) The characterisation of phenylalanine ammonia-lyase from several plant species. *Phytochemistry* **131**, 765–68.

Los, M. (1984). *Pesticide Synthesis Through Rational Approaches* ACS Chapter 3. o-(5-oxo-2-imidazolin-2-yl) aryl carboxylates: a new class of herbicides. 29–44.

Makepeace, R.J. (1986). Agric. Chem. Approvals Scheme, Pesticides 1986, HMSO, MAFF.

Menn, J.J. (1980). Contemporary frontiers in chemical pesticide research. *J. Agric. Food Chem.* **28**, 2–8.

Menn, J.J. & Henrick, C.A. (1981). Rational and biorational design of pesticides. *Phil. Trans. Roy. Soc. London*, **B295**, 57–71.

NACA (1984). (National Agricultural Chemicals Association) Industry Profile Study.

Registry of Toxic Effects of Chemical Substances. NIOSH (National Institute of Occupational Safety and Health) 1976 Edition, Rockville, Maryland.

Sauers, R.F. & Levitt, G. (1984). *Pesticide Synthesis Through Rational Approaches* ACS, Chapter 2.

Shurtleff, J.L. & Coble, H.D. (1985). Interference of certain broadleaf weed species in soybeans (*Glycine max*). *Weed Science* **33**, 654–7.

Storck, W.J. (1984). Liquidity proves tough for chemical firms to rebuild. *Chem. and Eng. News*, **62** (2) 36–7.

Templeton, G.E. (1985). Specific weed control with mycoherbicides. *Proc. British Crop Protection Conference – Weeds*, 601–7.

Walker, R.H., Patterson, M.G., Hauser, E., Isienhour, D.J., Todd, J.W. & Buchanan, G.A. (1984). Effects of insecticide, weed free period and row spacing on soybean (*Glycine max*) and Sicklepod (*Cassia obtusifolia*) growth. *Weed Science* **32**, 702–6.

Wilson, B.J., Thornton, M.E. and Lutman, P.J. (1985). Yields of winter cereals in relation to the timing of control of black grass and broad leaved weeds. *Aspects of Applied Biology* **9** The biology and control of weeds in cereals.

Wood, Mackenzie & Co Agrochemical Service. (1985, 1985, 1984, 1983, 1982, 1981) Reference Volume. Agrochemical Products Section. Pub. Wood Mackenzie & Co Ltd, 74–77 Queen Street, Edinburgh, EH2 4NS.

Wright, B.J., Dowsett, J.R., Rubery, P.H., Baillie, A.C. & Corbett, J.R. (1973). Rational herbicide design by synthesis of inhibitors for the peroxidase catalysed oxidation of indolyl-3-acetic acid. *Pestic. Sci.* **4**, 785–94.
Wright, B.J., Baillie, A.C., Wright, K., Dowsett, J.R. & Sharpe, T.M. (1980) Synthesis of potential herbicides designed to uncouple photophosphorylation. *Phytochemistry* **19** 61–5.

J.J.S. VAN RENSEN

Herbicides interacting with photosystem II

Introduction

Weed control using organic chemicals commenced just over half a century ago, when in 1932, 4,6-dinitro-*o*-cresol (DNOC) was first used as a weed-controlling agent. The phenoxyacetic acids such as 2,4-D followed in the 1940s. Chemical weed control was widely accepted when the ureas (1951), the triazines (1955) and the bipyridiniums (1960) became available. The latter three groups of herbicides act via the photosynthetic process. The ureas and triazines effectively block photosynthetic electron transport at the level of the Photosystem II acceptor site. Many reviews are now available on the effects of herbicides on Photosystem II, for example: Van Rensen (1982), Pfister & Urbach (1983), Sandmann & Böger (1986) and Renger (1986).

Research on the action of herbicides inhibiting photosynthesis has yielded much detailed information about their mechanisms of action. Furthermore, our understanding of the photosynthetic process has been greatly enhanced by the use of these chemicals as specific inhibitors. This chapter highlights important events of the research on Photosystem II herbicides and surveys some recent developments.

The photosynthetic electron transport pathway

The light energy conversion processes of photosynthesis are located in the grana of the chloroplasts, while the reduction of carbon dioxide occurs within the stroma. Grana consist of stacks of thylakoids, i.e. vesicle-like structures having an internal space surrounded by a membrane. The grana are interconnected by unappressed stroma thylakoids. The thylakoid membranes contain the electron and proton translocating components (Fig. 1).

Photosynthesis is initiated by the absorption of light energy by the chlorophylls of both photosystems (PSII and PSI). Excitation energy is then transferred to the reaction centres: P680 in Photosystem II, and P700 in Photosystem I. P680 and P700 are specialized chlorophyll *a* molecules which are able to accomplish a charge separation, resulting in $P680^+$ and $Pheo^-$ in light reaction II, and $P700^+$ and FeS^- in light reaction I (Pheo represents pheophytin; FeS are several iron sulphur centres). The charge separations are followed by electron and proton translocating reactions.

The electron hole of $P680^+$ is filled via various steps, denoted in Figure 1 as S, by an electron which is ultimately derived from water. This water splitting not only yields electrons, but also protons and oxygen. From $Pheo^-$, the electron is transported to the first stable quinone electron acceptor Q_A and then to the secondary acceptor Q_B. Because of its microenvironment, Q_A acts as a one-electron carrier and it can only be reduced as far as the semiquinone state. Q_B acts as a two-electron gate: it accumulates two electrons and transfers them in pairs to the plastoquinone pool (PQ). The reduction of plastoquinone is accompanied by protonation to a fully reduced hydroquinone (PQH_2) utilizing two protons from the stroma. From PQH_2, electrons are transferred via the cytochrome b_6/f complex to plastocyanin (PC). During this reaction two protons from PQH_2 are liberated into the internal space of the thylakoid. PC is the primary electron donor to $P700^+$. From FeS^- the electron is transferred to Fd (soluble ferredoxin), and via ferredoxin $NADP^+$ reductase (FP) to $NADP^+$. Under certain conditions a cyclic electron flow is possible by transfer of electrons from Fd to the cytochrome b_6/f complex. The protons which have been accumulated in the internal space of the thylakoid can flow back to the stroma through the ATP-ase, which phosphorylates ADP to ATP. A more detailed review of energy transduction in photosynthesis has been written by Ort (1986).

The electron and proton transporting components are located within protein structures which are embedded in a lipid bilayer. The thylakoid membrane contains four major functional polypeptide complexes. The Photosystem II complex consists of a reaction centre, the attendant antenna chlorophyll proteins and, in addition, the polypeptides involved in water oxidation. The Photosystem I complex contains a reaction centre and the additional antenna chlorophyll proteins. The intersystem

Fig. 1. Electron and proton transport pathways in photosynthetic electron flow. Explanation in text.

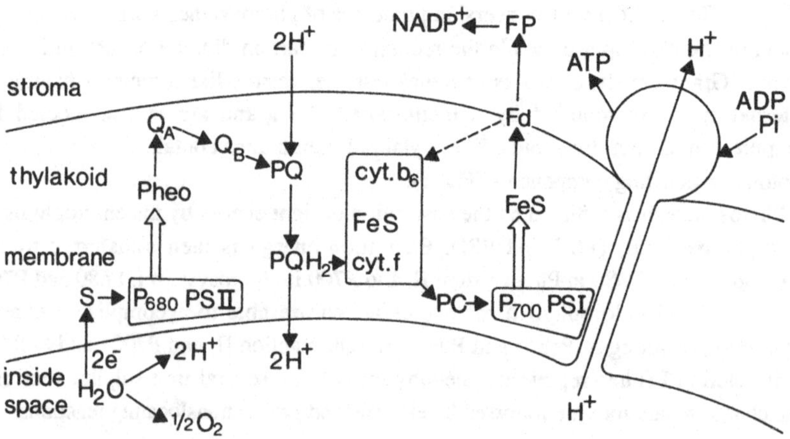

electron transport is catalysed by the cytochrome b_6/f complex and ATP-synthesis by the coupling factor complex. Of special interest for this chapter is one of the proteins of the Photosystem II complex: the so-called Q_B-protein (Kyle, 1985). The secondary quinone electron acceptor Q_B is reversibly associated with the 32 kD Q_B-protein. This polypeptide of the Photosystem II core is also the binding site of Photosystem II herbicides.

About 50% of all commercially available herbicides are inhibitors of photosynthesis. Moreland (1980) classified these herbicides into five groups: electron transport inhibitors, inhibitory uncouplers, electron acceptors, uncouplers, and energy charge inhibitors. The electron transport inhibitors include the urea and triazine herbicides; the inhibitory uncouplers include the phenol-type inhibitors. This chapter will concentrate on electron transport inhibitors, phenol-type herbicides and quinone inhibitors. The relation of bicarbonate with herbicide action will also be discussed.

Action of diuron-type herbicides

The most important inhibitors of this class are the urea and triazine herbicides. Both groups contain large numbers of active chemicals, with a common chemical structure of a sp^2 hybrid bound to N, O, or =CH and attached to a lipophilic substituent (Trebst & Draber, 1979). The site and mode of action are the same for all the members of these groups, but the differences in activity are caused by the various lipophilic side chains. In photosynthesis research, the urea 3-(3,4-dichlorophenyl)-1,1-dimethylurea (diuron = DCMU) and the triazine 2-chloro-4-ethylamino-6-isopropylamino-1,3,5-triazine (atrazine) are the best known.

Wessels & Van der Veen (1956) were the first to show that diuron inhibited the Hill reaction in isolated chloroplasts. The stimulation of fluorescence by diuron was explained by Duysens & Sweers (1963) by assuming that this inhibitor prevented the reoxidation of Q_A. By studying various parts of the electron transport pathway the site of inhibition of these herbicides was located between Q_A and the PQ pool. Photosystem I-dependent electron transport and cyclic electron flow was inhibited only at very high concentrations (Van Rensen, 1969).

The study of the mode of action of these herbicides was greatly stimulated by the introduction of a new technique by Tischer & Strotmann (1977). After a radiolabelled herbicide had been bound to isolated chloroplasts, other (not-radiolabelled) herbicides were added to see if they could replace the labelled compound. By this replacement technique they showed that phenylureas, triazines, pyridazinones and biscarbamates compete for the same binding site. Moreover, the relative concentration of specific binding sites was found to be 1 per 300–500 chlorophyll molecules, i.e. about 1 per electron transfer chain.

A characteristic of these herbicidal inhibitors is the reversibility of their effects. Van Rensen & Van Steekelenburg (1965) found that the inhibition of oxygen evolution in algae by diuron and by 2-methoxy-4,6-bis(ethylamino)-1,3,5-triazine (simeton) could be removed easily by washing. Izawa & Good (1965) showed that diuron was reversibly bound to chloroplasts. This implied that only weak bonds were involved in the interaction of these herbicides and the receptor molecule in the thylakoid membrane. Tischer & Strotmann (1979) measured the ΔH for binding of 4-amino-6-isopropyl-3-methylthio-1,2,4-triazine-5-one (metribuzin) and found it to be –50 kJ/mol. According to Shipman (1981) this binding energy is much too small for covalent binding in a protein, and it is also not consistent with hydrogen binding. He suggested that polar components of the herbicides bind via coulombic interactions at or near a highly polar protein site, probably a protein salt bridge or the terminus of an α-helix on the Q_B-protein.

The proteinaceous nature of the binding site of these herbicides was confirmed by experiments with the water-soluble enzyme trypsin. Regitz & Ohad (1975) and Renger (1976) demonstrated that diuron sensitivity of the Hill reaction was removed by proper treatment of chloroplasts with trypsin. These observations were extended by Böger and Kunert (1979) and by Van Rensen & Kramer (1979) to the triazine and phenolic herbicides. Trypsin treatment lowers the binding affinity of these herbicides (Trebst, 1979; Tischer & Strotmann, 1979; Steinback, Pfister & Arntzen, 1981). At the same time Q_A becomes accessible to ferricyanide. Therefore Renger (1976) postulated the presence of a protein shield above the quinone acceptors. He assumed that this protein acted as an allosteric regulator for the electron transport between Q_A and the PQ pool. Moreover, this protein was assumed to contain the binding sites for diuron-type herbicides, inhibition occurring by an allosteric mode of action.

Schemes visualizing binding sites of herbicides to the Q_B-protein were presented by Trebst & Draber (1979), Pfister & Arntzen (1979) and Johanningmeier, Neumann & Oettmeier (1983). Vermaas & Van Rensen (1981) incorporated, in addition, binding sites for bicarbonate. From comparative resistance studies with mutants, Böger, Sandmann & Miller (1981) concluded that there is a minimum number of two binding sites for each herbicide class. The basic principles are shown in Figure 2. It suggests two binding sites each for the ureas, the triazines, for bicarbonate, and one site for phenolic herbicides. The urea and triazine herbicides have one binding site in common.

Interaction of diuron-type herbicides with plastoquinone was first proposed by Van Rensen (1969, 1971). In 1974, Velthuys & Amesz suggested that diuron lowered the midpoint potential of Q_B (i.e. diuron increases the difficulty of Q_B-reduction). It is now widely accepted that the mechanism of herbicide action is a displacement of Q_B from its binding site at the Q_B-protein. This was independently and simultaneously proposed by Velthuys (1981) for the Photosystem II complex of plant thylakoids, and by Wraight (1981) for the reaction centre complex of purple photosynthetic bacteria. The action of the herbicides appears to be competitive with plastoquinone; the rate of release of the inhibitor from the site is many times slower than the rate of release for plastoquinone (Vermaas, Dohnt & Renger, 1984). The herbicides cannot be reduced by Q_A and electron transfer beyond this point is thereby prevented. Instead of a direct physical removal of Q_B from its binding site by the inhibitors, an allosteric action of the herbicides was suggested (Renger, 1976; Van Rensen, 1982; Renger, 1986). By allosteric action the binding of a herbicide and also the absence of bicarbonate causes a conformational change of the Q_B-protein. This change in conformation has two consequences, 1) the binding of another herbicide or bicarbonate is impaired, and 2) electron transport from Q_A to Q_B is inhibited. Several data are in favour of this allosteric model, which includes the idea of spatially separated herbicide binding sites with the Q_B-protein. First, the different chemical nature of molecules interacting with the Q_B-protein: diuron-type herbicides, phenol-type inhibitors, quinones, bicarbonate and formate. Secondly, plastoquinone not only interacts with the Q_B-protein, but also with the cytochrome b_6/f complex; most inhibitors of the Q_B-protein seem not to affect the cyt. b_6/f complex. Thirdly, there is evidence from herbicide resistance studies that show that thylakoids from triazine-resistant plants remain sensitive to urea herbicides. Experiments with resistant mutants of a cyanobacterium showed that loss of the diuron binding capacity did not necessarily affect the binding of atrazine and Q_B (Astier, Boussac & Etienne, 1984). In *Chlamydomonas* mutants the covalent binding of azidomonuron and azidoatrazine appeared to occur at different positions within the primary structure of the Q_B-protein (Boschetti, Tellenbach & Gerber, 1985).

Fig. 2. Visualization of the binding sites of the Q_B-quinone, herbicides and bicarbonate to a suggested D-1/D-2 complex of the reaction centre of Photosystem II. Urea herbicides could bind to sites 1 and 2, the triazines to sites 2 and 3, phenolic herbicides to site 4 and bicarbonate to sites 4 and 5.

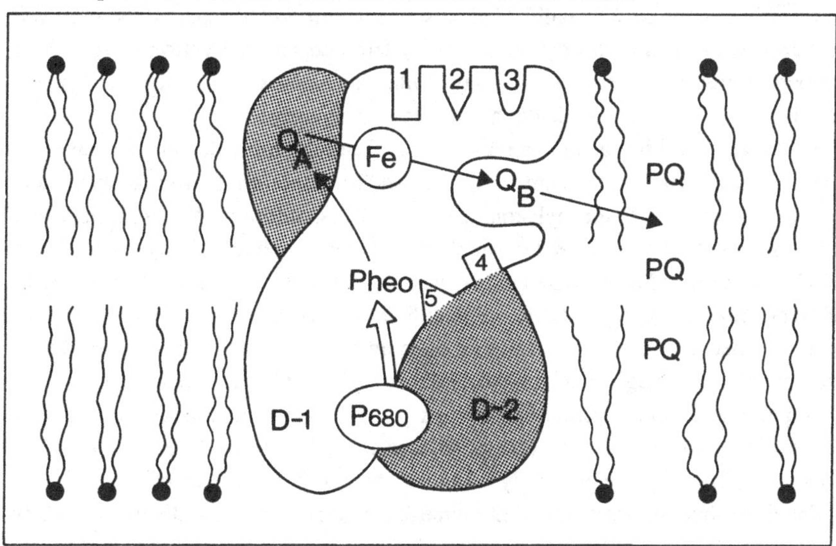

How can a conformational alteration of the Q_B-protein, caused by binding of a herbicide or the absence of bicarbonate, affect electron flow from Q_A to Q_B? Shipman (1981) rationalized that the Q_B-protein must provide a stabilization of Q_B^- in its semiquinone form since Q_B^- is stable in a hydrophobic environment for at least some seconds. This stabilization could be affected by protonation of the protein, by the generation of a strong electric field across the protein, or by relaxation of the protein conformation when Q_B is reduced. Hence inhibitor binding could affect the $Q_A^- Q_B \Leftrightarrow Q_A Q_B^-$ equilibrium by affecting the means of stabilization of Q_B^- by the Q_B-protein. Renger, Hagemann & Dohnt (1981) suggested, by analogy with electron transport between the redox-active haem groups of cytochrome c and cytochrome c peroxidase, that the Q_B-protein mediates electron transport between Q_A and Q_B via an 'electron channel', established by a specific arrangement of functional amino acid residues. Binding of an inhibitor could thus cause a structural modification of the protein units forming the electron tunnel. In this way the kinetics of electron transport between Q_A and Q_B could be changed.

Action of phenol-type herbicides and quinone inhibitors

The phenol-type herbicides include DNOC, 2,4-dinitro-6-s-butylphenol (dinoseb), 3,5-dibromo-4-hydroxybenzonitrile (bromoxynil) and 3,5-diiodo-4-hydroxy-benzonitrile (ioxynil). Because of its similarity to 2,4-dinitrophenol, DNOC was long considered to act only by uncoupling oxidative phosphorylation. However, in 1964, Kerr & Wain reported that it also inhibited the Hill reaction. Van Rensen, Van der Vet & Van Vliet (1977) and Van Rensen & Hobé (1979) demonstrated that at low concentrations DNOC inhibited photosynthetic electron transport at the same site as diuron, while at high concentration Photosystem I-dependent electron transport was uncoupled. Because of this dual effect of inhibition and uncoupling, Moreland (1980) classified these phenolic herbicides as inhibitory uncouplers.

Although inhibiting at the same site, the phenolic herbicides do not have the common basic chemical structure of the ureas and triazines, furthermore the interaction with the receptor site in the thylakoid membrane is also different (Van Rensen, Wong & Govindjee, 1978). Trebst & Draber (1979) studied structure–activity relationships of a large number of halogenated nitro- and dinitrophenols and found that the activity of ureas and triazines was related to lipophilicity and electronic parameters, whereas, with the phenolic herbicides, activity was more related to steric parameters. These herbicides have a lag-phase for the inhibition of Photosystem II activity, which disappears after preillumination or mild trypsin digestion, indicating hindered accessibility to the binding site (Böger & Kunert, 1979; Thiel & Böger, 1986). While ureas interfere non-competitively with the specific binding site of the phenolic herbicide iso-butyl-dinoseb, the phenol-type herbicides do so competitively (Oettmeier & Masson, 1980). Azido-atrazine and azido-dinoseb label different polypeptides (Oettmeier, Masson & Johanningmeier, 1980), azido-atrazine a 32 kD polypeptide

(Q_B-protein), while azido-dinoseb binds to a polypeptide of 40–50 kD (Johanningmeier, Neumann & Oettmeier, 1983).

Recently, a number of phenol analogues have been described as potent inhibitors of Photosystem II; they include benzoquinones (Oettmeier, Reimer & Link, 1978), naphthoquinones (Pfister *et al.*, 1981), pyridones, quinolones, pyrones, dioxobenzthiazoles and cyanoacrylates (Phillips & Huppatz, 1984). More details are given by Trebst & Draber (1986). These inhibitors displace each other as well as the diuron-type herbicides from the membrane. Trebst & Draber (1986) discussed a common essential element for the phenol- and quinone-type inhibitors.

Interaction of herbicides with bicarbonate

The Hill reaction in chloroplasts, that are depleted of CO_2 in the presence of formate, requires the addition of bicarbonate for reactivation. The major site of bicarbonate action is located between Q_A and plastoquinone. For reviews, see Govindjee & Van Rensen (1978), Vermaas & Govindjee (1981) and Van Rensen & Snel (1985). Because the sites of action of bicarbonate and diuron-type herbicides are located close to each other a relationship between bicarbonate effect and the action of herbicides was anticipated and indeed demonstrated (Van Rensen & Vermaas, 1981; Snel & Van Rensen, 1983, and Van Rensen, 1984). It was shown that the system thylakoid membrane versus bicarbonate has a Michaelis–Menten behaviour and that it can be treated like a system enzyme versus substrate. From double reciprocal plots of the rate of the Hill reaction as a function of the bicarbonate concentration the apparent dissociation constant (K_d) could be calculated, both in the absence and in the presence of a herbicide. The result of such an experiment with diuron is illustrated in Figure 3. It was found that, in the presence of many urea, triazine, and phenol-type herbicides the K_d-value for bicarbonate was increased by at least two-fold. This means that these herbicides decrease the apparent affinity of the thylakoid membrane for bicarbonate. Herbicide binding studies using radiolabelled atrazine (Khanna et al., 1981) or ioxynil (Vermaas, Van Rensen & Govindjee, 1982) revealed that the number of binding sites of these herbicides was the same in control, CO_2-depleted, and bicarbonate-reactivated chloroplasts. However, the binding constant for both herbicides was increased three-fold in the depleted chloroplasts compared to control and bicarbonate-reactivated chloroplasts.

With respect to their interaction with bicarbonate, the phenol-type herbicides appear to behave somewhat differently from the urea and triazine herbicides. The interaction of bicarbonate (or formate) and herbicide binding probably is of an allosteric nature. Both the binding of a herbicide and the absence of bicarbonate may cause a conformational alteration of the Q_B-protein, which changes the affinity for another herbicide or for bicarbonate, and causes inhibition of electron transport.

Herbicide binding kinetics

As is already described, the Q_B-quinone is not permanently bound to the Q_B-protein. In its full reduced state, Q_BH_2, it is only weakly bound and can easily leave its binding site. The semiquinone, Q_B^- is more tightly bound than the oxidized form, Q_B. Owing to the equilibrium between Q_A and Q_B, an electron shared between the two molecules spends about 95% of the time on Q_B and the remainder of the time on Q_A. Also herbicides bind and release at their binding site to the Q_B-protein. These exchange parameters can be measured using a method, developed by Vermaas, Dohnt

Fig. 3. Double reciprocal plot of the Hill reaction rate as a function of bicarbonate concentration in CO_2-depleted pea chloroplasts in the absence and in the presence of DCMU (diuron). Reaction medium: 50 mM Na-phosphate, 100 mM Na-formate, 100 mM NaCl, 5 mM $MgCl_2$, 0.5 mM ferricyanide, 33 μg chlorophyll/ml, (pH 6.5).

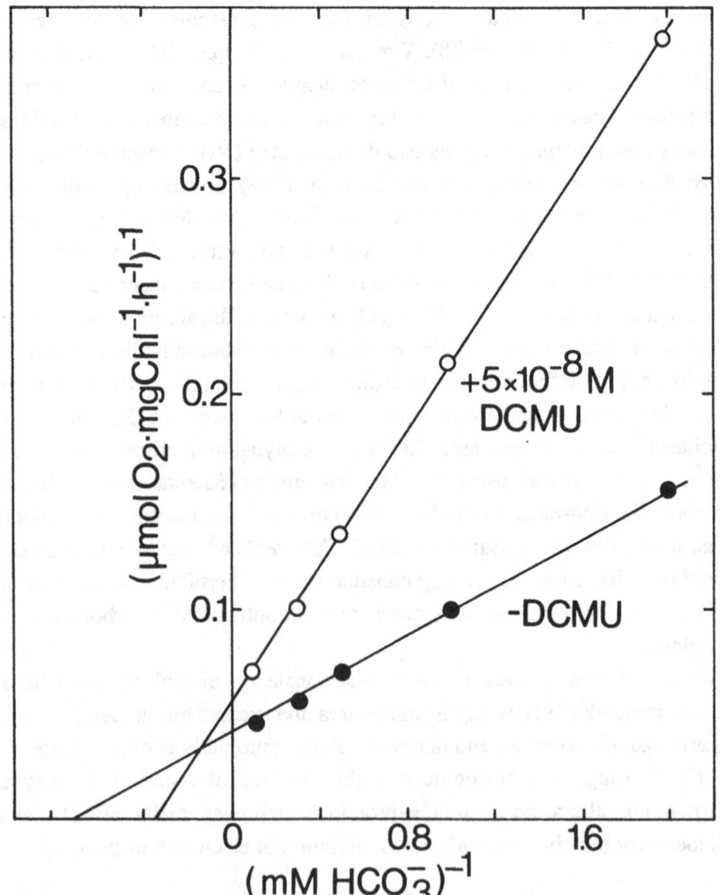

& Renger (1984). It is based on the flash-induced oxygen evolution patterns of isolated broken chloroplasts, which are measured in the absence and in the presence of herbicides. The exchange parameters are obtained by fitting experimental data to those calculated with a kinetic model. This model is derived from the following equations:

$$S_n \cdot Q_A \cdot Q_B + I \underset{E_3}{\overset{E_1}{\Leftrightarrow}} S_n \cdot Q_A \cdot I + Q_B$$

$$S_n \cdot (Q_A \cdot Q_B)^- + I \underset{E_4}{\overset{E_2}{\Leftrightarrow}} S_n \cdot (Q_A \cdot I)^- + Q_B$$

In these equations, S_n (where $n = 0,1,2,3$) represents the redox state of the oxygen evolving complex. In the presence of slowly exchanging herbicides, having residence times on the Q_B-protein of the same order of magnitude as the duration of the flash train or longer, the oscillation is hardly damped compared to the control. In this case only the amplitude of the signal is diminished. However, when the herbicide exchange is occurring with the same or higher frequency than the firing of the flashes, the damping of the oscillation is considerably stronger. This is caused by the fact that then reaction centres are blocked for a certain time span, and start making turnovers at the moment the herbicide is displaced by a PQ-molecule. Thus, centres can get out of phase with each other, and produce O_2 at different flashes. By comparing flash patterns obtained with different flash frequencies and herbicide concentrations, the exchange parameters E1 to E_4 can be calculated.

Using this method Naber & Van Rensen (1987) found for the herbicides atrazine, diuron and dinoseb that the E_1-parameters are 20–200-fold higher than those for E_2. This means that these herbicides can bind much easier to the Q_B-protein when Q_B is in its oxidized form than when it is in its semiquinone form. This result is in agreement with the hypothesis that Q_B^- has a much higher affinity for its binding environment than Q_B and Q_BH_2.

The mechanism of triazine resistance

In the last decade resistance against triazine herbicides has developed in many weed species. This resistance is specific for all triazine herbicides; the concentration of triazine herbicide causing 50% inhibition of the Hill reaction in isolated chloroplasts from resistant biotypes is often about 1000 times higher than that in chloroplasts from the sensitive biotypes. There is little difference in sensitivity for urea herbicides, while resistant biotypes are somewhat more sensitive to phenol-type herbicides (Pfister & Arntzen, 1979; Oettmeier *et al.*, 1982; Jansen *et al.*, 1986). The resistance is caused by lack of binding of triazine herbicides (Pfister & Arntzen, 1979). In the model of Figure 2 this could be visualized by the disappearance of binding site number 3. The resistance could be correlated with an alteration of the Q_B-protein. After it was found that the chloroplast *psbA* gene codes for the Q_B-protein, Hirschberg *et al.* (1984) demonstrated a change from adenine in the susceptible to guanine in the resistant biotype. This leads to a substitution of serine at position 264 for glycine within the Q_B-protein.

The alteration of the Q_B-protein did not only result in triazine resistance, but also in a ten-fold decrease in the rate of electron flow from Q_A to Q_B (Pfister & Arntzen, 1979; Bowes, Crofts & Arntzen, 1980). Vermaas & Arntzen (1983) related this decreased rate with a large effect on the semiquinone equilibrium between Q_A and Q_B. The impaired rate of electron transport between Q_A and Q_B was suggested to be the cause of a significantly lowered rate of light- and CO_2-saturated photosynthesis (Holt, Stemler & Radosevich, 1981; Ahrens & Stoller, 1983). However, an influence of impaired electron transport between Q_A and Q_B on the overall electron flow may be questioned. The lower rate of electron transport from Q_A to Q_B in the resistant biotype is still about ten times faster than the oxidation of reduced plastoquinone. The latter reaction remains the rate determining step having a half-time of about 20 ms (Stiehl & Witt, 1969). Both Ort *et al.* (1983) and Jansen *et al.* (1986) measured a lower rate of electron flow between Q_A and plastoquinone in resistant chloroplasts; the whole chain electron transport activity was, however, not significantly different in the two biotypes.

Triazine resistance initiated the study of the molecular biology of the Q_B-protein. By growing algae in the presence of sublethal concentrations of a herbicide, many mutants resistant to various herbicides have been obtained. These mutants have been of great value in the elucidation of the molecular architecture of the herbicide binding site.

Architecture of the herbicide binding site

The core complex of Photosystem II consists of five major integral membrane-spanning polypeptide subunits of 47, 44, 34, 32 and 10 kD molecular weight. This complex contains the reaction centre chlorophyll P680, core antenna chlorophylls, pheophytin, Fe, cytochrome b559, two acceptor plastoquinones Q_A and Q_B and a primary electron donor, possibly also a plastoquinone. Of special interest for this chapter are the 32 and 34 kD proteins. In earlier studies these proteins were designated as D-1 and D-2, where D referred to a diffuse peak (Ellis, 1977; Chua & Gilham, 1977). The protein responsible for herbicide binding and for the trypsin sensitive shield was identified by photoaffinity labelling. Azidoatrazine binds to the 32 kD polypeptide (Pfister *et al.*, 1981) that is rapidly turning over and is coded for by a photogene *psbA* (Mattoo *et al.*, 1981). Also azidomonuron (Boschetti, Tellenbach & Gerber, 1985) and an azidotriazinone (Oettmeier *et al.*, 1984) bind covalently to this polypeptide. Since then it is called the 32 kD (or D-1) herbicide and Q_B-binding protein. This Q_B-protein is probably one of the most well-studied proteins in thylakoid membranes. A detailed discussion is available in the recent review by Kyle (1985).

The role of the other protein, 34 kD D-2, has long been uncertain. It differs from the D-1 protein in containing lysine; the D-1 protein being lysine-free. From experiments with a lysine specific protease Renger, Hagemann & Vermaas (1984) concluded that binding of atrazine is additionally affected by a lysine containing polypeptide. This may very well be the D-2 protein. Further, they indicated that interaction of herbicides with their binding sites is more complex and allosteric effects may be also of functional relevance. In Figure 2 a model is suggested, in which the D-1 and D-2 proteins are closely related in charge separation, primary electron flow, and transfer of electrons between the primary and secondary quinone acceptors. Several binding sites for herbicides are indicated, allosterically affecting the Q_B-binding site and thereby electron flow between Q_A and Q_B.

Further indications for a role of the D-2 protein come from the close analogy in the organization of electron transport at the acceptor side of purple bacteria and Photosystem II of green plants. Inspection of the hydropathy index plots and of the amino acid sequence showed a striking structural similarity between D-1 and the L-subunit of bacteria as well as between D-2 and the M-subunit. The data obtained led to the conclusion that the transmembrane folding of D-1 and D-2 is basically the same as for L- and M-subunits in purple bacteria, especially with respect to the existence of five membrane-spanning helices (Deisenhofer *et al.*, 1985; Trebst, 1986).These similarities indicate that D-1 and D-2 may form the apoenzyme of the Photosystem II reaction centre in analogy with purple bacteria, where the L- and M-subunits form the bacterial reaction centre. The structure of the bacterial reaction centres of *Rhodopseudomonas viridis* has been recently resolved by X-ray-diffraction analysis (Deisenhofer *et al.*, 1985). It appeared that the reaction centre apoprotein of purple bacteria is formed by an intermingled complex of the L- and M-subunits. If analogously the D-1 and D-2 would establish the matrix for the prosthetic groups of the Photosystem II reaction centre complex, the well-known structure of purple bacterial reaction centres could be a guideline for further studies about the structural organization of Photosystem II reaction centres. For a detailed discussion of the D-1/D-2 model, see Renger (1986) and Trebst & Draber (1986).

The study of herbicide resistant organisms is of great value for our information of the architecture of the herbicide binding sites. In higher plants triazine resistance is intensively investigated. At present, many algal and cyanobacterial mutants resistant to various herbicides are available. A few recent examples are described by Erickson *et al.* (1984), Astier *et al.* (1986), and Johanningmeier, Bodner & Wildner (1987). The molecular basis for this resistance was in each case a change in the *psbA* gene coding for the Q_B-protein. Several amino acid substitutions in this D-1 protein were reported: Ser 264, Val 219, Phe 255, Leu 275 and Ala 251. All these amino acid residues are located at the stroma side of helix 4 and 5 of the D-1 polypeptide and have to be of functional significance for herbicide binding (Trebst & Draber, 1986).

Through the years there have been reports of herbicide action at the donor side of Photosystem II. A few recent reports are Mathis & Rutherford (1984) and Metz, Bricker & Seibert (1985). Metz *et al.* (1986) suggested a role of the D-1 protein in affecting also the oxidizing side of Photosystem II. Labelling experiments with ^{125}I lead Ikeuchi & Inoue (1987) to suggest that the D-1 protein bears the photooxidation site or has a domain very close to the photooxidation site on the donor side of Photosystem II, in addition to the well established binding site for Q_B and herbicides on the acceptor side of Photosystem II. An action of the Q_B-protein at the donor side of Photosystem II could explain effects of herbicides at this site.

Conclusion

This chapter reviews aspects of research on the mechanisms of action of herbicides inhibiting Photosystem II. The site of their action is between the quinone acceptors Q_A and Q_B. Considerations are presented which led to a D-1/D-2 model for the structural organization of the Photosystem II reaction centre complex (D-1 being the 32 kD Q_B-protein). The herbicides bind to the D-1 polypeptide and in an allosteric manner decrease the affinity of Q_B for its binding site. In this way the herbicides inhibit electron transport between Q_A and Q_B.

References

Ahrens, W.H. & Stoller, E.W. (1983). Competition, growth rate and CO_2 fixation in triazine-susceptible and resistant smooth pigweed (*Amaranthus hybridus*). *Weed Sci.*, **31**, 438–44.

Astier, C., Boussac, A. & Etienne, A.L. (1984). Evidence for different binding sites on the 33 kDa protein for DCMU, atrazine and Q_B. *FEBS Lett.*, **167**, 321–6.

Astier, C., Meyer, I., Vernotte, C. & Etienne A.L. (1986). Photosystem II electron transfer in highly herbicide resistant mutants of *Synechocystis* 6714. *FEBS Lett.*, **207**, 234–8.

Böger, P. & Kunert, K.-J. (1979). Differential effects of herbicides upon trypsin-treated chloroplasts. *Z. Naturforsch.*, **34c**, 1015–20.

Böger, P., Sandmann, G. & Miller, R. (1981). Herbicide resistance in a mutant of the microalgae *Bumilleriopsis filiformis*. *Photosynthesis Res.*, **2**, 61–74.

Boschetti, A., Tellenbach, M. & Gerber, A. (1985). Covalent binding of 3-azido-monuron to thylakoids of DCMU-sensitive and resistant strains of *Chlamydomonas reinhardii*. *Biochim. Biophys. Acta*, **810**, 12–19.

Bowes, J., Crofts, A.R. & Arntzen, C.J. (1980). Redox reactions on the reducing side of photosystem II in chloroplasts with altered herbicide binding properties. *Arch. Biochem. Biophys.*, **200**, 303–8.

Chua, N.H. & Gilham, N.W. (1977). The sites of synthesis of the principal thylakoid membrane polypeptides in *Chlamydomonas reinhardii*. *J. Cell. Biol.*, **77**, 441–52.

Deisenhofer, J., Epp, O., Miki, K., Huber, R. & Michel, H. (1985). Structure of the protein subunit in the reaction centre of *Rhodopseudomonas viridis* at 3 Å resolution. *Nature*, **318**, 618–24.

Duysens, L.N.M. & Sweers, H.E. (1963). Mechanism of two photochemical reactions in algae as studied by means of fluorescence. In: *Studies on Microalgae and Photosynthetic Bacteria*. ed. *Jpn. Soc. Plant Physiol.*, pp. 353–72. Tokyo, Univ. of Tokyo Press.

Ellis, R.J. (1977). Protein synthesis by isolated chloroplasts. *Biochim. Biophys. Acta*, **463**, 185–218.

Erickson, J.M., Rahire, M., Bennoun, P., Depelaire, P., Diner, B. & Rochaix, J.D. (1984). Herbicide resistance in *Chlamydomonas reinhardii* results from a mutation in the chloroplast gene for the 32 kilodalton protein of photosystem 2. *Proc. Natl. Acad. Sci. USA*, **81**, 3617–21.

Govindjee & Van Rensen, J.J.S. (1978). Bicarbonate effects on the electron flow in isolated broken chloroplasts. *Biochim. Biophys. Acta*, **505**, 183–213.

Hirschberg, J., Bleeker, A., Kyle, D.J., McIntosh, L. & Arntzen, C.J. (1984). The molecular basis of triazine-herbicide resistance in higher-plant chloroplasts. *Z. Naturforsch.*, **39c**, 412–20.

Holt, J.S., Stemler, A.J. & Radosevich, S.R. (1981). Differential light responses of photosynthesis by triazine-resistant and triazine-susceptible *Senecio vulgaris* biotypes. *Plant Physiol.*, **67**, 744–8.

Ikeuchi, M. & Inoue, Y. (1987). Specific [^{125}I]-labeling of D 1 (herbicide-binding protein). An indication that D 1 functions on both the donor and acceptor sides of photosystem II. *FEBS Lett.*, **210**, 71–6.

Izawa, S. & Good, N.E. (1965). The number of sites sensitive to 3-(3,4-dichlorophenyl)-1,1-dimethylurea, 3-(4-dichorophenyl)-1,1-dimethylurea and 2-chloro-4-(2-propylamino)-6-ethylamino-s-triazine in isolated chloroplasts. *Biochim. Biophys. Acta*, **102**, 20–38.

Jansen, M.A.K., Hobé, J.H., Wesselius, J.C. & Van Rensen, J.J.S. (1986). Comparison of photosynthetic activity and growth performance in triazine-resistant and susceptible biotypes of *Chenopodium album*. *Physiol. Vég.*, **24**, 475–84.

Johanningmeier, U., Neumann, E. & Oettmeier, W. (1983). Interaction of a phenolic inhibitor with photosystem II particles. *J. Bioenerg. Biomembr.*, **15**, 43–66.

Johanningmeier, U., Bodner, U. & Wildner, G.F. (1987). A new mutation in the gene coding for the herbicide-binding protein in Chlamydomonas. *FEBS Lett.*, **211**, 221–4.

Kerr, M.W. & Wain, R.L. (1964). Inhibition of the ferricyanide Hill reaction of isolated bean leaf chloroplasts by 3,5-diiodo-4-hydroxybenzonitrile (ioxynil) and related compounds. *Ann. Appl. Biol.*, **54**, 447–50.

Khanna, R., Pfister, K., Keresztes, A., Van Rensen, J.J.S. & Govindjee (1981). Evidence for a close spatial location of the binding sites for CO_2 and for photosystem II inhibitors. *Biochim. Biophys. Acta*, **634**, 105–16.

Kyle, D. (1985). The 32,000 dalton Q_B-protein of photosystem II. *Photochem. Photobiol.*, **41**, 107–16.

Mathis, P. & Rutherford, A.W. (1984). Effect of phenolic herbicides on the oxygen-evolving side of photosystem II. Formation of the carotenoid cation. *Biochim. Biophys. Acta*, **767**, 217–22.

Mattoo, A.K., Pick, U., Hoffmann-Falk, H. & Edelman, M. (1981). The rapidly metabolized 32,000-dalton polypeptide of the chloroplast is the 'proteinaceous shield' regulating photosystem II electron transport and mediating diuron herbicide sensitivity. *Proc. Natl. Acad. Sci. USA*, **78**, 1572–6.

Metz, J.G., Bricker, T.M. & Seibert, M. (1985). The azido[^{14}C]-atrazine photoaffinity technique labels a 34-kDa protein in *Scenedesmus* which functions on the oxidizing side of photosystem II. *FEBS Lett.*, **185**, 191–6.

Metz, J.G., Pakrasi, H.B., Seibert, M. & Arntzen, C.J. (1986). Evidence for a dual function of the herbicide-binding D 1 protein in photosystem II. *FEBS Lett.*, **205**, 269–74.

Moreland, D.E. (1980). Mechanisms of action of herbicides. *Ann. Rev. Plant Physiol.*, **31**, 597–638.

Naber, J.D. & Van Rensen, J.J.S. (1987). Determination of the exchange parameters of herbicides on the Q_B-protein of photosystem II. In: *Progress in Photosynthesis Research*, Vol III, ed. J. Biggins, pp. 767–70. Dordrecht; Martinus Nijhoff Publishers.

Oettmeier, W., Reimer, S. & Link, K. (1978). Quantitative structure–activity relationship of substituted benzoquinones as inhibitors of photosynthetic electron transport. *Z. Naturforsch.*, **33c**, 695–703.

Oettmeier, W. & Masson, K. (1980). Synthesis and thylakoid membrane binding of the radioactively labeled herbicide dinoseb. *Pestic. Biochem. Physiol.*, **14**, 86–97.

Oettmeier, W., Masson, K. & Johanningmeier, U. (1980). Photoaffinity labeling of the photosystem II herbicide binding protein. *FEBS Lett.*, **118**, 267–70.

Oettmeier, W., Masson, K., Fedtke, K., Konze, J. & Schmidt, R.R. (1982). Effect of different photosystem II inhibitors on chloroplasts isolated from species either susceptible or resistant toward s-triazine herbicides. *Pestic. Biochem. Physiol.*, **18**, 357–67.

Oettmeier, W., Masson, K., Soll, H. & Draber, W. (1984). Herbicide binding at photosystem II. A new azido-triazinone photoaffinity label. *Biochim. Biophys Acta*, **767**, 590–5.

Ort, D.R., Ahrens, W.H., Martin, B. & Stoller, E.W. (1983). Comparison of photosynthetic performance in triazine resistant and susceptible biotypes of *Amaranthus hybridus*. *Plant Physiol.*, **72**, 925–30.

Ort, D.L. (1986). Energy transduction in oxygenic photosynthesis; an overview of structure and mechanism. In: *Encyclopedia of plant physiology*, N.S., Photosynthesis III, ed. L.A. Staehelin and C.J. Arntzen, pp. 143–96. Berlin; Springer-Verlag.

Pfister, K. & Arntzen, C.J. (1979). The mode of action of photosystem II-specific inhibitors in herbicide-resistant weed biotypes. *Z. Naturforsch.*, **34c**, 996–1009.

Pfister, K., Lichtenthaler, H.K., Burger, G., Musso, H. & Zahn, M. (1981). The inhibition of photosynthetic light reactions by halogenated naphthoquinones. *Z. Naturforsch.*, **36c**, 645–55.

Pfister, K., Steinback, K.E., Gardner, G. & Arntzen, C.J. (1981). Photoaffinity labeling of a herbicide receptor protein in chloroplast membranes. *Proc. Natl. Acad. Sci. USA*, **78**, 981–5.

Pfister, K. & Urbach, W. (1983). Effects of biocide and growth regulators; physiological basis. In *Encyclopedia of plant physiology*, N.S., 12D, Plant Ecology, IV, ed. O.L. Lange *et al.*, pp. 329–91. Berlin; Springer-Verlag.

Phillips, J. & Huppatz, J. (1984). Cyanoacrylate inhibitors of photosynthetic electron transport. Nature of the interaction with the receptor site. *Z. Naturforsch.*, **39c**, 335–7.

Regitz, G. & Ohad, I. (1975). Changes in the protein organization in developing thylakoids of *Chlamydomonas reinhardii* Y-1 as shown by sensitivity to trypsin. In: *Proc. IIIrd Int. Congress Photosynthesis*, ed. M. Avron, pp. 1615–25. Amsterdam; Elsevier.

Renger, G. (1976). Studies on the structural and functional organization of system II of photosynthesis; the use of trypsin as a structurally selective inhibitor at the outer surface of the thylakoid membrane. *Biochim. Biophys. Acta*, **440**, 287–300.

Renger, G., Hagemann, R. & Dohnt, G. (1981). Properties of the proteinaceous component acting as apoenzyme for the functional plastoquinone groups on the acceptor side of system II. *Biochim. Biophys. Acta*, **636**, 17–26.

Renger, G., Hagemann, R. & Vermaas, W.F.J. (1984). Studies on the functional mechanism of system II herbicides in isolated chloroplasts. *Z. Naturforsch.*, **39c**, 362–7.

Renger, G. (1986). Herbicide interaction with photosystem II, recent developments. *Physiol. Vég.*, **24**, 509–21.

Sandmann, G.& Böger, P. (1986). Sites of herbicide inhibition at the photosynthetic apparatus. In: *Encyclopedia of plant physiology*, N.S., Photosynthesis III, ed. L.A. Staehelin & C.J. Arntzen, pp. 595–602. Berlin; Springer-Verlag.

Shipman, L.L. (1981). Theoretical study of the binding site and mode of action for photosystem II herbicides. *J. Theor. Biol.*, **90**, 123–48.

Snel, J.F.H. & Van Rensen, J.J.S. (1983). Kinetics of the reactivation of the Hill reaction in CO_2-depleted chloroplasts by addition of bicarbonate in the absence and in the presence of herbicides. *Physiol. Plant.*, **57**, 422–7.

Steinback, K.E., Pfister, K. & Arntzen, C.J. (1981). Trypsin-mediated removal of herbicide binding sites within the photosystem II complex. *Z. Naturforsch.*, **36c**, 98–108.

Stiehl, H.H. & Witt, H.T. (1969). Quantitative treatment of the function of plastoquinone in photosynthesis. *Z. Naturforsch.*, **24b**, 1588–98.

Thiel, A. & Böger, P. (1986). Binding of ioxynil to photosynthetic membranes. *Pestic. Biochem. Physiol.*, **25**, 270–8.

Tischer, W. & Strotmann, H. (1977). Relationship between inhibitor binding and inhibition of photosynthetic electron transport. *Biochim. Biophys. Acta*, **460**, 113–25.

Tischer, W. & Strotmann, H. (1979). Some properties of the DCMU-binding site in chloroplasts. *Z. Naturforsch.*, **34c**, 992–5.

Trebst, A. (1979). Inhibition of photosynthetic electron flow by phenol and diphenylether herbicides in control and trypsin-treated chloroplasts. *Z. Naturforsch.*, **34c**, 986–91.

Trebst, A. & Draber, W. (1979). Structure activity correlations of recent herbicides in photosynthetic reactions. In: *Advances in Pesticide Science*, part 2, ed. H. Geissbüler, pp. 223–34. Oxford; Pergamon Press.

Trebst, A. (1986). The topology of the plastoquinone and herbicide binding peptides of photosystem II in the thylakoid membrane. *Z. Naturforsch.*, **41c**, 240–5.

Trebst, A. & Draber, W. (1986). Inhibitors of photosystem II and the topology of the herbicide and Q_B binding polypeptide in the thylakoid membrane. *Photosynthesis Res.*, **10**, 381–92.

Van Rensen, J.J.S. & Van Steekelenburg, P.A. (1965). The effect of the herbicides simetone and DCMU on photosynthesis. *Meded. Landbouwhogeschool Wageningen*, **65 (13)**, 1–8.

Van Rensen, J.J.S. (1969). Polyphosphate formation in *Scenedesmus* in relation to photosynthesis. In: *Progress in Photosynthesis Research*, Vol. III, ed. H. Metzner, pp. 1769–76. Tübingen.

Van Rensen, J.J.S. (1971). Action of some herbicides in photosynthesis of *Scenedesmus*, as studied by their effects on oxygen evolution and cyclic photophosphorylation. *Meded. Landbouwhogeschool*, **71–9**, 1–80.

Van Rensen, J.J.S., Van der Vet, W. & Van Vliet, W.P.A. (1977). Inhibition and uncoupling of electron transport in isolated chloroplasts by the herbicide 4,6-dinitro-*o*-cresol. *Photochem. Photobiol.*, **25**, 579–83.

Van Rensen, J.J.S., Wong, D. & Govindjee (1978). Characterization of the inhibition of photosynthetic electron transport in pea chloroplasts by the herbicide 4,6-dinitro-*o*-cresol by comparative studies with 3-(3,4-dichlorophenyl)-1,1-dimethylurea. *Z. Naturforsch*, **33c**, 413–20.

Van Rensen, J.J.S. & Hobé, J.H. (1979). Mechanism of action of the herbicide 4,6-dinitro-*o*-cresol in photosynthesis. *Z. Naturforsch.*, **34c**, 1021–3.

Van Rensen, J.J.S. & Kramer, H.J.M. (1979). Short-circuit electron transport insensitive to diuron-type herbicides induced by treatment of isolated chloroplasts with trypsin. *Plant Sci. Lett.*, **17**, 21–7.

Van Rensen, J.J.S. & Vermaas, W.F.J. (1981). Action of bicarbonate and photosystem II inhibiting herbicides on electron transport in pea grana and in thylakoids of a blue–green alga. *Physiol. Plant.*, **51**, 106–10.

Van Rensen, J.J.S. (1982). Molecular mechanisms of herbicide action near photosystem II. *Physiol. Plant.*, **54**, 515–21.

Van Rensen, J.J.S. (1984). Interaction of photosystem II herbicides with bicarbonate and formate in their effects on photosynthetic electron flow. *Z. Naturforsch.*, **39c**, 374–7.

Van Rensen, J.J.S. & Snel, J.F.H. (1985). Regulation of photosynthetic electron transport by bicarbonate, formate and herbicides in isolated broken and intact chloroplasts. *Photosynthesis Res.*, **6**, 231–46.

Velthuys, B.R. & Amesz, J. (1974). Charge accumulation at the reducing side of system 2 of photosynthesis. *Biochim. Biophys. Acta*, **333**, 85–94.

Velthuys, B.R. (1981). Electron dependent competition between plastoquinone and inhibitors for binding to photosystem II. *FEBS Lett.*, **126**, 277–81.

Vermaas, W.F.J. & Govindjee (1981). Unique role(s) of carbon dioxide and bicarbonate in the photosynthetic electron transport system. *Proc. Indian Natl. Sci. Acad.*, **B47**, 581–605.

Vermaas, W.F.J. & Van Rensen, J.J.S. (1981). Mechanism of bicarbonate action on photosynthetic electron transport in broken chloroplasts. *Biochim. Biophys. Acta*, **636**, 168–74.

Vermaas, W.F.J., Van Rensen, J.J.S. & Govindjee (1982). The interaction between bicarbonate and the herbicide ioxynil in the thylakoid membrane and the effects of amino acid modification on bicarbonate action. *Biochim. Biophys. Acta*, **681**, 242–7.

Vermaas, W.F.J. & Arntzen, C.J. (1983). Synthetic quinones influencing herbicide binding and photosystem II electron transport; the effects of triazine-resistance on quinone binding properties in thylakoid membranes. *Biochim. Biophys. Acta*, **725**, 483–91.

Vermaas, W.F.J., Dohnt, G & Renger, G. (1984). Binding and release kinetics of inhibitors of Q_A^- oxidation in thylakoid membranes. *Biochim. Biophys. Acta*, **765**, 74–83.

Wessels, J.S.C. & Van der Veen, R. (1956). The action of some derivatives of phenylurethan and of 3-phenyl-1,1-dimethylurea on the Hill reaction. *Biochim. Biophys. Acta*, **19**, 548–9.

Wraight, C.A. (1981). Oxidation–reduction physical chemistry of the acceptor quinone complex in bacterial photosynthetic reaction centers; evidence for a new model of herbicide activity. *Isr. J. Chem.*, **21**, 348–54.

ALAN D. DODGE

Herbicides interacting with photosystem I

Introduction

Photosystem I (PSI), like Photosystem II (PSII) considered in the previous chapter, is an integral part of the chloroplast electron transport system. Although PSI may act independently of PSII in cyclic flow, both are required for the continuous maintenance of electron flow from water to $NADP^+$ for CO_2 incorporation. Whereas with inhibitors of PSII (Chapter 2) interaction with a protein component resulted in the indirect cessation of electron flow, herbicides considered in connection with PSI intercept electrons directly.

In the generation of PSI, electrons expelled from P700 are raised to a negative potential (possibly –900 mV) to electron acceptors A_0 and A_1, that are forms of chlorophyll *a*. The donor and acceptors are associated with 70 kD pigment proteins that span the thylakoid membrane lipid bilayer. The subsequent electron acceptor is the hypothetical X, possibly an iron–sulphur centre with a potential of around –700 mV, and then two iron–sulphur centres A and B (Fig. 1) with potentials of around –590 mV and –530 mV. These are associated with two subunits of 18 and 16 kD and are probably identical to the former acceptor $P430^-$. These two centres function either in series or in parallel, and are almost certainly bound ferredoxin, and link to soluble ferredoxin and ferredoxin-$NADP^+$ reductase and hence $NADP^+$ (Fig. 1) (Haehnel 1984).

Bipyridiniums

The most important herbicides to interact in this part of the chloroplast are bipyridiniums, paraquat and diquat. These compounds, discovered in the late 1950s,

Fig. 1 Sequence of electron acceptors at Photosystem I. For details see text.

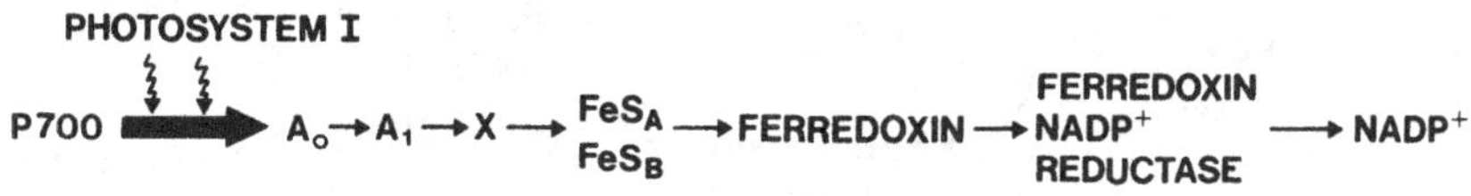

PHOTOSYSTEM I

P700 ⟶ A_0 → A_1 → X → FeS_A / FeS_B → FERREDOXIN → FERREDOXIN NADP$^+$ REDUCTASE → NADP$^+$

have been widely used as total kill herbicides, where rapid phytotoxic symptoms of chlorophyll bleaching are seen. Experiments with isolated chloroplasts (Zweig, Shavit & Avron, 1965) showed that they compete for electron flow with ferredoxin, thus, as a consequence *in vivo*, the failure of $NADP^+$ reduction results in the rapid cessation of carbon dioxide incorporation (Harris & Dodge, 1972*b*). The redox potentials of these compounds, paraquat –446 mV, and diquat –349 mV is similar to that of soluble ferredoxin (–420 mV) and experiments by Warden & Bolton (1974) suggested that electron diversion was from $P430^-$, equivalent to the iron–sulphur centres A and B.

Experiments by Mees (1960) showed that both light and oxygen were essential for the rapid action of these herbicides, the light requirement being initially to generate electron flow, and the oxygen for the production of toxic oxygen radicals. Early work suggested that the initial toxic agent was hydrogen peroxide (Davenport, 1963), that could give rise to damaging hydroxyl radicals (Dodge, 1971). We now have convincing evidence that the reoxidation of reduced paraquat leads to the generation of superoxide (O_2^-). In the presence of oxygen, the paraquat radical will react with a reaction rate constant of around $7.7 \times 10^8 M^{-1}s^{-1}$, to give a μM concentration of superoxide within the chloroplast grana (Farrington *et al.*, 1973). Spin trapping experiments by Harbour & Bolton (1975) using the trap 5,5-dimethyl-1-pyrroline-1-oxide (DMPO), also confirmed the bipyridinium catalysed production of superoxide in what was probably the first utilization of this technique in biological systems. The reduction of paraquat by one electron gives, in the absence of oxygen, a stable highly coloured radical cation (Zweig *et al.*, 1965), however, in oxygen, the herbicide acts as a cyclic catalyst (Fig. 2). This is illustrated in Figure 3, which shows that the inhibitory action of 0.05 μM paraquat on $NADP^+$ reduction is enhanced in air, by virtue of the catalytic reoxidation of the reduced cation.

Fig. 2. Sequence of the electron flow from iron sulphur centres A/B to either bipyridiniums (BP) or ferredoxin.

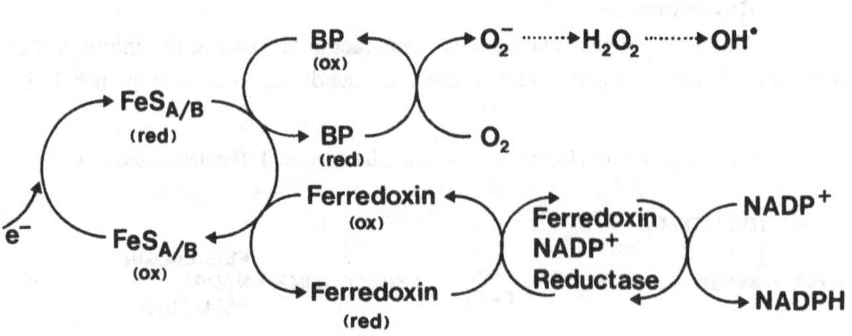

Figure 4(a) shows an electron spin resonance trace obtained with isolated chloroplasts illuminated in the presence of oxygen and paraquat. The definitive superoxide signal was replaced by a hydroxyl radical signal on the addition of superoxide dismutase (Fig. 4b) which converted O_2^- to hydrogen peroxide. This experiment provides a good model for the subsequent *in vivo* events that follow the endogenous generation of superoxide. Figure 5 shows the interaction of O_2^- and H_2O_2 in a Fenton-type reaction mediated by an iron catalyst. Under normal circumstances plant cell iron that is not utilized in iron–sulphur centres, ferredoxin, cytochromes, and other molecules is stored within phytoferritin. O_2^- may promote the release of iron from this store (Saito, Thomas & Aust, 1985), thus enhancing the deleterious iron-catalysed reactions.

Hydroxyl radicals (OH•) generated in this way rapidly react with membrane unsaturated fatty acids, deoxyribose sugar and thymine of DNA, a number of protein amino acids including methionine and histidine, and also numerous aromatic compounds. The immediate consequence is that cellular destruction will follow (Harris & Dodge, 1972*a*). Thus a cascade of events is initiated, not only involving

Fig. 3. The effect of 0.05 μM paraquat on $NADP^+$ reduction by isolated thylakoids under air ○ or nitrogen ● with electron donation from ascorbate. Additions in μmoles: ascorbate, 20; dichlorophenol indophenol 0.5; $NADP^+$ 0.5; monuron 0.015; Tris-HCl buffer pH 8.0, 150: ferredoxin 150 μg. From Dodge A.D. & Davenport H.E. (1965 unpublished).

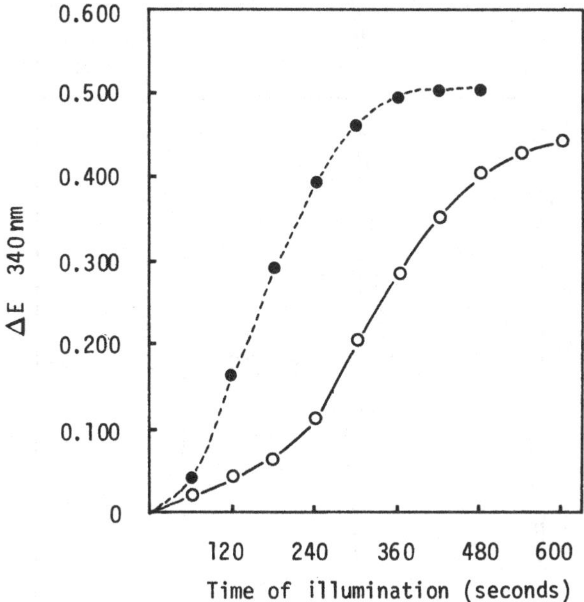

radical damage, but also that promoted by osmotic and pH changes, and the release of toxic compounds from the cell vacuole.

The generation of O_2^- via electron flow from ferredoxin to oxygen may occur to a limited extent under normal circumstances, possibly determined by the content of

Fig. 4. Electron spin resonance traces of isolated chloroplast thylakoids illuminated with 0.5 μM paraquat. Additions 100 mM DMPO: 1 mM DTPA. (A) Spectrum of DMPO-superoxide spin adduct. (B) Spectrum of DMPO-hydroxyl spin adduct after addition of superoxide dismutase. From Dodge, A.D., Thronalley, P.J. & Bannister, J.V. (1982 unpublished).

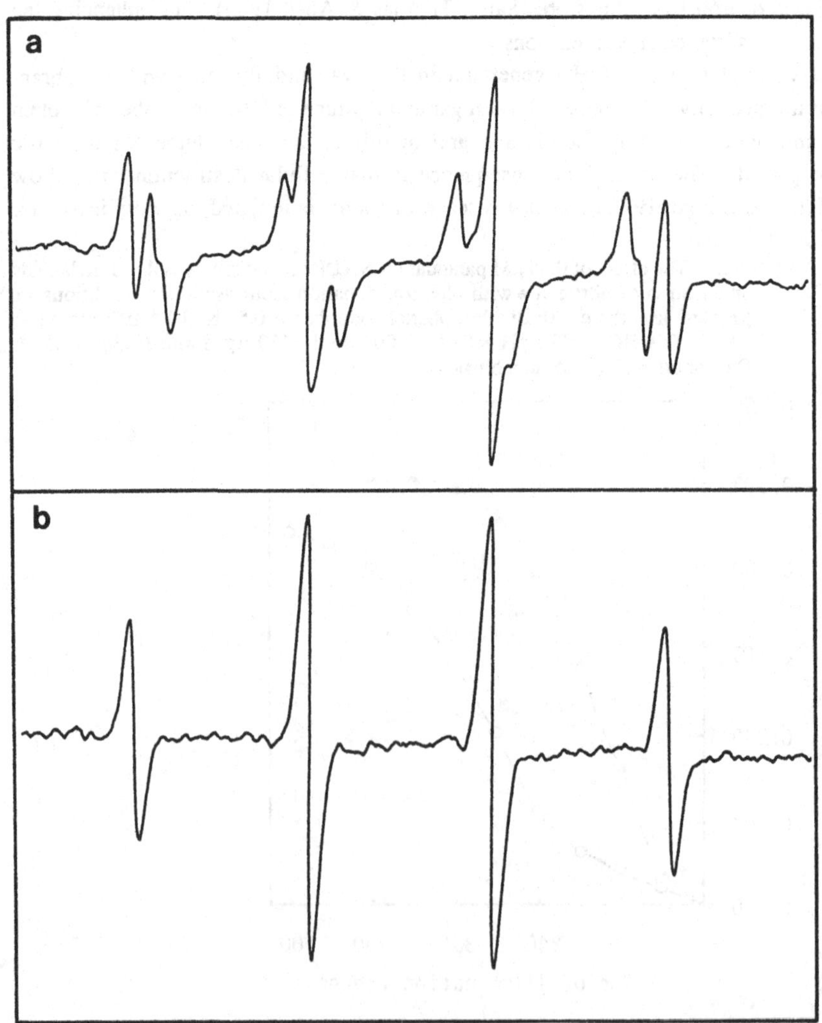

reduced ferredoxin that has not been reoxidized by ferredoxin–NADP+ reductase (Asada & Nakano, 1978). Thus, an endogenous Mehler reaction could account for up to 10–20% of electron flow (Asada *et al.*, 1977; Furbank, 1984), and involve not only the generation of O_2^- by a one-electron donation, but also H_2O_2 by a two-step reaction (Allen, 1975). O_2^- and H_2O_2 are efficiently scavenged, however, if formed under these circumstances by a series of enzymes and quenching systems that exist in the chloroplast (Fig. 6).

The initial event in O_2^- scavenging involves superoxide dismutase enzymes (SOD) that catalyse reaction (i).

$$O_2^- + O_2^- + 2H^+ \xrightarrow{SOD} H_2O_2 + O_2 \qquad \text{(i)}$$

The predominant chloroplast SOD is a Cu–Zn enzyme that is partially bound to the chloroplast thylakoids (Jackson *et al.*, 1979). This enzyme occurs in a ratio of approximately one molecule to 3000 chlorophylls, and is sufficient to maintain the O_2^- concentration at a steady state level of around 6.0×10^{-9} M. The interaction of the bipyridinium herbicides with chloroplast electron flow, probably results in the elevation of superoxide by an uncontrollable 10–20 fold (Asada *et al.*, 1977). O_2^- is also scavenged directly by ascorbate at a rate constant of 2.7×10^5 M^{-1}/s^{-1}, that is present within the chloroplast stroma in a concentration of around 20 mM. Although chloroplasts contain no catalase to remove hydrogen peroxide, there is a series of

Fig. 5. Reaction of superoxide and hydrogen peroxide in the presence of an iron catalyst; a Fenton reaction.

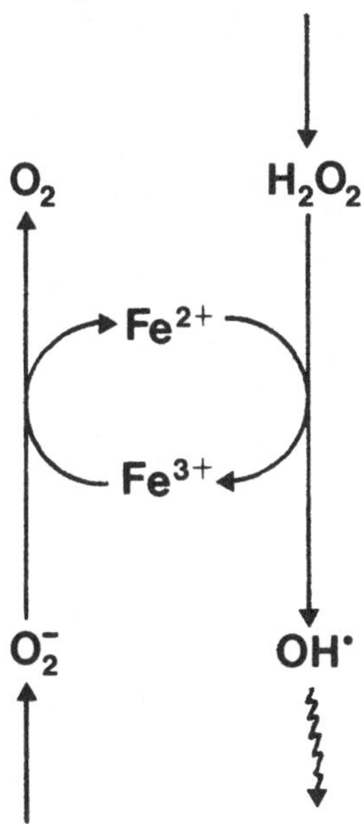

linked enzymes involving the cooperation of both ascorbate and glutathione that is also present in millimolar amounts (Law, Charles & Halliwell, 1983; Asada & Badger, 1984).

This sequence, the 'ascorbate–glutathione' cycle (Fig. 6), involves the enzymes ascorbate peroxidase, dehydroascorbate reductase, and glutathione reductase. If OH• should arise, it is scavenged not only by ascorbate and glutathione, but also by

Fig. 6. Chloroplast scavenging enzymes for superoxide and hydrogen peroxide, incorporating the 'ascorbate–glutathione' cycle. (i) superoxide dismutase (ii) ascorbate peroxidase (iii) dehydroascorbate reductase (iv) glutathione reductase (v) ferredoxin–NADP+ reductase.

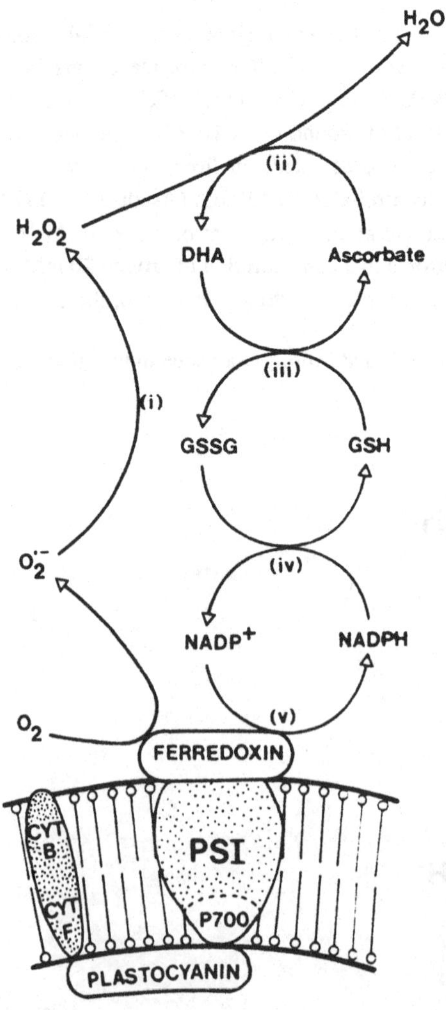

thylakoid membrane bound α-tocopherol. This is present in a ratio of around one molecule to 24 chlorophylls. Other scavenging systems may involve flavonols (Takahama, 1983) and polyamines (Drolet *et al.*, 1986).

In a number of experiments, the activity of paraquat has been shown to be retarded by O_2^- or H_2O_2 scavenging systems, but also enhanced by scavenging enzyme inhibitors. Youngman *et al.* (1979) showed that the action of paraquat was limited in whole leaves if also treated with a superoxide scavenger complex, copper penicillamine. Enzyme induction was demonstrated by sub-toxic levels of paraquat. In *Chlorella sorokiniana* SOD, (Rabinowitch *et al.*, 1983) and ascorbate peroxidase and glutathione reductase in pea leaves (Gillham & Dodge, 1984). In both instances a greater tolerance to higher concentrations of paraquat was achieved. Asada *et al.* (1977), working with maize, showed that SOD activity was higher in younger leaves, and this was correlated with increased tolerance to paraquat. Gillham & Dodge (1987*d*) showed, furthermore, that pea leaves pretreated under higher light intensities contained higher activities of ascorbate peroxidase, dehydroascorbate reductase, glutathione reductase and ascorbate, and this was also linked with an increased tolerance to paraquat. Work by Schmidt & Kunert (1987) showed that it was possible to elevate the endogenous level of leaf ascorbate by exogenous feeding of *Ipomoea purpurea* cotyledons with sodium ascorbate or the precursor L-galactono-1,4-lactone. The elevated level of ascorbate leads to partial protection to paraquat.

Figure 7 shows the enhancing effect of the SOD inhibitor diethyl dithocarbamate

Fig. 7. Chlorophyll bleaching in pea leaf discs incubated on water ●; 1 μM DDTC O; 1 μM paraquat Δ; or DDTC plus paraquat □; (from Gillham D.J., 1986).

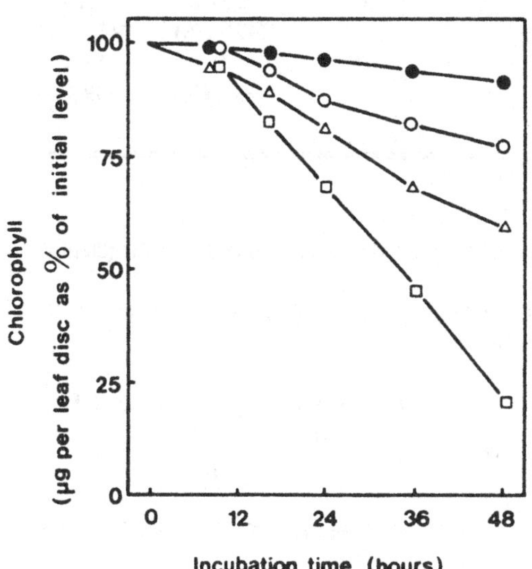

Table 1. *Inhibition of chloroplast electron transport (Hill reaction) by nitro-DPE herbicides and photosynthetic inhibitors, chloroxuron, monuron and atrazine. (From Gillham D.J. & Dodge, A.D., 1987a.)*

Herbicide	I_{50} value for inhibition of the Hill reaction (µM)
Nitrofen	50
Nitrofluorfen	16
Acifluorfen	150–380
Chloroxuron	0.03–0.08
Monuron	0.15–0.63
Atrazine	0.25

Table 2. *The effect of acifluorfen[a] and oxyfluorfen[b] on NADPH oxidation by the ferredoxin:ferredoxin–NADP+ reductase system[c]. (From Gillham D.J. & Dodge, A.D., 1987a.)*

Addition to basic reaction mixture	Rate of NADPH oxidation (nmol min^{-1})		
	Control	Acifluorfen	Oxyfluorfen
Ferredoxin–NADP$^+$ reductase (0.5 unit)	3.78	4.4	4.6
Ferredoxin (150 µg)	1.79	1.62	1.62
Ferredoxin–NADP$^+$ reductase + ferredoxin	13.32	16.01 (+20.2%)	15.86 (+19.1%)
Ferredoxin–NADP$^+$ reductase + ferredoxin + thylakoid membranes	13.25	18.4 (+38.9%)	19.8 (+49.4%)

[a]+25 µM
[b]+25 µM
[c]The basic 3.0 ml reaction mixture contained 0.33 M Tris–HCl buffer and NADPH (80 µM)

Fig. 8. Pathway of NADPH oxidation by ferredoxin and ferredoxin–NADP$^+$ reductase, and the site of DPE interaction.

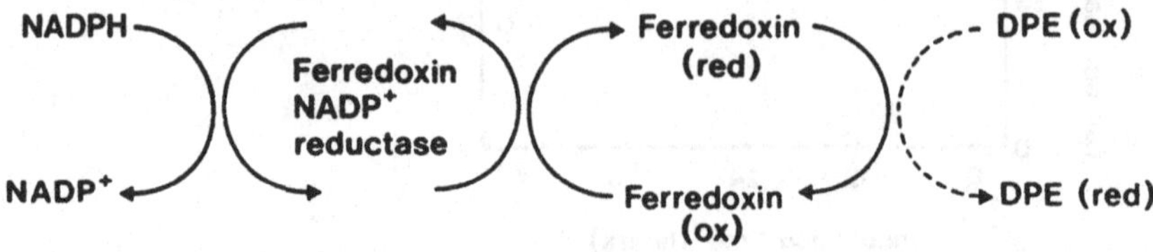

(DDTC) on the action of paraquat on pea leaf discs. DDTC was also shown to be an effective inhibitor of *in vivo* SOD activity (Gillham, 1986). Work with intact isolated chloroplasts showed that on illumination in the presence of paraquat there was a rapid oxidation of both ascorbate (Foyer, Rowell & Walker, 1983) and glutathione (Law, Charles & Halliwell, 1983).

Diphenylether herbicides

The *p*-nitro diphenyl ether herbicides are commonly known as the diphenyl ethers (DPE). Their structure, activity and mode of action has been reviewed recently (Gillham & Dodge, 1987*a*). The first important molecule in this class was nitrofen, discovered in the 1960s and shown by Matsunaka (1969) to be light activated. Although this compound and a number of other related molecules have been shown to inhibit photosynthetic electron flow, their I_{50} is such that this is unlikely to be the primary effect (Table 1). Like the bipyridinium herbicides, oxygen has been shown to enhance herbicidal action (Gillham & Dodge, 1987*b*) and also, as in the case of nitrofluorfen and fomesafen to promote a PSI oxygen uptake (Ridley, 1983), but at a considerably reduced rate. Gillham & Dodge (1987*c*), found that the DPE herbicides acifluorfen and oxyfluorfen failed to catalyse PSI induced oxygen uptake and O_2^- generation.

Although a number of mechanisms have been put forward to account for the light activation of these molecules (Orr & Hess, 1982; Ridley, 1983) there is good evidence that in the initial stages, at least, chloroplast electron flow is required (Böger, 1984; Gillham and Dodge, 1987*b*).

Gillham & Dodge (1987*c*) showed that, in a reconstituted chloroplast system with the addition of ferredoxin and ferredoxin–$NADP^+$ reductase, the addition of either acifluorfen or oxyfluorfen promoted the reoxidation of reduced $NADP^+$ in a dark, diaphorase-type reaction (Fig. 8). Table 2 shows that maximal rates were achieved in the presence of chloroplast thylakoid membranes, even though there was no photochemical reaction. Further experiments (Gillham & Dodge, 1987*c*) showed that lipid peroxidation induced by acifluorfen in isolated thylakoids was enhanced if washed-out ferredoxin was returned. Collectively these results indicate that DPE activation occurs within the vicinity of ferredoxin at the terminal end of PSI and that binding to chloroplast membranes, unlike the bipyridiniums, is a pre-requisite of action. Work by Ridley (1983) showed that the DPE fomesafen attached to chloroplast membranes by ionic interaction, at a site closely associated with the CF_1 coupling factor. Although such interaction did not affect photosynthetic phosphorylation, it could be physically in close proximity to PSI.

The idea of binding to membranes was also implicit in the studies of Draper & Casida (1985) with nitrofen. It is possible that this could follow the initial reduction to a nitro-anion radical (Kunert & Böger, 1981; Lambert, Kroneck & Böger, 1985). According to Draper & Casida (1985) the nitro-anion (nitroso) could bind to a membrane unsaturated fatty acid, giving rise to an alkenylarylhydroxylamine and hence a damaging nitroxide radical (Fig. 9). These workers found convincing evidence for the generation of such radicals using electron spin resonance studies in nitrofen-treated beet leaves.

A role for radicals in the action of DPE herbicides has been implicated in experiments with radical scavengers and in particular α-tocopherol (Orr & Hess, 1982) or α-tocopherol plus ascorbate (Finckh & Kunert, 1985). Work by Tappel (1980) suggested that the effectiveness of α-tocopherol was maintained as a free radical scavenger if the tocopherol radical, produced by radical reactions, was maintained in the reduced state by a pool of ascorbate. Although α-tocopherol is

Fig. 9. Sequence of events following nitro-DPE reduction. Covalent binding to carbon–carbon double bonds of unsaturated lipids and subsequent formation of nitroxide radicals (From Draper W.M. & Casida J.E., 1985).

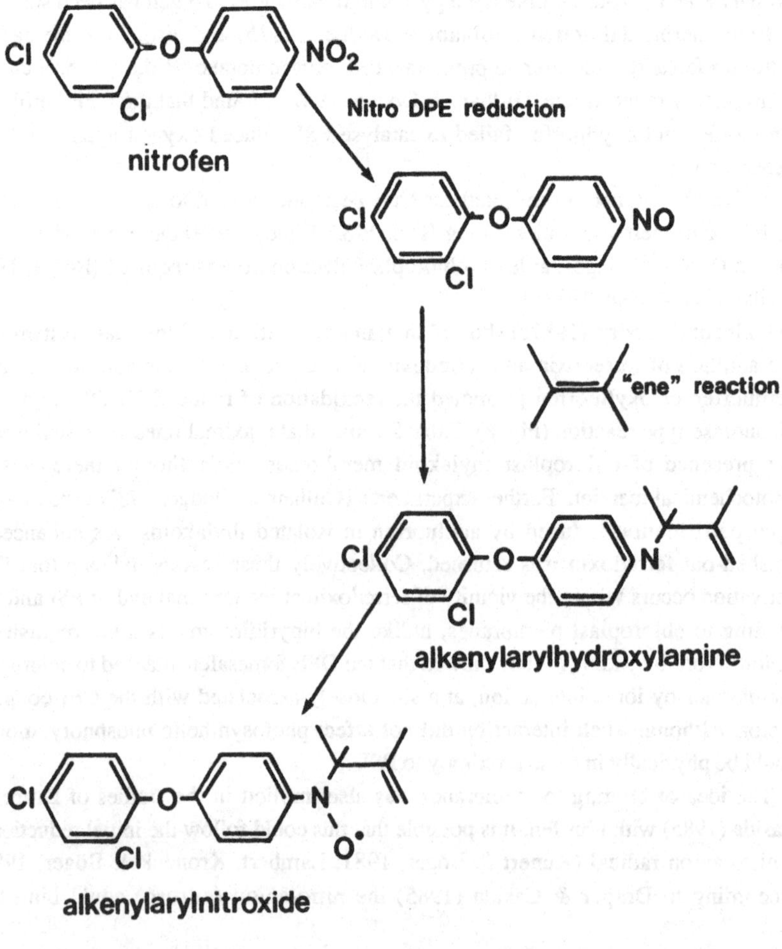

membrane bound, and ascorbate located within the stroma, it is assumed that a thylakoid surface interaction could occur, and that, for optimal scavenging, a ratio of one α-tocopherol to around 15 ascorbate would be most effective (Finckh & Kunert, 1985). In addition, ascorbate itself is an effective radical scavenger (Bodannes & Chan, 1979).

The effect of DPE herbicide treatment on endogenous scavenging systems was demonstrated by Kenyon & Duke (1985). They incubated cucumber cotyledons on acifluorfen, and found a rapid decrease in scavenging enzymes, including superoxide dismutase, dehydroascorbate reductase and glutathione reductase, as well as of ascorbate and glutathione.

General damaging events

The bipyridinium and DPE herbicides mentioned above promote leaf bleaching, as do the photosynthetic electron transport inhibitors as outlined in Chapter 2 (Pallett & Dodge, 1980) and also glutamine synthetase inhibitors such as glufosinate–ammonia (Chapter 8; Köcher, 1983). Whereas with bipyridiniums and DPE herbicides, damage is almost certainly initiated by free radicals, with the other two classes, damage follows the cessation of photosynthesis, and could involve the action of triplet chlorophyll (^{3}Chl), and singlet oxygen ($^1O_2$). These two potentially toxic agents will almost certainly be involved in the later stages of chloroplast thylakoid membrane breakdown, when the photosynthetic machinery has been inactivated.

Photosynthetically inactive chlorophyll, which is nevertheless still capable of absorbing light energy, is sensitized to the singlet state, with a lifetime of 10^{-6}–10^{-8}/s. If unquenched, inter-system crossing, involving spin inversion may occur, to yield the longer lived triplet state (lifetime 10^{-3}/s). Molecular oxygen (triplet oxygen $^3O_2$) can interact with the triplet sensitizer, giving rise to singlet oxygen (equation (ii)) with a lifetime of around 2–4 μs in water.

$$^3\mathrm{Chl} + {}^3O_2 \rightarrow \mathrm{Chl} + {}^1O_2 \qquad \text{(ii)}$$

Both ^{3}Chl and $^1O_2$ can instigate damage in biological molecules, and in particular unsaturated fatty acids of membranes, certain protein amino acids such as histidine, methionine and tryptophan, and guanine of nucleic acids. $^1O_2$ is scavenged by ascorbate and glutathione, and together with ^{3}Chl by carotenoids pigments that are located within the chloroplast thylakoid membranes. These are the subject of the next chapter.

References

Allen, J.F. (1975). A two-step mechanism for the photosynthetic reduction of oxygen by ferredoxin. *Biochem. Biophys. Res. Commun.*, **66**, 36–43.

Asada, K., Takahashi, M.-A., Tanaka, K. & Nakano, Y. (1977). Formation of active oxygen and its fate in chloroplasts. In *Biochemical and Medical Aspects of*

Active Oxygen. Hayaishi, O. & Asada, K. Eds., University Park Press, Baltimore, 45–63.

Asada, K. & Nakano, Y. (1978). Affinity for oxygen in photoreduction of molecular oxygen and scavenging of hydrogen peroxide in spinach chloroplasts. *Photochem. Photobiol.*, **28**, 917–20.

Asada, K. & Badger, M. (1984). Photoreduction of $^{18}O_2$ and $H_2{}^{18}O_2$ with concomitant evolution of $^{16}O_2$ in intact spinach chloroplasts. *Plant Cell Physiol.*, **25**, 1169–79.

Bodannes, R.S. & Chan, P.C. (1979). Ascorbic acid as a scavenger of singlet oxygen. *FEBS Lett.*, **105**, 195–6.

Böger, P. (1984). Multiple modes of action of diphenylethers. *Z. Naturforsch.* Teil C, **39**, 468–75.

Davenport, H.E. (1963). The mechanism of cyclic phosphorylation by illuminated chloroplasts. *Proc. Roy. Soc. B.*, **157**, 332–45.

Dodge, A.D. (1971). The mode of action of the bipyridinium herbicides, paraquat and diquat. *Endeavour*, **30**, 130–5.

Draper, W.M. & Casida, J.E. (1985). Nitroxide radical adducts of nitrodiphenyl ether herbicides and other nitroaryl pesticides, with unsaturated cellular lipids. *J. Agric. Food Chem.*, **33**, 103–8.

Drolet, G., Dumbroff, E.B., Legge, R.L. & Thompson, J.E. (1986). Radical scavenging properties of polyamines. *Phytochem.*, **25**, 367–71.

Farrington, J.A., Ebert, M., Land, E.J. & Fletcher, K. (1973). Bipyridinium quaternary salts and related compounds. V. Pulse radiolysis studies of the reaction of paraquat radical with oxygen. Implications for the mode of action of bipyridinium herbicides. *Biochem. Biophys. Acta*, **314**, 372–81.

Finckh, B.F. & Kunert, K.J. (1985). Vitamins C & E, antioxidative systems against herbicide-induced lipid peroxidation in higher plants. *J. Agric. Food Chem.*, **33**, 574–7.

Foyer, C., Rowell, J. & Walker, D.A. (1983). Measurement of the ascorbate content of spinach leaf protoplasts and chloroplasts during illumination. *Planta*, **157**, 239–44.

Furbank, R.T. (1984). Photoreduction of oxygen in higher plants: *What's new. Plant Physiol.*, **15**, 33–6.

Gillham, D.J. & Dodge, A.D. (1984). Some effects of paraquat on plants. *Plant Physiol.*, **75**, Suppl., 51.

Gillham, D.J. (1986). Aspects pf chloroplast protection against photo-oxidative damage. Ph.D. thesis. University of Bath.

Gillham, D.J. & Dodge, A.D. (1987*a*). The mode of action of nitro-diphenylether herbicides. *Prog. Pestic. Biochem & Toxicol.*, **6**, 147–67.

Gillham, D.J. & Dodge, A.D. (1987*b*). Studies into the action of the diphenyl ether herbicides acifluorfen and oxyfluorfen. 1. Activation by light and oxygen in leaf tissue. *Pestic. Sci.*, **19**, 19–24.

Gillham, D.J. & Dodge, A.D. (1987*c*). Studies into the action of the diphenyl ether herbicides acifluorfen and oxyfluorfen . II. The interaction with photosynthetic electron transport reactions. *Pestic. Sci.*, **19**, 25–34.

Gillham, D.J. & Dodge, A.D. (1987*d*). Chloroplast superoxide and hydrogen peroxide scavenging systems from pea leaves: seasonal variations. *Plant Sci.*, **50**, 105–9.

Haehnel. W. (1984). Photosynthetic electron transport in higher plants. *Ann. Rev. Plant Physiol.*, **35**, 659–93.

Harbour, J.R. & Bolton, J.R. (1975). Superoxide formation in spinach chloroplasts: electron spin resonance detection by spin trapping. *Biochem. Biophys. Res. Comm.*, **64**, 803–7.

Harris, N. & Dodge, A.D. (1972*a*). The effect of paraquat on flax cotyledon leaves: changes in fine structure. *Planta*, **104**, 201–9.

Harris, N. & Dodge, A.D. (1972*b*). The effect of paraquat on flax cotyledon leaves: physiological and biochemical changes. *Planta*, **104**, 210–9.

Jackson, C., Dench, J., Moore, A.D., Halliwell, B., Foyer, C., & Hall, D.O. (1978). Subcellular localisation and identification of superoxide dismutase in the leaves of higher plants. *Eur. J. Biochem.*, **91**, 339–44.

Kenyon, W.H. & Duke, S.O. (1985). Effects of acifluorfen on endogenous antioxidants and protective enzymes in cucumber (*Cucumis sativus* L.) cotyledons. *Plant Physiol.*, **79**, 862–6.

Köcher, H. (1983). Influence of the light factor on physiological effects of the herbicide HOE 39866. *Aspects Appl. Biol.*, **4**, 227–34.

Kunert, K.J. & Böger, P. (1981). The bleaching effect of the diphenylether oxyfluorfen. *Weed Sci.*, **29**, 169–73.

Lambert, R., Kroneck, P.M. & Böger, P. (1985). Radical formation and peroxidative action of phytotoxic diphenyl ethers. *Z. Naturforsch.* Teil C, **39**, 486–91.

Law, M.W., Charles, S.A. & Halliwell, B. (1983). Glutathione and ascorbic acid in spinach (*Spinaca oleracea*) chloroplasts. *Biochem. J.*, **210**, 899–903.

Matsunaka, S. (1969). Acceptor of light energy in photoactivation of diphenylether herbicides. *J. Agric. Food Chem.*, **17**, 171–5.

Mees, G.C. (1960). Experiments on the herbicidal action of 1,1'-ethylene-2, 2'-dipyridylium dibromide. *Ann. Appl. Biol.*, **48**, 601–12.

Orr, G.L. & Hess, F.D. (1982). Mechanism of action of the diphenyl ether herbicide acifluorfen-methyl in excised cucumber (*Cucumis sativus* L.) cotyledons. *Plant Physiol.*, **69**, 502–7.

Pallett, K.E. & Dodge, A.D. (1980). Studies into the action of some photosynthetic inhibitor herbicides. *J. Exp. Bot.*, **31**, 1051–66.

Rabinowitch, H.D., Clare, D.A., Crapo, J.D. & Fridovich, I. (1983). Positive correlation between superoxide dismutase and resistance to paraquat toxicity in the green algae *Chlorella sorokiniana*. *Arch. Biochem. Biophys.*, 225, 640–8.

Ridley, S. (1983). Interaction of chloroplasts with inhibitors. *Plant Physiol.*, **72**, 461–8.

Saito, M., Thomas, C.E. & Aust, S.D. (1985). Paraquat and ferritin-dependent lipid peroxidation. *J. Free Rad. Biol. Med.*, **1**, 179–85.

Schmidt, A. & Kunert, K.J. (1987). Antioxidative systems, defense against oxidative damage in plants, in *Molecular Strategies for Crop Protection* Vol 48, UCLA Symposia on Molecular and Cellular Biology, Arntzen, C. & Ryan C. Eds., Alan R. Liss, New York.

Takahama, U. (1983). Suppression of lipid photoperoxidation by quercetin and its glycosides in spinach chloroplasts. *Photochem. Photobiol.*, **38**, 363–7.

Tappel, A.L. (1980). Measurement of and protection from *in vivo* lipid peroxidation, in *Free Radicals in Biology*, 4 (Pryor, W.A. ed) Academic Press, New York, 1–47.

Warden, J.T. & Bolton, J.R. (1974). Flash photolysis electron spin resonance studies of the dynamics of photosystem I in green plant photosynthesis. I. Effects of acceptors and donors in subchloroplast particles. *Photochem. Photobiol.*, **20**, 251–62.

Youngman, R.J., Dodge, A.D., Lengfelder, E. & Elstner, E.F. (1979). Inhibition of paraquat phytotoxicity by a novel copper chelate with superoxide dismutating activity. *Experientia*, **35**, 1295–6.

Youngman, R.J. & Dodge, A.D. (1979). Mechanism of paraquat action: inhibition of the herbicidal effect by a coper chelate with superoxide dismutating activity. *Z. Naturforsch.* Teil C, **34**, 1032–5.

Zweig, G., Shavit, N. & Avron, M. (1965). Diquat in photoreactions of isolated chloroplasts. *Biochem. Biophys. Acta*, **109**, 332–46.

G. BRITTON, P. BARRY & A.J. YOUNG

Carotenoids and chlorophylls: herbicidal inhibition of pigment biosynthesis

Introduction: bleaching herbicides

Many structurally unrelated bleaching herbicides cause either white or yellow chlorosis of leaves, which is obviously a consequence of the total or partial absence of the normal chloroplast pigments, i.e. chlorophylls and carotenoids. The chlorosis may result from the inhibition of pigment biosynthesis or from the destruction of existing pigment. As a general rule, biosynthesis inhibitors only give rise to chlorosis in newly developing leaves, and are most effective as herbicides when given as a pre-emergence treatment. On the other hand, the inhibition of photosynthesis and photosynthetic electron transport by other herbicides frequently leads to the destruction of existing chloroplast pigments. Although this chapter is intended primarily to discuss the inhibitory effects of herbicides on pigment biosynthesis, these effects can be difficult to distinguish from pigment destruction.

Chlorophylls and carotenoids in photosynthesis

To appreciate the importance of pigment biosynthesis as a target for herbicide activity it is necessary to understand the fundamental roles that the pigments play in normal photosynthesis. The chlorophylls and carotenoids are located specifically in the pigment–protein complexes (PPC) within the thylakoid membranes. Each functional PPC has its own distinctive pigment composition. Thus the reaction centre core complexes of photosystems I and II (PSI and II) have chlorophyll *a* and are enriched in β-carotene whereas the light-harvesting chlorophyll-proteins (LHCP) associated with PSI and PSII contain both chlorophylls *a* and *b* together with the xanthophylls, lutein, violaxanthin and neoxanthin (Siefermann-Harms, 1985). The chlorophylls of antennae or LHCP are obviously essential as the main light-harvesting pigments which pass their excitation energy on to the special chlorophyll-dimers of the reaction centres where charge separation occurs to drive photosynthetic electron transport. The carotenes associated with the reaction centres may play a peripheral role in electron transport (Mathis & Schenck, 1982), but they are particularly important as protective agents against chlorophyll-sensitized photo-oxidation. The carotenoids of the LHCP may also play a similar protective role, in addition to their importance as accessory light-harvesting pigments.

Protective role of carotenoids

A scheme which summarizes in simple terms the protective role of carotenoids is given in Figure 1 (Krinsky, 1971). At high light intensity, the light-harvesting chlorophylls may receive and absorb more light energy than can be dissipated through the normal channels into reaction centres and the electron transport system. Under these saturating conditions the excitation energy of the chlorophylls can be given up in a number of ways, including fluorescence, but part of the excited chlorophyll will undergo intersystem crossing to the triplet state, ^{3}CHL. This longer-lived species can pass its excitation energy on to oxygen to produce the highly reactive singlet oxygen $^1O_2$, which, if unchecked, can bring about the destruction of lipids, membranes and tissues. Carotenoid molecules protect against this photodynamic effect by quenching the excitation energy of ^{3}CHL, and provide a second line of defence by

Fig. 1. Scheme illustrating the photosensitized production of singlet oxygen, $^1O_2$, via triplet chlorophyll, ^{3}CHL and the protective quenching action of carotene (CAR).

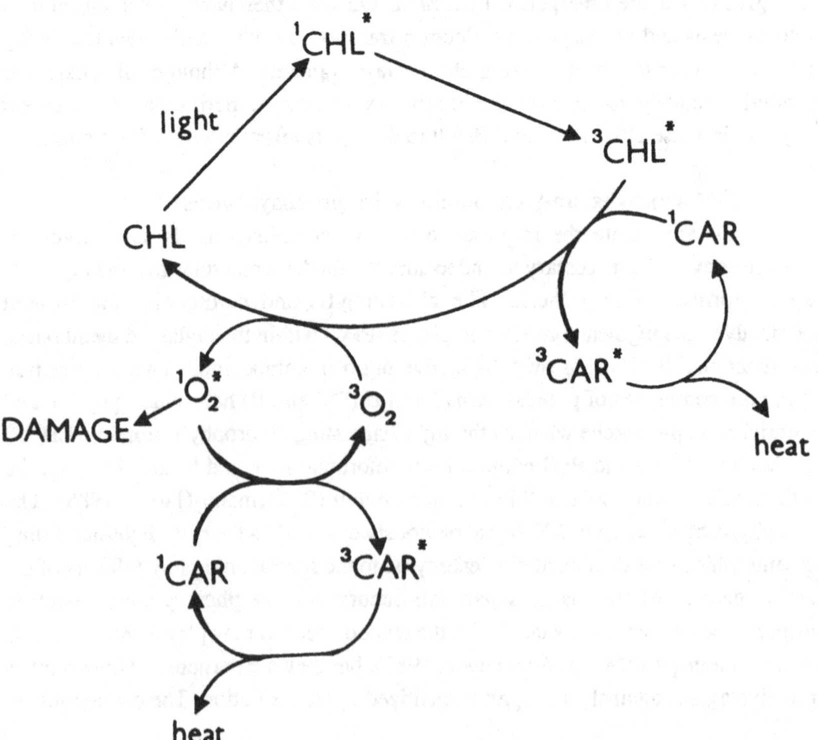

their ability to quench the energy of $^1O_2$ if any should be produced. To be an effective quencher, the carotenoid molecule must have a chromophore of eight or more conjugated double bonds. If the amount of carotenoid is reduced, either by inhibition of its biosynthesis or by destruction, the efficiency of the protective mechanism will be diminished, the probability of damaging amounts of $^1O_2$ being produced will increase, and tissue damage may follow, especially at high light intensity.

Chlorophyll biosynthesis

The main features of the long and complex pathway of chlorophyll biosynthesis (Castelfranco & Beale, 1983) are summarized in Figure 2. Apart from the inhibitory effect of amitrole (Rüdiger & Benz, 1979) and *S*–ethyl dipropylthiocarbamate (Wilkinson, 1985) on the hydrogenation of the C_{20} isoprenoid side–chain, which causes chlorophyll with a geranylgeranyl sidechain to accumulate in place of the normal phytyl ester, virtually all examples of the chemical inhibition of chlorophyll biosynthesis involve effects on the formation or utilization of the key committed intermediate δ-aminolaevulinic acid (ALA). ALA-dehydratase, the enzyme which converts ALA into porphobilinogen, is strongly and competitively inhibited by the structurally related compounds laevulinic acid (Beale, 1978), 4,6-dioxoheptanoic acid (Meller & Gassman, 1981*a, b*) and 4,5-dioxovaleric acid (Porra & Meisch, 1984; Shioi, Doi & Sasa, 1985). These inhibitors have been widely used in studies of chlorophyll biosynthesis and its regulation. Recently a pyridoxal phosphate antagonist, gabaculin (5-amino-1,3-cyclohexadienylcarboxylic acid or 3-amino-2,3-dihydrobenzoic acid) has been reported to inhibit the formation of ALA from glutamate (Jenkins, Rogers & Kerr, 1983; Flint 1984). Gabaculin inhibited chlorophyll formation very effectively in illuminated segments of etiolated barley leaf and in intact plants when supplied via the roots. Carotenoid synthesis was also affected to some extent, probably indirectly. Barley *(Hordeum vulgare),* wheat *(Triticum aestivum),* oats *(Avena sativa),* and rye *(Secale cereale)* were all similarly susceptible to gabaculin, but maize *(Zea mays)* was markedly more tolerant (Hill *et al.*, 1985, 1986; Caiger *et al.*, 1986). In barley, gabaculin was 100-fold more potent than 4,6-dioxoheptanoic acid and 500-fold more potent than laevulinic acid as an inhibitor of chlorophyll biosynthesis.

None of these compounds has found use as a herbicide. All are general inhibitors of porphyrin biosynthesis and, therefore, affect haem, etc as well as chlorophyll.

Reports that other bleaching herbicides inhibit chlorophyll biosynthesis have usually not been substantiated. The decreases seen in chlorophyll levels are most likely to be secondary effects occurring as a consequence of the inhibition of carotenoid biosynthesis and as part of general pigment destruction.

The strategy of applying exogenous ALA to plants is an interesting one. By this means, an important regulatory step is overcome, and high levels of protochlorophyllide accumulate in the dark without concomitant production of apoproteins of the chloroplast PPC. When such plants are then irradiated, the accumulated protochlorophyllide can sensitize photodynamic damage (Rebeiz *et al.*, 1984).

Carotenoid biosynthesis

Pathways, reactions and enzymes

The overall pathway of carotenoid biosynthesis (Fig. 3) is well established (Britton, 1976, 1983) but the enzymes have proved very difficult to study. It is generally believed that carotenoid biosynthesis takes place on a membrane-bound multi-enzyme complex which contains at least the phytoene synthetase, desaturase, cyclase and hydroxylase enzymes. This would explain why cell and tissue disruption so rapidly destroys carotenogenic activity.

Formation of phytoene. In its early stages, carotenoid biosynthesis follows the

Fig. 2 Scheme illustrating chlorophyll biosynthesis; A, summary of overall pathway;

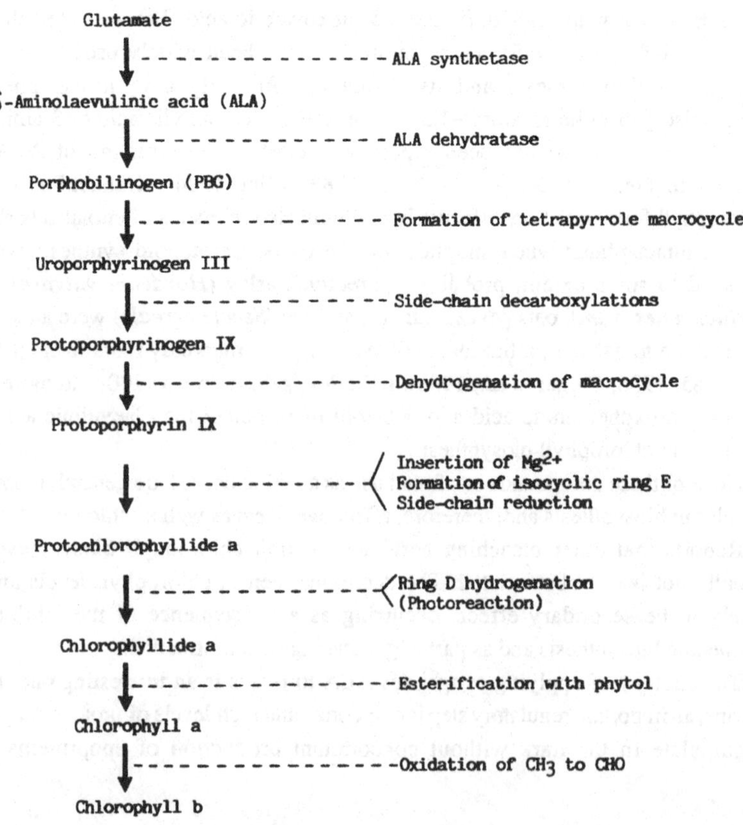

Fig. 2 (contd). B, formation of δ-aminolaevulinic acid (ALA) and porphobilinogen (PBG).

B

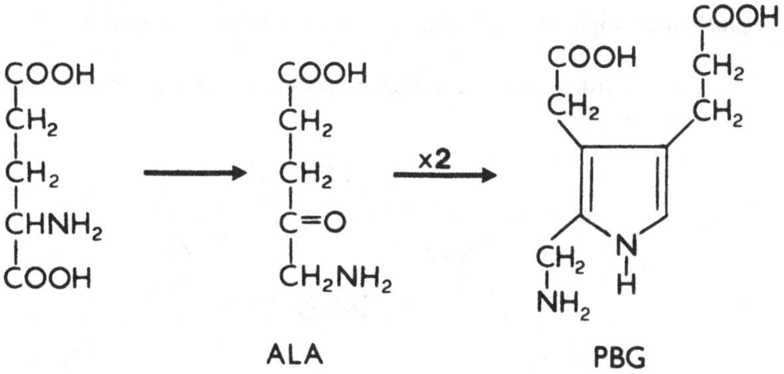

Fig. 3. Summary of the overall pathway of carotenoid biosynthesis.

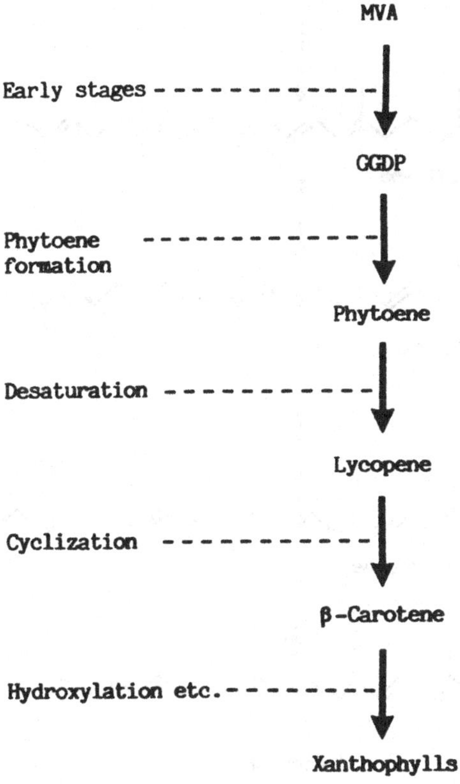

common isoprenoid pathway from mevalonate (MVA) via isopentenyl diphosphate (IDP) to the C_{20} geranylgeranyl diphosphate (GGDP). The first step, which is specific to carotenoids, is the formation of phytoene from two molecules of GGDP, via prephytoene diphosphate. In plants, any phytoene that may accumulate is normally

Fig. 4. Sequence of desaturation reactions in the formation of lycopene from (15-*cis*)-phytoene.

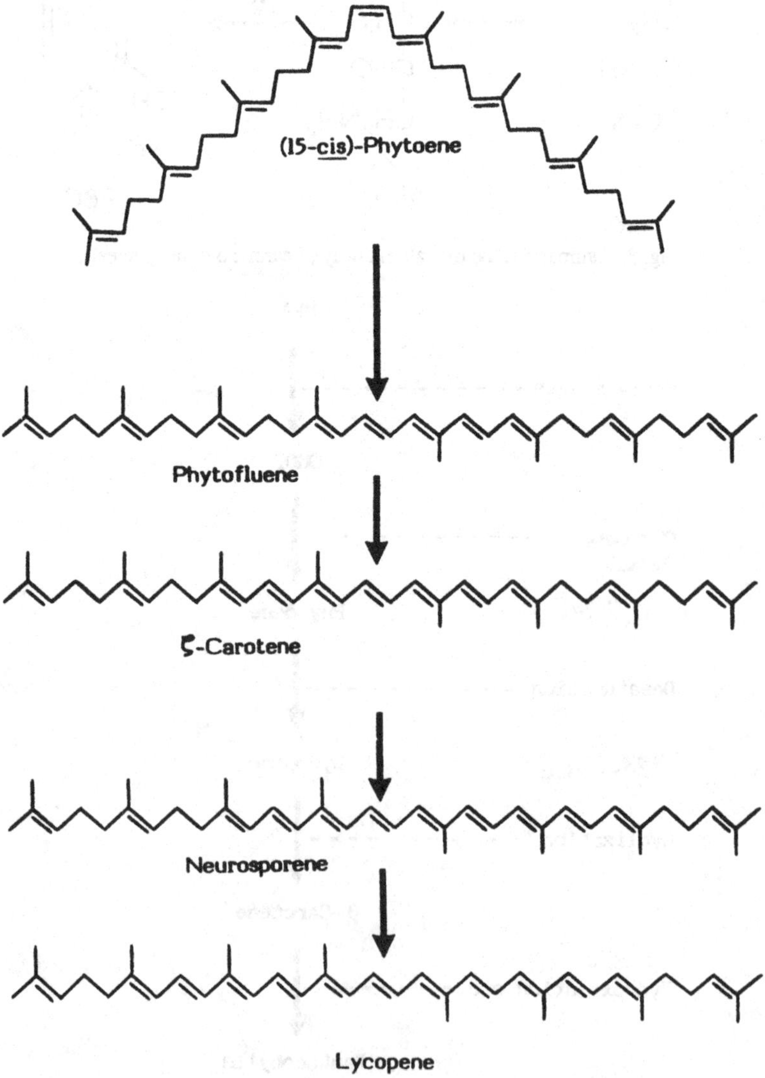

(15-cis)-Phytoene

Phytofluene

ζ-Carotene

Neurosporene

Lycopene

the (15-*cis*) or (15Z)-isomer. Cell-free preparations capable of very efficient synthesis of phytoene from MVA, IDP or GGDP have been obtained from a number of plant and microbial sources. Phytoene synthetase is a peripheral component of the multienzyme complex and is easy to extract by mild detergent solubilization. Metal ions (Mg^{2+}or Mn^{2+}) are the only requirement for its action, though ATP and detergent can be strongly stimulatory (Islam *et al.*, 1977).

Desaturation. The conversion of phytoene into the normal coloured carotenoids requires a series of desaturation reactions. In each of these, a further double bond is introduced by the *trans* elimination of two hydrogen atoms (McDermott, Britton & Goodwin, 1973), and the chromophore is extended by two conjugated double bonds. In higher plants, the desaturations occur alternately in the two halves of the molecule and give successively, phytofluene, ζ-carotene, neurosporene and finally lycopene (Figure 4). It is assumed that isomerization about the C-15,15' double bond takes place early in this sequence. Although crude cell-free preparations can be obtained which can incorporate early precursors (MVA, IDP or GGDP) into lycopene, β-carotene, etc, and must therefore retain desaturase activity, it is very difficult to isolate the individual enzymes from these preparations, and direct desaturation has been very difficult to achieve. There are, however, early reports of the incorporation of radioactivity from phytoene, phytofluene and ζ-carotene into later unsaturated carotenes by crude preparations from tomato chromoplasts (Porter & Spurgeon, 1979). The involvement of a closely associated simple electron transport system of the cytochrome P450 type in the desaturations is suspected (Britton, 1979).

Cyclization. Lycopene is assumed to be the normal immediate acyclic precursor which is cyclized (Figure 5), but any carotenoid in which one half of the molecule has reached the lycopene level of desaturation is a potential substrate. The normal chloroplast carotenoids have cyclic β- and ε-end-groups. The mechanism and stereochemistry of the formation of these rings are illustrated in Figure 6 (Britton, 1985). The β-ring and ε-ring are formed independently and cannot be interconverted. The direct cyclization of lycopene to give γ-carotene, β-carotene, etc can be achieved comparatively easily with crude, detergent-solubilized preparations of fungi (Bramley, Aung Than & Davies, 1977), bacteria (M. Browne and G. Britton, unpublished results) and plant chromoplasts (Kushwaha *et al.*, 1969; Camara & Dogbo, 1986). There appear to be no specific cofactor requirements.

Introduction of oxygen functions. Hydroxylation and epoxidation occur as the final stages of xanthophyll biosynthesis, e.g. lutein is formed by hydroxylation of α-carotene (Figure 7); zeaxanthin, violaxanthin and neoxanthin are derived from β-carotene. The hydroxylation requires molecular O_2 and apparently occurs by a mixed-function oxidase reaction. Direct hydroxylation of carotenes is difficult but has been achieved with crude membrane preparations of a *Flavobacterium* (Brown, Britton &

Fig. 5. Pathway for the formation of cyclic carotenes from lycopene.

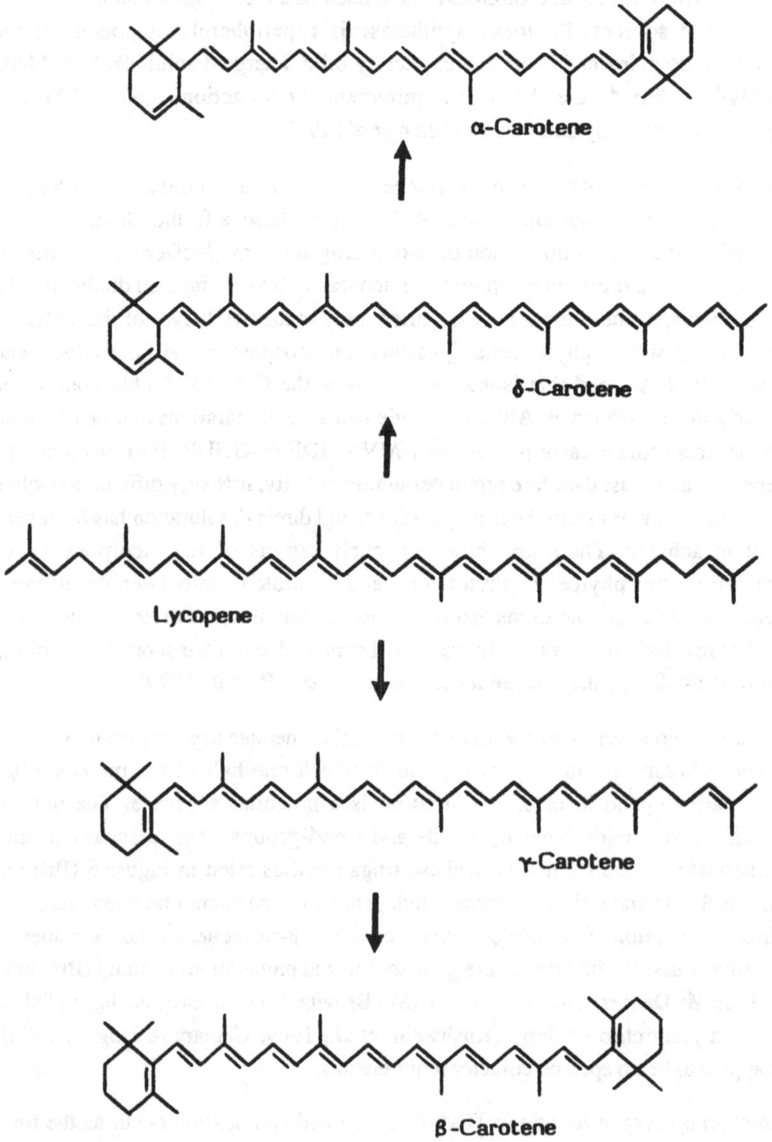

α-Carotene

δ-Carotene

Lycopene

γ-Carotene

β-Carotene

Fig. 6. Mechanism and stereochemistry of formation of the β-and ε-rings in carotenoid biosynthesis.

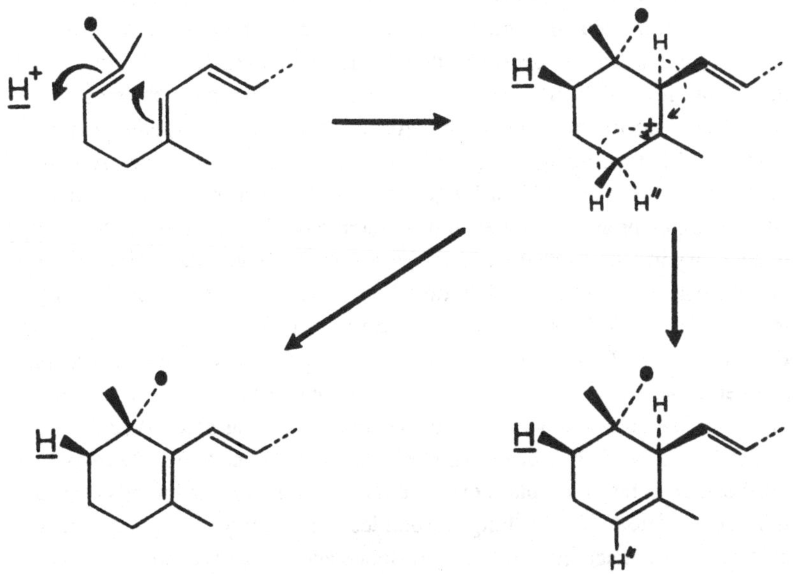

Fig. 7. Formation of the xanthophyll, lutein, by direct hydroxylation of α-carotene.

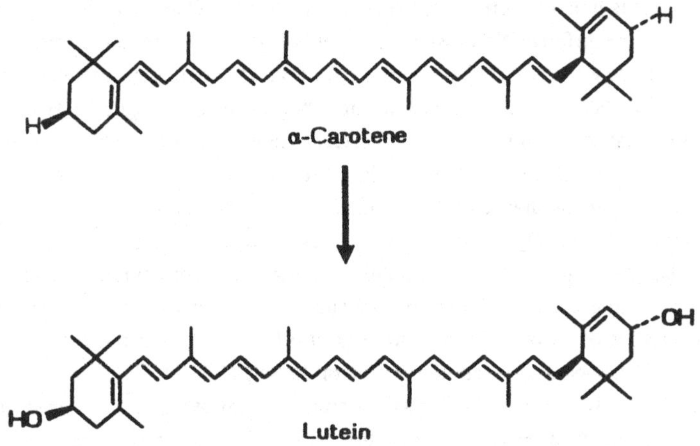

Goodwin, 1975) and a cyanobacterium, *Aphanocapsa* (Sandmann & Bramley, 1985*a*).

Carotenoid biosynthesis in the chloroplast

Carotenoid biosynthesis is an integral part of the construction of the pigment–protein complexes in the thylakoids, and its regulation is closely linked to that of the formation of other components, e.g. chlorophyll, proteins, lipids. The carotenoids that accumulate in the chloroplast are certainly biosynthesized within that organelle. There are conflicting views, however, about whether early biosynthetic intermediates such as MVA or IDP need to be imported (Kreuz & Kleinig, 1981) or whether the chloroplast is completely autonomous (Grumbach & Bach, 1979). There is also dispute about where within the chloroplast the later stages of carotenoid biosynthesis take place. Thus in radish (*Raphanus sativus*) carotenoid biosynthesis, including phytoene formation, was restricted to the thylakoid fraction (Grumbach & Britton, 1984). This contrasts with the situation in spinach, where phytoene synthetase and desaturase activities were found in the envelope fraction but the thylakoid fraction was devoid of activity (Lütke-Brinkhaus *et al.*, 1982).

Very little work has been reported on the carotenogenic enzymes and their organization in the chloroplast, even though this is the system directly relevant to the inhibitory action of bleaching herbicides. The carotenogenic enzymes of the chloroplast are much more difficult to isolate and study than those of chromoplasts, presumably because they are more tightly associated with membrane structures.

Regulation. The overall story of carotenoid biosynthesis and its regulation in chloroplasts is thus very complicated (Britton, 1986). Carotenoids are biosynthesized in at least three different circumstances: (i) in bulk during the initial construction of the photosynthetic apparatus, (ii) in mature chloroplasts, when some synthesis continues as part of turnover, and (iii) as part of a response to changes in environmental conditions, especially light intensity. Different regulatory factors may be involved in these different situations. Also, when dark-grown plants are illuminated, the enzymes that were responsible for carotenoid biosynthesis in the etiolated plants may continue to function, especially in the early stages, before new enzyme synthesis and chloroplast development become established. Nor is chloroplast development likely to follow the same course universally. Substantial differences may be expected in the characteristics of carotenoid biosynthesis in relation to chloroplast development in different plant species or even in varieties of the same species. Factors such as the age of the plant or tissue, and the growth conditions employed, are very important, and cotyledons and true leaves, even of the same individual plant, may behave quite differently. It is also clear that the course of chloroplast development, and therefore its associated carotenoid biosynthesis, is not the same in a greening etiolated system as it is in plants growing normally in the light, so extrapolations and comparisons must be made with extreme caution.

Because the carotenoids are localized and function in different pigment–protein complexes in the chloroplast, the factors which determine the different carotenoid compositions of these complexes must also be understood. The time-course of xanthophyll synthesis appears not to be the same as that of bulk carotene synthesis in the same developing system (Grumbach, 1979). It is known that β- and ε-rings are not interconverted (Williams, Britton & Goodwin, 1967), so in the final stages lutein synthesis must be independent of that of β-carotene and the other xanthophylls. The stage at which the two pathways diverge, however, is not known, i.e. whether lutein and β-carotene are made *de novo* by two distinct enzyme assemblies, or whether the divergence involves only the final cyclization and hydroxylation reactions, after phytoene synthesis and desaturation have taken place on a common enzyme assembly. Double-labelling experiments suggest, however, that the carotenoid molecules destined for the PPC of stacked and unstacked thylakoids, for example, are not made at separate sites (Sergeant & Britton, 1984). It is not known what determines how a carotenoid molecule is directed towards its specific PPC site within the thylakoid.

Carotenogenic enzymes as a target for herbicide action

Any substance that affects the early stages of carotenoid biosynthesis (before phytoene synthesis) would presumably have an unspecific effect on the formation of all isoprenoids, both inside and outside the chloroplast. Phytoene synthetase is a likely control point in the biosynthetic pathway, but its specific inhibition would probably be difficult to achieve, since the reaction mechanism is so similar to that of the formation of the sterol precursor, squalene. The desaturase enzymes appear to be the best target for herbicide action. The desaturation is specific to carotenoid biosynthesis and relatively easy to disrupt. Its inhibition results in the accumulation of phytoene and other intermediates which have short chromophores and cannot protect against photo-oxidation, so the plants will rapidly be killed by light and oxygen. Inhibition of cyclization causes lycopene to accumulate in place of the normal cyclic carotenoids. In principle, the long chromophore of lycopene should protect efficiently against photo-oxidative damage, but lycopene does not do so *in vivo*; presumably it cannot be fitted into the thylakoid PPC at the correct sites and in the same orientation as β-carotene and lutein. Thus seedlings treated with CPTA [2-(4-chlorophenylthio)-triethylammonium chloride] accumulate lycopene and are initially red but later they bleach and die. Herbicides which block carotenoid cyclization specifically might be slower to act than the desaturation inhibitors, but they would give the same eventual result and could be very specific. If hydroxylation were inhibited the corresponding carotenes would accumulate but it is likely that they would not be able to occupy the same specific sites as the xanthophylls and maintain the PPC structures. However, any substance that inhibits carotenoid hydroxylation would probably not be specific and would also affect other oxidative processes such as the formation of squalene epoxide and sterols.

Herbicides that affect carotenoid compositions

Inhibitors of carotenoid biosynthesis

Some herbicides which act by blocking carotenoid biosynthesis have been studied extensively. Chief among these are the pyridazinones metflurazon (SAN 6706) and norflurazon (SAN 9789) which, when given as pre-emergence treatments, inhibit carotenoid biosynthesis in the developing seedlings. These compounds specifically block desaturation and cause the accumulation of phytoene (Bartels & McCullough, 1972; Ben-Aziz & Koren, 1974; Eder, 1979).

Similar effects have been reported for other herbicidal compounds from widely differing structural groups, e.g. fluridone (Bartels & Watson, 1978), difunone (Urbach, *et al.*, 1976), substituted 4-hydroxypyridines (Ridley, 1982) and substituted diphenyl ethers (Lambert & Böger, 1983).

Other substances, including dichlormate (Burns *et al.*, 1971) the substituted 6-methylpyrimidines J334, J739 and J852 (Ridley, 1982) and a dihydropyrone, LS76-1243, (Vial & Borrod, 1984) also inhibit desaturation, but their action is rather different because it is ζ-carotene and not phytoene that accumulates. Since all four desaturations occur by the same reaction mechanism, this difference in behaviour is presumably related to the organization of the multienzyme complex. The widely used aminotriazole (amitrole) also inhibits carotenoid biosynthesis but its action seems not to be uniform. In wheat it has been reported to inhibit desaturation so that phytoene accumulates, together with some phytofluene and ζ-carotene (Burns *et al.*, 1971). Another report, however, shows that in wheat (Rüdiger *et al.*, 1976) as well as in maize and radish seedlings (Guillot-Salomon, Douce & Signol, 1967; Grumbach, 1981) the main effect is on cyclization so that lycopene accumulates. However, as shown below (Figure 10) a mixture of intermediates is normally present in amitrole-treated plants, indicating a less specific effect than that of other herbicides.

Many other compounds, especially amine derivatives, have been shown to have significant effects on the carotenoid compositions of fruit, especially grapefruit. They cause, for example, the accumulation of lycopene or formation of poly*cis*-carotenoids (Yokoyama *et al.*, 1982). Apart from CPTA these compounds, described as chemical regulators, do not seem to have been examined for effects on green plants or for possible herbicidal activity.

Strategies for evaluating carotenoid biosynthesis herbicides

Model enzyme systems. It is obviously important to understand the mechanisms by which bleaching herbicides inhibit the biosynthesis of the normal chloroplast carotenoids. The logical conclusion is that the herbicides directly inhibit the activity of the carotenogenic enzymes. Bramley, Sandmann & coworkers have therefore adopted the strategy of studying the inhibitory effects of herbicides on carotenogenesis in cell-free enzyme systems, and have used crude cell extracts of the mould *Phycomyces blakesleeanus* and the cyanobacterium *Aphanocapsa* as models. Fluridone, difunone, J852, and a series of N-alkylphenoxybenzamides were among the compounds that caused the incorporation of IDP or MVA by cell extracts of *Phycomyces* strain C115 to be diverted into phytoene instead of into β-carotene, thereby confirming a direct inhibitory effect on the desaturase enzymes (Bramley *et al.*, 1984). Similarly, [^{14}C]-GGDP was incorporated into phytoene instead of into the normal cyclic carotenoids when it was incubated with a cell-free system from *Aphanocapsa* in the presence of several phenoxybenzamides and substituted 2-phenylpyridazinones (Sandmann *et al.*, 1984; Clarke *et al.*, 1985). Kinetic studies showed that 3-(2,5-dimethylphenoxy)-N-ethylbenzamide is a non-competitive inhibitor of the phytoene desaturase complex. This work has now been extended to include incubations with a coupled enzyme system containing a crude cell extract of an appropriate mutant strain of *Phycomyces*, to provide the enzymes for the synthesis of phytoene or β-carotene, together with membranes of *Aphanocapsa* which were able to metabolize these intermediates further to xanthophylls (Bramley & Sandmann, 1985; Sandmann & Bramley, 1985 *a,b*). The hydroxylation was blocked to some extent by CO and by the mono-oxygenase inhibitors tetcyclacis and LAB117682 (Sandmann & Bramley, 1985*a*).

Even in these model systems, however, carotenoid biosynthesis is a complicated process, requiring a properly organized enzyme system. There are many different facets of each reaction which could be affected by a herbicide and the same eventual symptoms of inhibition could be produced by completely different mechanisms. Thus, as an alternative to blocking directly the active site, an inhibitor could prevent synthesis of the carotenogenic enzymes or their organization into a multienzyme complex in the membrane, interfere with the organization or functioning of the electron transport system that is believed to be associated with the desaturase enzymes, prevent the isomerization that would be required in the conversion of (15*Z*)-phytoene into the normal (all-*E*)-carotenes, or restrict the availability of substrates or cofactors. Carotenoid biosynthesis in the developing chloroplast would be even more complicated; additional factors such as uptake of the substrates into the chloroplast, the incorporation of the carotenoids into the thylakoid PPC, and the availability of the other components of the PPC could also be affected.

A screening programme with model carotenogenic enzyme systems must therefore be flexible enough to detect the inhibitory properties of any compound, even though its action may not be directly on the active site of an enzyme.

Screening by hplc. A powerful alternative strategy is to employ a rapid, reliable and sensitive method to screen the biochemical effects of the herbicide on the plants themselves, especially if this procedure is versatile enough to distinguish immediately direct effects on biosynthesis from pigment destruction (see below) and to provide information about the biochemical mechanisms of either. Advances in high-performance liquid chromatography (hplc) instrumentation and procedures, coupled with increased knowledge and understanding of the analysis of carotenoids, and their distribution, roles and possible transformations within the chloroplast, now make such screening feasible as a routine procedure. The amount of information that can be obtained is enormous, though care and experience are needed in its interpretation. When correctly applied, this procedure will not only reveal immediately any direct inhibitory effect on carotenoid biosynthesis but also identify which stage in the biosynthesis is inhibited, indicate the specificity of the inhibition and even provide information about the effects on individual carotenoids and the different functional PPC, thereby allowing the physiological and biochemical consequences of the effect to be assessed and predicted.

Figure 8 shows a chromatogram obtained by reversed phase hplc of an extract of normal green barley seedlings. Chromatograms illustrated in Figures 9–11 were obtained in a similar way from extracts of seedlings that had been treated with known inhibitors of carotenoid biosynthesis, namely norflurazon, amitrole and CPTA, respectively. The differences are striking. In the case of the norflurazon-treated plants (Figure 9), only small amounts of chlorophylls were present and the normal cyclic

Fig. 8. Reversed phase hplc chromatogram of a normal leaf extract. Column: Zorbax-ODS, 5 µm, 25 x 0.46 cm. Solvent: linear gradient 0–100% ethyl acetate in acetonitrile-water (9:1). Flow rate: 1 ml/min. Monitoring wavelength 445 nm. Identifications: A-neoxanthin; B-violaxanthin; C-lutein-5,6-epoxide; D-antheraxanthin; E-lutein; F-zeaxanthin; G,G'-chlorophyll *b;* H,H'-chlorophyll *a;* J-α-carotene; K-β-carotene.

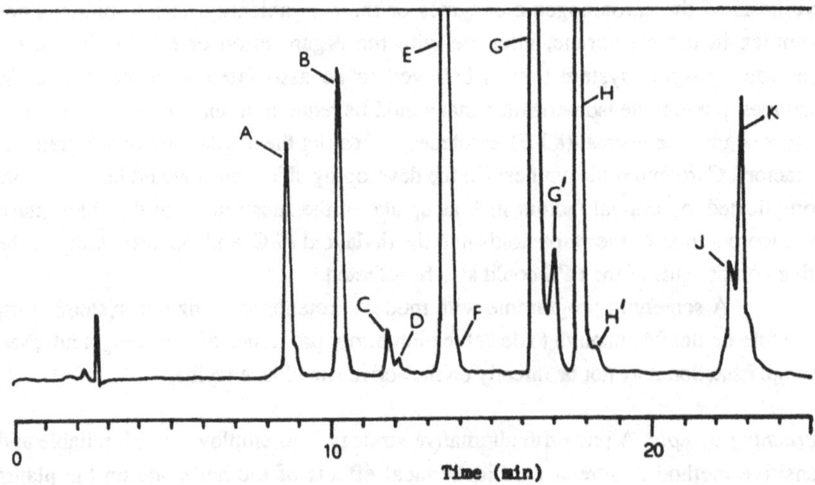

carotenoids were completely absent, but phytoene was present in large amounts, accompanied by a trace of phytofluene. This is the typical pattern expected for a compound which inhibits the desaturation of phytoene very specifically. Interestingly, two derivatives of phytoene with the polarity of monohydroxy and dihydroxy compounds were also detected, the former in substantial quantities. The presence of unusual substances such as these may be particularly valuable for distinguishing subtle differences in the modes of action of different herbicidal compounds. In the case of amitrole, the inhibition was incomplete and less specific. Several intermediates, especially phytoene and lycopene, were present in substantial quantities (Figure 10), showing that desaturation and cyclization were both affected.

Fig. 9. Reversed phase hplc chromatogram of an extract of leaves of barley grown in the presence of norflurazon. Hplc conditions as in Figure 8. Monitoring wavelengths: a, 287 nm; b, 445 nm. Identifications: A–K, as in Figure 8; L-phytoene; M–N-oxy-derivatives of phytoene.

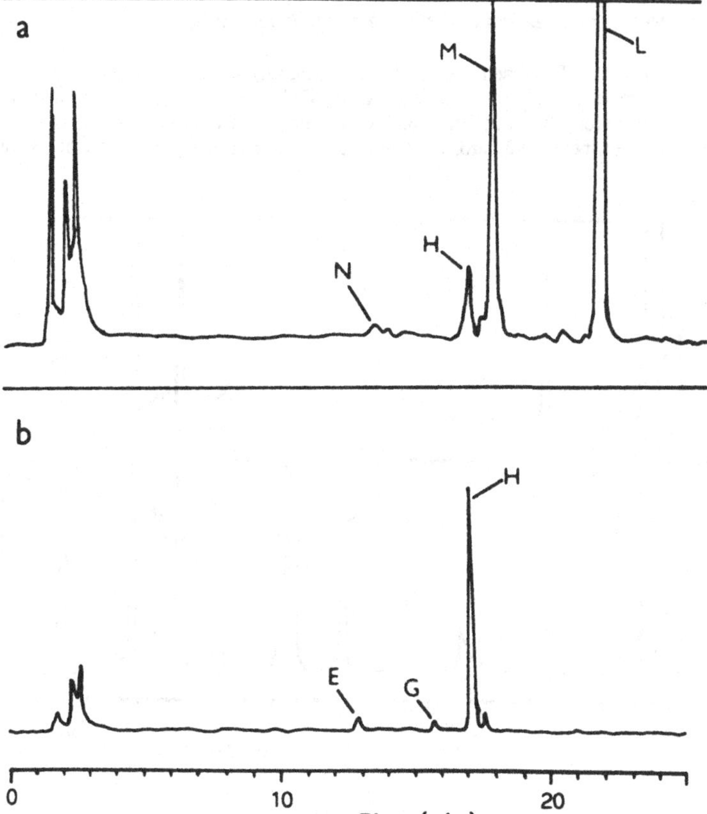

Although CPTA is not used as a herbicide, its effects are typical of those expected of any inhibitor of carotenoid cyclization (Figure 11). The amounts of cyclic carotenes and xanthophylls were greatly reduced, and lycopene was the main pigment. However, several other interesting carotenoids were present in substantial amounts. One of these, δ-carotene, contains an ε-ring and must therefore have accumulated in place of the normal ε-ring carotenoid, lutein. Only traces of the corresponding β-ring carotenoid, γ-carotene, were detected, showing that cyclization to form a β-ring was more strongly inhibited than that to form an ε-ring, and also that inhibition of the biosynthesis of β-carotene and its xanthophyll derivatives violaxanthin and neoxanthin, was greater than that of lutein. No oxygenated derivatives of the monocyclic hydrocarbons γ-carotene and δ-carotene were detected, showing that the hydroxylation and epoxidation reactions that occur after cyclization could not proceed unless both rings were present. The presence of substantial amounts of hydroxy-derivatives of lycopene, however, shows that, if cyclization is inhibited, hydroxylation of accumulated acyclic intermediates can occur, and may provide valuable information relevant to the organization of the carotenogenic enzymes.

Fig. 10. Reversed phase hplc chromatogram of an extract of leaves of barley grown in the presence of amitrole. Hplc conditions as in Figure 8. Monitoring wavelengths: a, 287 nm; b, 450–470 nm. Identifications: A–N, as in Figures 8,9; P-lycopene; Q-δ-carotene; R-γ-carotene; S-phytofluene (detected at 350 nm).

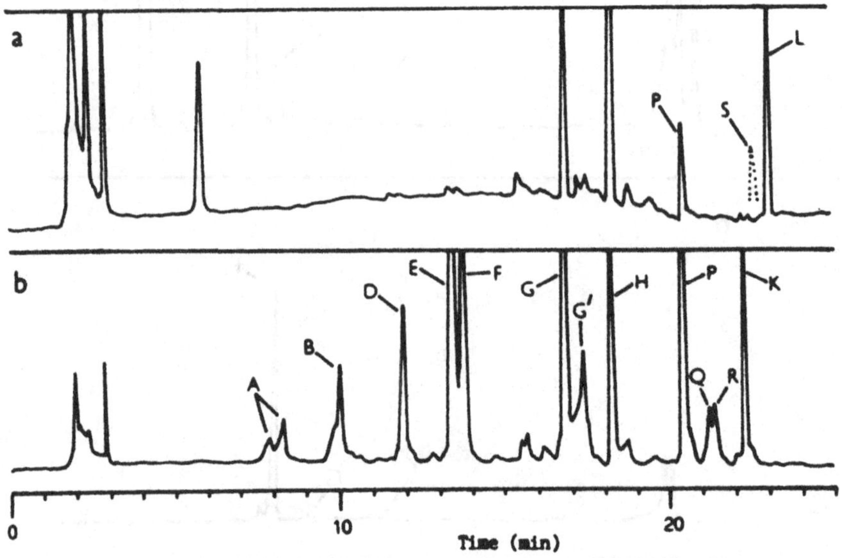

These examples illustrate how clearly a single hplc assay reveals the effect of a herbicide and also the extent, complexity and subtlety of the information that can be obtained about its action.

Pigment destruction as a consequence of herbicide action

Although this discussion deals primarily with pigment biosynthesis and its inhibition by herbicides, similar symptoms of chlorosis can be produced by other herbicides whose action leads to the destruction of existing pigments. For example, bleaching is a long-term result of the action of PSII herbicides such as diuron and monuron (Stanger & Appleby, 1972; Ridley, 1977, 1982; Pallett & Dodge, 1979) and of bipyridylium herbicides, e.g. paraquat, which affect photosynthetic electron transport around PSI (Harris & Dodge, 1972; Youngman & Dodge, 1979). Similar effects have been described more recently for diphenylether herbicides, including nitrofen and oxyfluorfen (Kunert & Böger, 1981) (See chapter 3).

Pigment destruction of this kind is not solely a feature of herbicide action, however. Similar effects can be obtained simply by exposing isolated chloroplasts or thylakoids to strong light. It appears that many factors, including the tissue disruption necessary in the preparation of chloroplasts and thylakoids, render the pigments much more susceptible to oxidative destruction either in the presence or absence of herbicides. A possible explanation is that, in the intact plant, all the protective

Fig. 11. Reversed phase hplc chromatogram of an extract of leaves and cotyledons of soya bean grown in the presence of CPTA. Hplc conditions as in Figure 8, monitoring wavelength 470 nm. Identifications: A–S, as in Figures 8–10; T-oxy-derivatives of lycopene.

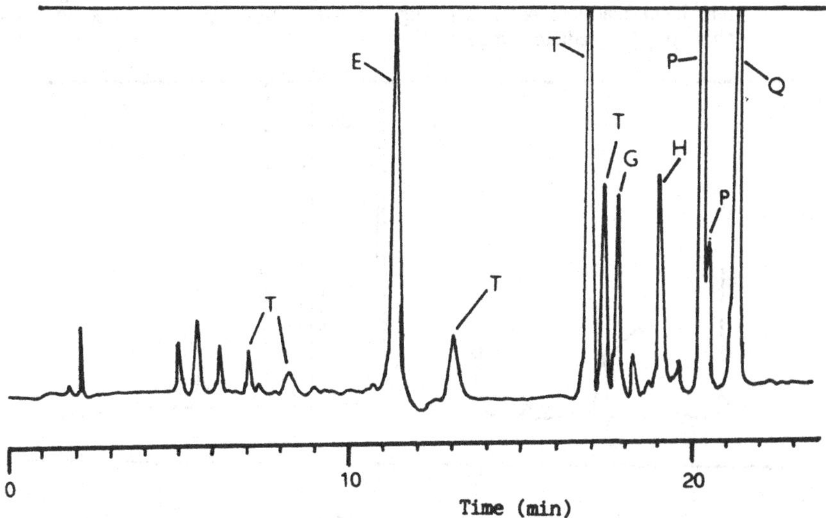

defences, including carotenoids, tocopherol, ascorbate, superoxide dismutase and catalase, act in concert to deal with the whole range of oxidizing species ($^1O_2$, OH•, O_2^-, H_2O_2). When the tissue is disrupted, this concerted action is no longer possible, so that the oxygen radicals become available to initiate lipid peroxidation and carotenoid destruction. It is therefore risky to extrapolate effects of herbicides from isolated chloroplasts or thylakoids to intact plants. It is worth noting that any herbicide or other factor that affects the lipid composition of thylakoids or PPC may alter the susceptibility of the thylakoid or plant to pigment destruction by photosynthesis inhibitor herbicides. Much work is needed before the mechanisms of these herbicidal effects which indirectly cause pigment bleaching are understood.

However, the hplc screening procedure described earlier will immediately distinguish between bleaching of existing pigments and inhibition of biosynthesis, and can give much important information about the bleaching events. When a typical chromatogram of an extract from barley cotyledons that have been treated with paraquat (Figure 12) is compared with that of an untreated leaf extract (Figure 8), substantial and characteristic differences can be seen, particularly the large decrease in the β-carotene content and the appearance of β-carotene-5,6-epoxide. It is particularly useful to follow the time-course of pigment changes, since the same eventual symptoms of chlorosis follow from different primary events and mechanisms. Changes in the pigment compositions of an intact plant or leaf after treatment with these photosynthesis inhibitors occur more slowly and may not be the same as those seen with isolated chloroplasts or thylakoids.

Fig. 12. Reversed phase hplc chromatogram of an extract of barley chloroplasts treated with paraquat. Hplc conditions as in Figure 8. Monitoring wavelength 445 nm. Identifications: A–T, as in Figures 8–11; U-β-carotene-5,6-epoxide; V-chlorophyll degradation product.

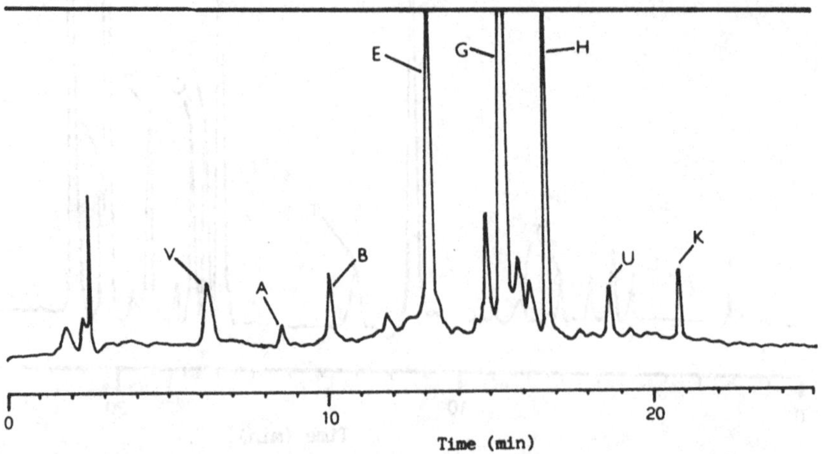

Conclusions

Increasing knowledge about carotenoid biosynthesis and about the protective role of carotenoids in photosynthesis and the way this can be overcome by oxidizing conditions, together with the precise analytical methods now available make it possible for effects of herbicides on carotenoid biosynthesis and carotenoid destruction easily to be distinguished and characterized. The proper understanding of the effects or mechanism of action of any bleaching herbicide is a goal that could now be achieved.

Acknowledgements

Much of the information in this paper, especially about hplc analysis and the effects of herbicides on carotenoid compositions, is taken from our own unpublished work. The financial support of SERC and BAT Industries Ltd, Southampton, is gratefully acknowledged.

References

Bartels, P.G. & McCullough, C. (1972). A new inhibitor of carotenoid synthesis in higher plants: 4-chloro-5-(dimethylamino)-2-(α,α,α-trifluoro-*m*-tolyl)-3(2H)-pyridazinone (Sandoz 6706). *Biochemical and Biophysical Research Communications*, **48**, 16–22.

Bartels, P.G. & Watson, C.W. (1978). Inhibition of carotenoid synthesis by fluridone and norflurazone. *Weed Science*, **26**, 198–203.

Beale, S.I. (1978). δ-Aminolaevulinic acid in plants: its biosynthesis, regulation and role in plastid development. *Annual Review of Plant Physiology*, **29**, 95–120.

Ben-Aziz, A. & Koren, E. (1974). Inteference on carotenogenesis as a mechanism of action of the pyridazinone herbicide Sandoz 6706: accumulation of C_{40} carotenoid precursors, inhibition of β-carotene synthesis and enhancement of phytoene epoxidation. *Plant Physiology*, **54**, 916–920.

Bramley, P.M., Aung Than & Davies, B.H. (1977). Alternative pathways of carotene cyclization in *Phycomyces blakesleeanus*. *Phytochemistry*, **16**, 235–8.

Bramley, P.M., Clarke, I.E., Sandmann, G. & Böger, P. (1984). Inhibition of carotene biosynthesis in cell extracts of *Phycomyces blakesleeanus*., *Zeitschrift für Naturforschung*, **39c**, 460–3.

Bramley, P.M. & Sandmann, G. (1985). *In vitro* and *in vivo* biosynthesis of xanthophylls by the cyanobacterium *Aphanocapsa*. *Phytochemistry*, 24, 2919–22.

Britton, G. (1976). Biosynthesis of carotenoids. In *Chemistry and Biochemistry of Plant Pigments*, 2nd edn, vol. 1, ed. T.W. Goodwin, pp. 262–327. London: Academic Press.

Britton, G. (1979). Carotenoid biosynthesis – a target for herbicide activity. *Zeitschrift für Naturforschung*, **34c**, 979–985.

Britton, G. (1983). *The Biochemistry of Natural Pigments*. Cambridge: Cambridge University Press.

Britton, G. (1985). Stable isotopes in carotenoid biochemistry. *Pure and Applied Chemistry*, **57**, 701–8.

Britton, G. (1986). Biosynthesis of chloroplast carotenoids. In *Regulation of Chloroplast Differentiation*, ed. G. Akoyunoglou & H. Senger, pp. 125–34. New York: Alan R. Liss, Inc.

Brown, D. J., Britton, G. & Goodwin, T. W. (1975). Carotenoid biosynthesis by a cell-free preparation from a *Flavobacterium* species. *Biochemical Society Transactions*, **3**, 149–50.

Burns, E.R., Buchanan, G. A. & Carter, M.D. (1971). Inhibition of carotenoid synthesis as a mechanism of action of amitrole, dichlormate and pyriclor. *Plant Physiology*, **47**, 144–8.

Caiger, D. P., Pearson, S.A., Smith, A.J. & Rogers, L.J. (1986). Differential effects of gabaculine and laevulinic acid on protochlorophyllide regeneration. *Plant Cell and Environment*, **9**, 495–9.

Camara, B. & Dogbo, O. (1986). Demonstration and solubilization of lycopene cyclase from *Capsicum* chromoplast membranes. *Plant Physiology*, **80**, 172–4.

Castelfranco, P.A. & Beale, S.I. (1983). Chlorophyll biosynthesis: recent advances and areas of current interest. *Annual Review of Plant Physiology*, **34**, 241–78.

Clarke, I.E., Sandmann, G., Bramley, P.M. & Böger, P. (1985). Phytotoxicity of *m*-phenoxybenzamides: inhibition of cell-free phytoene desaturation. *Pesticide Biochemistry and Physiology*, **23**, 335–40.

Eder, F. A. (1979). Pyridazinones, their influence on the biosynthesis of carotenoids and the metabolism of lipids in plants. *Zeitschrift für Naturforschung*, **34c**, 1052–4.

Flint, D.H. (1984). Gabaculine inhibits δ-ALA synthesis in chloroplasts. *Plant Physiology*, **75** supplement, 170.

Grumbach, K.H. (1979). Evidence for the existence of two β-carotene pools and two biosynthetic β-carotene pathways in the chloroplast. *Zeitschrift für Naturforschung*, **34c**, 1205–8.

Grumbach, K.H. (1981). Effect of SAN 6706, amitrole and photosystem-II herbicides on chloroplast development. In *Proceedings of the Fifth International Photosynthesis Congress*, ed. G. Akoyunoglou, pp. 625–36. Philadelphia: Balaban.

Grumbach, K.H. & Bach, T. J. (1979). The effect of PSII herbicides, amitrol and SAN 6706 on the activity of 3-hydroxy-3-methylglutaryl coenzyme-A-reductase and the incorporation of [2-^{14}C] acetate and [2-^{3}H] mevalonate into chloroplast pigments of radish seedlings. *Zeitschrift für Naturforschung*, **34c**, 941–3.

Grumbach, K.H. & Britton, G. (1984). Carotenoid biosynthesis in radish (*Raphanus sativus*) and the effect of bleaching herbicides. In *Advances in Photosynthesis Research*, Vol. 4, ed.. C. Sybesma, pp. 69–72. Dordrecht, Netherlands: Martinus Nijhoff/Dr. W. Junk.

Guillot-Salomon, T., Douce, R. & Signol, M. (1967). Relations between alterations in plastid structure, pigment level, and polar lipid composition in aminotriazole-treated corn leaves. *Bulletin de la Société Francaise de Physiologie Végétale*, **13**, 63–79.

Harris, N. & Dodge, A.D. (1972). The effect of paraquat on flax cotyledon leaves: physiological and biochemical changes. *Planta*, **104**, 210–19.

Hill, C.M., Pearson, S.A., Smith, A.J. & Rogers, L.J. (1985). Inhibition of chlorophyll synthesis in *Hordeum vulgare* by 3-amino-2,3-dihydrobenzoic acid (gabaculin). *Bioscience Reports*, **5**, 775–81.

Hill, C.M., Caiger, D. P., Pearson, S.A., Smith, A.J. & Rogers, L.J. (1986). Mode of action of gabaculin, an inhibitor of chlorophyll biosynthesis. *Biochemical Society Transactions*, **14**, 36–7.

Islam, M.A., Lyrene, S.A., Miller, E.M. & Porter, J.W. (1977). Dissociation of prelycopersene pyrophosphate synthetase from phytoene synthetase complex of tomato fruit plastids. *Journal of Biological Chemistry*, **252**, 1523–5.

Jenkins, C.L.D., Rogers, L.J. & Kerr, M.W. (1983). Inhibition of glycolate metabolism by aminooxyacetate: effects on pigment formation in higher plants. *Phytochemistry*, **22**, 347–52.

Kreuz, K. & Kleinig, H. (1981). On the compartmentation of isopentenyl diphosphate synthesis and utilization in plant cells. *Planta*, **153**, 578–81.

Krinsky, N.I. (1971). Function. In *Carotenoids*, ed. O. Isler, pp. 669–716. Basel: Birkhäuser.

Kunert, K.J. & Böger, P. (1981). The bleaching effect of the diphenylether oxyfluorfen. *Weed Science*, **29**, 169–73.

Kushwaha, S.C., Subbarayan, C., Beeler, D.A. & Porter, J.W. (1969). Conversion of lycopene-[15, 15'-3H] to cyclic carotenes by soluble extracts of higher plant plastids. *Journal of Biological Chemistry*, **244**, 3635–42.

Lambert, R. & Böger, P. (1983). Inhibition of carotenogenesis by substituted diphenylethers of the *m*-phenoxybenzamide type. *Pesticide Biochemistry and Physiology*, **20**, 183–7.

Lütke-Brinkhaus, F., Liedvogel, B., Kreuz, K. & Kleinig, H. (1982). Phytoene synthase and phytoene dehydrogenase associated with envelope membranes from spinach chloroplasts. *Planta*, **156**, 176–80.

McDermott, J.C.B., Britton, G. & Goodwin, T.W. (1973). Carotenoid biosynthesis in a *Flavobacterium* species: stereochemistry of hydrogen elimination in the desaturation of phytoene to lycopene, rubixanthin and zeaxanthin. *Biochemical Journal*, **134**, 1115–7.

Mathis, P. & Schenck, C.C. (1982). The functions of carotenoids in photosynthesis. In *Carotenoid Chemistry and Biochemistry*, ed. G. Britton & T.W. Goodwin, pp. 339–51. Oxford: Pergamon.

Meller, E. & Gassman, M.L. (1981*a*). The effects of levulinic acid and 4,6-dioxoheptanoic acid on the metabolism of etiolated and greening barley leaves. *Plant Physiology*, **67**, 728–32.

Meller, E. & Gassman, M.L. (1981*b*). Studies with 4,6-dioxoheptanoic acid on etiolated and greening barley leaves. *Plant Physiology*, **67**, 1065–8.

Pallett, K.E. & Dodge, A.D. (1979). The role of light and oxygen in the action of the photosynthetic inhibitor herbicide monuron. *Zeitschrift für Naturforschung*. **34c**, 1058–61.

Porra, R. J. & Meisch, H.-U. (1984). The biosynthesis of chlorophyll. *Trends in Biochemical Science*, **9**, 99–104.

Porter, J.W. & Spurgeon, S.L. (1979). Enzymatic synthesis of carotenes. *Pure and Applied Chemistry*, **51**, 609–22.

Rebeiz, C.A., Montazer-Zouhoor, A., Hopen, H.J. & Wu, S.M. (1984). Photodynamic herbicides: 1. Concept and phenomenology. *Enzyme and Microbial Technology*, **6**, 390–401.

Ridley, S.M. (1977). Interaction of chloroplasts with inhibitors: induction of chlorosis by diuron during prolonged illumination *in vitro*. *Plant Physiology*, **59**, 724–32.

Ridley, S.M. (1982). Carotenoids and herbicide action. In *Carotenoid Chemistry and Biochemistry*, ed. G. Britton & T.W.Goodwin, pp. 353–69. Oxford: Pergamon.

Rüdiger, W. & Benz, J. (1979). Influence of aminotriazol on the biosynthesis of chlorophyll and phytol. *Zeitschrift für Naturforschung*. **34c**, 1055–7.

Rüdiger, W., Benz, J., Lempert, U., Schoch, S. & Steffens, D. (1976). Hemmung der Phytol-Akkumulation mit Herbiziden. Geranylgeraniol und Dihydrogeranylgeraniol-haltiges Chlorophyll aus Weizenkeimlingen. *Zeitschrift für Pflanzenphysiologie*, **80**, 131–43.

Sandmann, G. & Bramley, P.M. (1985*a*). The *in vitro* biosynthesis of β-cryptoxanthin and related xanthophylls with *Aphanocapsa* membranes. *Biochimica et Biophysica Acta*, **843**, 73–7.

Sandmann, G. & Bramley, P.M. (1985*a*). Carotenoid biosynthesis by *Aphanocapsa* homogenates coupled to a phytoene-generating system from *Phycomyces blakesleeanus*. *Planta*, **164**, 259–63.

Sandmann, G., Clarke, I.E., Bramley, P.M. & Böger, P. (1984). Inhibition of phytoene desaturase–the mode of action of certain bleaching herbicides. *Zeitschrift für Naturforschung*, **39c**, 443–9.

Sergeant, J.M. & Britton, G. (1984). Observations on the biosynthesis of chloroplast carotenoids. In *Advances in Photosynthesis Research*, Vol. 4, ed. C. Sybesma, pp. 779–82. The Hague: Nijhoff/Dr. W. Junk.

Shioi,Y., Doi, M. & Sasa, T. (1985). Inhibition of porphobilinogen synthase activity by 4,5-dioxovalerate. *Plant Cell Physiology*, **26**, 379–82.

Siefermann-Harms, D. (1985). Carotenoids in photosynthesis. 1. Location in photosynthetic membranes and light-harvesting function. *Biochimica et Biophysica Acta*, **811**, 325–55.

Stanger, C.E. & Appleby, A.P. (1972). A proposed mechanism for diuron-induced phytotoxicity. *Weed Science*, **20**, 357–63.

Urbach, D., Suchanka, M. & Urbach, W. (1976). Effect of substituted pyridazinone herbicides and of difunone (EMD-IT 5914) on carotenoid biosynthesis in green algae. *Zeitschrift für Naturforschung*, **31c**, 652–5.

Vial, J. & Borrod, G. (1984). Short note on dihydropyrones, a new herbicidal family with bleaching properties. *Zeitschrift für Naturforschung*, **39c**, 459.

Wilkinson, R.E. (1985). Inhibition of conversion of geranylgeranyl-chlorophyll to phytol-chlorophyll by S-ethyldipropylthiocarbamate (EPTC). *Pesticide Biochemistry and Physiology*, **23**, 289–93.

Williams, R.J.H., Britton, G. & Goodwin, T.W. (1967). The biosynthesis of cyclic carotenes. *Biochemical Journal*, **105**, 99–105.

Yokoyama, H., Hsu, W.J., Poling, S.M. & Hayman, E. (1982). Chemical regulation of carotenoid biosynthesis. In *Carotenoid Chemistry and Biochemistry*, ed. G. Britton & T.W. Goodwin, pp. 371–85. Oxford: Pergamon.

Youngman, R.J. & Dodge, A.D. (1979). Mechanism of paraquat action: inhibition of the herbicidal effect by a copper chelate with superoxide dismutating activity. *Zeitschrift für Naturforschung*, **34c**, 1032–5.

JOHN L. HARWOOD, STUART M. RIDLEY AND KEVIN A. WALKER

Herbicides inhibiting lipid synthesis

Introduction

A number of herbicides have been reported, over the years, to inhibit lipid metabolism. The effects of many of these compounds are believed to be secondary. However, three classes of herbicides may have a more specific effect on fatty acid (and lipid) synthesis. These groups of compounds are the substituted pyridazinones, thiocarbamates and a rather diverse class of graminaceae-selective herbicides which include the oxyphenoxy propionic acids and cyclohexanediones. These three groups form the substance of this brief review.

The overall topic of herbicides and lipid metabolism has been reviewed by several authors (Duke, 1985; Fedtke, 1982; Rivera & Penner, 1979; St John, 1982).

Substituted pyridazinones

The mode of action of pyridazinone herbicides has been reviewed recently and appears to involve several target sites in plants (St John, 1982; Duke, 1985). Depending on the exact structure of the compound, and the plant test species, the effects have been noted to include inhibition of photosynthetic O_2 evolution, of pigment synthesis and changes in fatty acid composition.

Two substituted pyridazinones which have excited particular interest with regard to their effects on fatty acid formation are San 9785, (BASF 13 338; 4-chloro-5-(dimethylamino)-2-phenyl-3(2H) pyridazinone) and San 6706 (metflurazon); 4-chloro-5-(dimethylamino)-2-(α,α,α-trifluoro-*m*-tolyl)-3(2H)-pyridazinone)(Figure 1). It was noted by St John (1976) that San 9785 altered the proportions of linoleic and α-linolenic acids in monogalactosyl- and digalactosyldiacylglycerol. In contrast, San 9774 merely reduced the proportion of α-linolenate in monogalactosyldiacyl-glycerol alone. In other experiments on the action of San 9785 on α-linolenate synthesis in wheat, it was noted that the herbicide specifically inhibited the acid's synthesis (Fig. 2) during the hardening period required for frost resistance (Willemot, 1977). The concurrent rise in linoleic acid seen during these studies suggested that San 9785 had a direct effect on the Δ15-desaturase. Similar experiments have been also reported by St John *et al.* (1979) who found that the reduced levels of α-linolenate were associated with early low temperature injury symptoms such as frost-banding. In addition, other species such as cotton and broad bean, were also found to show rises in linoleate and

Fig. 1. Structure of the substituted pyridazinones San 9785 and San 6706.

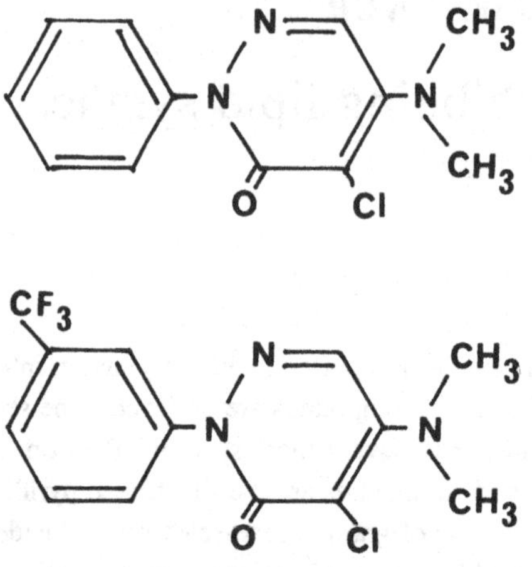

SAN 9785

SAN 6706

Fig. 2. Changes in cereal fatty acids following San 9785 treatment. Data from wheat taken from Willemot (1977) and for rye and barley from St John *et al.* (1979) where details can be found.

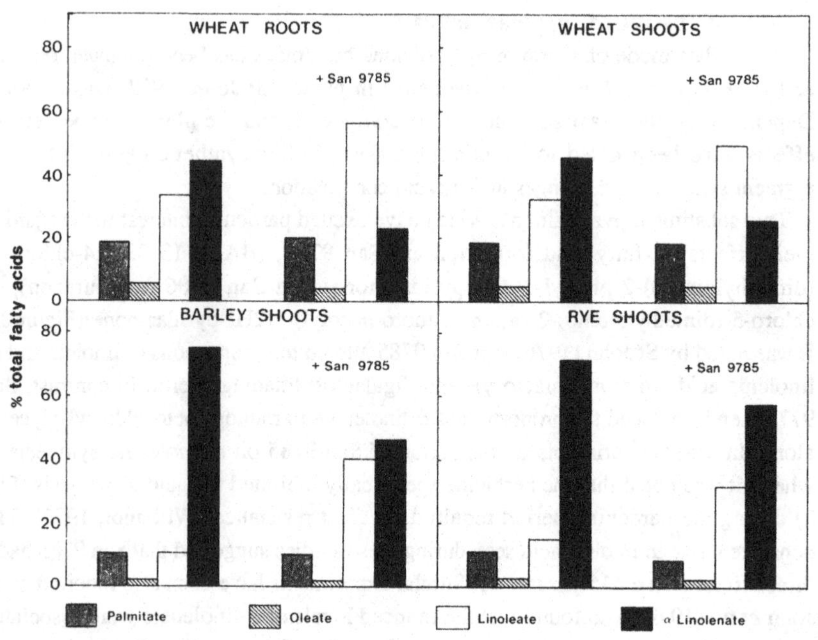

decreases in α-linolenate following San 9785 treatment (St John & Christiansen, 1976; Khan *et al.* 1979).

The hypothesis that San 9785 inhibited the Δ15-desaturase was tested more directly by using radiolabelled substrates. Murphy *et al.* (1980) showed that the susceptibility of different plants to the herbicide varied considerably. Labelling of α-linolenate from [^{14}C]-acetate was inhibited in cucumber and maize, severely inhibited in barley and rye grass and unaffected in pea and spinach leaves. Total lipid radioactivity was unaffected and the loss in [^{14}C]-linolenate was compensated by an increase in [^{14}C]-linoleate. The generally higher sensitivity of the moncotyledons in these studies (Murphy *et al.*, 1980) agreed with the *in vivo* studies of Khan *et al.* (1980) who found that linolenate formation in barley was affected more than that in broad bean at all herbicide concentrations. The inhibition of linoleate desaturation was tested more directly by using [^{14}C]-oleate and [^{14}C]-linoleate precursors. In susceptible species the conversion of linoleate to linolenate was severely inhibited (Murphy *et al.*, 1980). Interestingly, the accumulation of [^{14}C]-linoleate is almost exclusively confined to the monogalactosyldiacylglycerol and the labelling pattern of the highly radioactive phosphatidylcholine is hardly affected (Davies & Harwood, 1983; Murphy *et al.*, 1980; Willemot *et al.*, 1982). Data for several species labelled from [^{14}C]-acetate are shown in Fig. 3.

San 9785 appears to be more selective in its effects than other substituted pyridazinones. Although linoleate desaturation is strongly inhibited, the herbicide has only weak effects on photosynthesis, has a low phytotoxicity in most systems and has little action of pigment accumulation (St John, 1982; Frosch *et al.*, 1979). In barley leaves 10^{-4}M San 9785 reduced linolenate and increased linoleate labelling but had no effect on the labelling of other fatty acids nor on plant growth or lipid content. Minor changes were seen in the ultrastructure of developing chloroplasts which might have related to the diminished α-linoleate levels (Davies & Harwood, 1983; Khan *et al.*, 1979; Leech, Walton & Baker, 1985). Moreoever, the reductions in α-linolenate were mainly found in chloroplast lipids and were less obvious in the essentially non-chloroplastic phosphatidylcholine and phosphatidylethanolamine (Davies & Harwood 1983; Murphy *et al.*, 1985). Higher herbicide concentrations or longer incubation periods can cause a net loss in chloroplast galactosylglycerides and chlorophyll and also impair several photosynthetic functions (Laskay & Lehoczki, 1986; Laskay *et al.*, 1986; Lem & Williams, 1983; Willemot *et al.*, 1982). The selective nature of the inhibition by San 9785 of the Δ15-desaturase but not other desaturases is emphasised in the radiolabelling experiments of Murphy *et al.* (1980, 1985) cf. Table 1. Such data also emphasises the usual role of monogalactosyldiacylglycerol as substrate for linoleate desaturation (cf. Jones & Harwood, 1980; Harwood, 1980*a*). Detailed differences observed between the so-called '16:3' and '18:3' plant types are discussed by Willemot *et al.* (1982).

Table 1. *Changes in cereal fatty acids following San 9785 treatment.*

	Fatty acid composition (% total)				
	16:0	16:1	18:0	18:1	18:3
Wheat roots[a]					
Control	18.3	0.6	4.7	32.8	43.6
+ San 9785	19.0	1.0	3.3	54.8	22.0
Wheat shoots[a]					
Control	17.6	0.5	4.4	32.1	45.3
+ San 9785	18.8	1.1	4.8	49.4	25.9
Barley shoots[b]					
Control	10.0	1.3	1.7	10.0	75.4
+ San 9785	10.4	1.3	1.5	40.2	46.6
Rye shoots[b]					
Control	11.3	1.0	2.4	14.2	71.3
+ San 9785	8.8	0.8	1.7	30.8	57.6

Data taken from [a]Willemot (1977) and [b]St John *et al.* (1979) where details can be found.

Apart from San 9785, several other pyridazinone derivatives have been shown to affect fatty acid composition and synthesis. These include norflurazon (San 9789) and various different phenyl-pyridazinones (St John, 1982; Eder, 1979; St John & Hilton, 1976; St John *et al.*, 1984). Both norfluarazon and its dimethylamino analogue metflurazon (San 6706; Figure 1) inhibit the desaturation of phytoene to lycopene during carotene biosynthesis and hence cause bleaching of leaves. Several laboratories have shown that San 6706, in addition to inhibiting linoleate desaturation in some species, also prevents the conversion of palmitate to *trans* -Δ3-hexadecenoate (Khan *et al*, 1979; Davies & Harwood, 1983). The effect on hexadecenoate production is associated with the phosphatidylglycerol fraction where the *trans* -Δ3 isomer is exclusively located at its *sn*-2 position (Harwood, 1980*b*). As with San 9785, some plants such as barley were more sensitive to San 6706 than others (such as broad bean). Thus, in the former, San 6706 inhibited both *trans* -Δ3-hexadecenoate and linolenate synthesis whereas in *V. faba* only *trans* -Δ3-hexadecenoate formation was significantly affected (Khan *et al.*, 1979).

In summary, therefore, substituted pyridazinones are inhibitors of fatty acid desaturation, particularly the conversion of linoleate to α-linolenate. Of the compounds studied, San 9785 is the most specific and seems to have its primary mode of action on the Δ15-desaturase using monogalactosyldiacylglycerol substrate. The overall pathways for a fatty acid synthesis in plants are summarised in Fig. 4.

Thiocarbamates

Several thiocarbamate herbicides (Fig. 5) have been reported to inhibit lipid metabolism. Their effects have been reviewed (Rivera & Penner, 1979; Fedtke, 1982; Duke, 1985). S-Ethyl dipropylthiocarbamate (diallate) was shown to reduce epicuticular wax while 2-chloroallyl diethyldithiocarbamate (CDEC) altered the structure but not the amount of wax (Still, Davis & Zander, 1970). EPTC was reported to reduce petiolar thickness and will affect total fatty acid synthesis in sicklepod (Wilkinson & Hardcastle, 1969) while CDEC was shown to reduce ^{14}C-malonate incorporation into the lipids of excised hypocotyls (Mann & Pu, 1968). Mindful of the conversion of very long-chain fatty acids into the various plant wax components (cf. Harwood & Russell, 1984), Harwood & Stumpf (1971) tested EPTC, diallate and CDEC on germinating pea seeds – a system known to synthesise significant quantities of > C_{18} fatty acids. They found that, although the total incorporation of radioactivity from [^{14}C]-acetate into fatty acids was unaffected by 10^{-5} M herbicide, the proportion of very long-chain fatty acids made was significantly

Fig. 3. Effect of San 9785 on the incorporation of radioactivity from [^{14}C]-acetate into the fatty acids of monogalactosyldiacylglycerol (MGDG) and phosphatidylcholine (PC) in different plant leaf segments.

Data taken from Willemot *et al.* (1982). The fatty acids were separated by $AgNO_3$-tlc but evidence from our laboratory (see Wharfe & Harwood, 1978) showed that the four bands essentially correspond to palmitate, oleate, linoleate and α-linolenate as the only radioactive fatty acids. Legend see Figure 2.

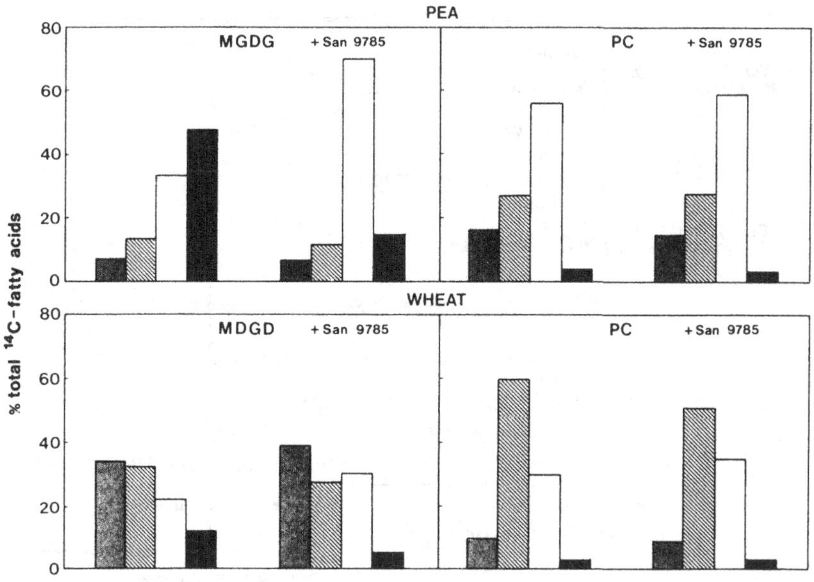

reduced (Table 4). This inhibition explained the defective wax formation induced by such herbicides.

Reduction in cuticular lipid formation by these and other thiocarbamates has been reviewed (Rivera & Penner, 1979). In the case of diallate the effects include increased leaf wettability which allows the leaf to be more effectively treated by foliage herbicides (Davis & Dusbabek, 1973). This increased leaf wettability is also found with agents such as trichloroacetic acid which have also been reported to inhibit fatty acid elongation (cf. Harwood & Stumpf, 1971).

After Harwood & Stumpf (1971) had found that EPTC and other thiocarbamates could inhibit very long-chain fatty acid synthesis rather specifically, their experiments were followed up by others. One of the typical cuticular components, alkanes, are not present in significant quantities in germinating peas but it was thought (at that time) that these compounds were made from very long-chain fatty acids by decarboxylation (Kollatukudy, 1980). Accordingly, leaf slices from young pea plants, which were capable of significant alkane synthesis were utilised in further work. Kollutukudy & Brown (1974) also used EPTC, CDEC and diallate and found that the action of the

Fig. 4. Simplified scheme for fatty acid synthesis in plant leaves, showing only the major pathways. The sites of action of substituted pyridazinones, oxyphenoxypropionic acids and thiocarbamates are shown.

(Abbreviations: Cond. 2 = β-ketoacyl-ACP synthetase II; PC = phosphatidylcholine; MGDG = monogalactosyldiacylglycerol) ACC = acetyl CoA carboxylase; ACP = acyl carrier protein.

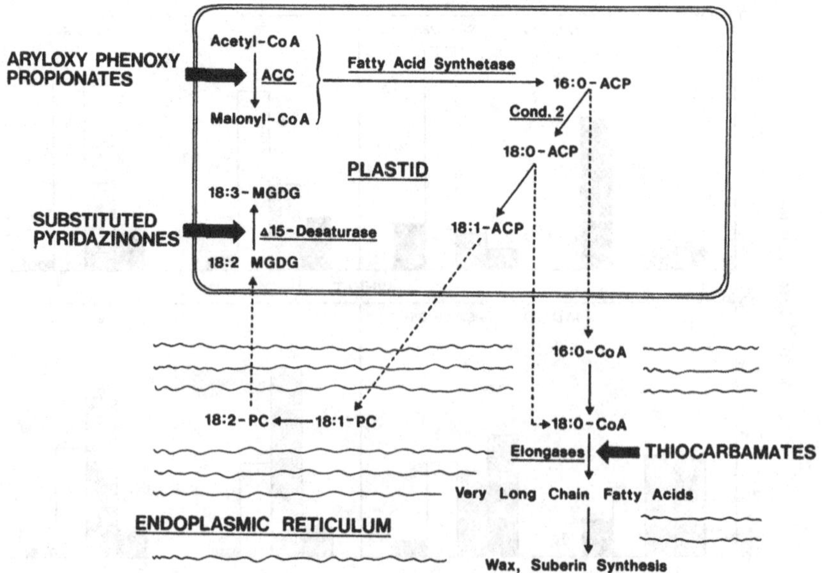

herbicides was mainly on the labelling of the surface rather than the internal lipids (Table 5). As a percentage of the untreated control, the labelling of the alkane and secondary alcohol fractions were reduced by 90% and the primary alcohols by about 50% with 10^{-5} M diallate. The alkanes and secondary alcohols were also particularly affected by CDEC and EPTC. When they examined the type of fatty acid labelled, Kollatukudy & Brown (1974) also found that EPTC had a selective effect on very long-chain (> C_{18}) fatty acid labelling (Table 6). Palmitate and stearate labelling was unaffected as previously noted for germinating pea (Harwood & Stumpf, 1971). The particular decrease in C_{26} acid and relative accumulation of label in C_{22} (although in net terms total very long-chain fatty acid labelling was reduced 50%) was considered

Fig. 5. Structures of some thiocarbamate herbicides (CDEC, diallate, triallate, EPTC).

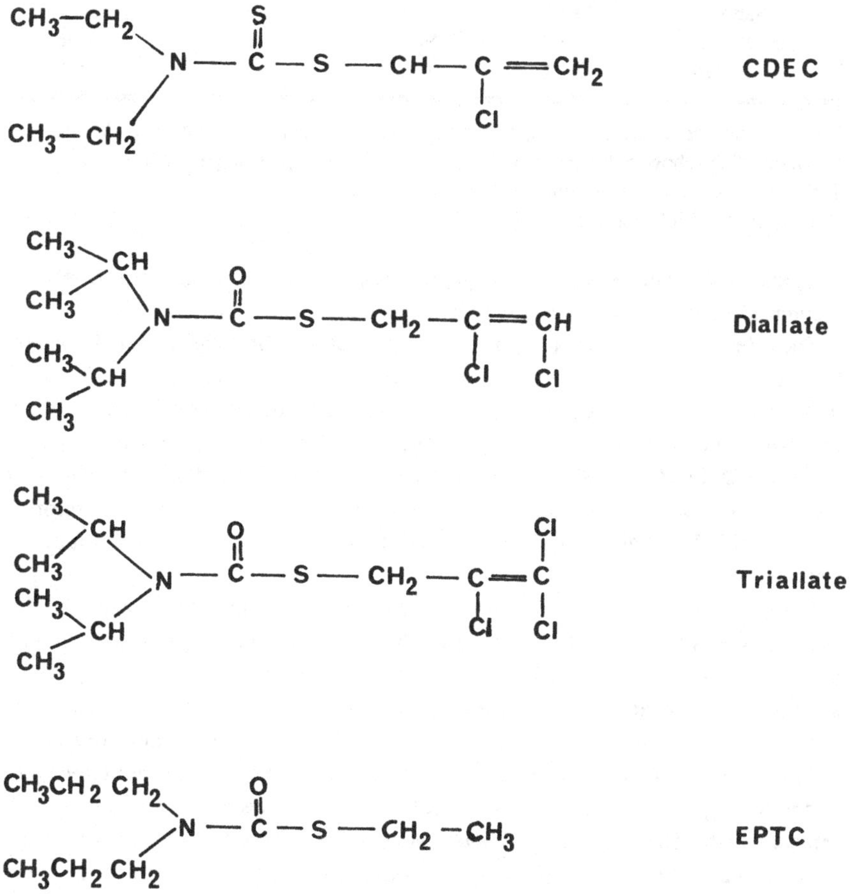

$(CH_3-CH_2)_2N-C(=S)-S-CH_2-C(Cl)=CH_2$ CDEC

$((CH_3)_2CH)_2N-C(=O)-S-CH_2-C(Cl)=CH(Cl)$ Diallate

$((CH_3)_2CH)_2N-C(=O)-S-CH_2-C(Cl)=C(Cl)_2$ Triallate

$(CH_3CH_2CH_2)_2N-C(=O)-S-CH_2-CH_3$ EPTC

Table 2. *Effect of San 9785 on the incorporation of radioactivity from [^{14}C]-acetate into the fatty acids of monogalactosyldiacylglycerol and phosphatidylcholine in different plant leaf segments.*

		Fatty acid composition (% total)			
		Saturated	Monounsaturated	Diunsaturated	Triunsaturated
Pea					
MGD	control	7 ± 1	13 ± 1	33 ± 2	47 ± tr
	+ SAN 9785	6 ± 1	11 ± tr	69 ± 2	14 ± 1
PC	control	16 ± 1	26 ± 2	55 ± 1	3 ± tr
	+ San 9785	14 ± 1	25 ± 1	59 ± 1	2 ± 1
Wheat					
MGD	control	34 ± 6	32 ± 3	22 ± 4	12 ± 5
	+ San 9785	39 ± 7	27 ± 1	30 ± 3	5 ± 3
PC	control	9 ± 1	59 ± 3	29 ± 3	3 ± tr
	+ San 9785	9 ± 1	51 ± 3	36 ± 2	3 ± 1

The fatty acids were separated by $AgNO_3$-tlc but evidence from our laboratory (cf. Wharfe & Harwood, 1978) showed that the four bands essentially correspond to palmitate, oleate, linoleate and α-linoleate as the only radioactive fatty acids.
Data taken from Willemot *et al.* (1982).

to be due to either more than one elongation system being present or, alternatively, to accumulation of the C_{22} acid as a partially elongated product.

The effects of thiocarbamates on wax components and lipids have been discussed by Ezra *et al.* (1983). Although as discussed above, the major effects of thiocarbamates in whole tissues seem to be confined to the wax components or their immediate precursors, some observations have also been made with isolated chloroplasts. EPTC, diallate, butylate, pebulate and vernolate inhibit incorporation of radioactivity from [^{14} C]-acetate and [^{14} C]-malonate in these fractions (Wilkinson & Smith, 1975, 1976) although the physiological significance of these observations is unclear.

Since thiocarbamates had been shown clearly in pea to inhibit the formation of the secondary alcohols and alkanes of wax by reducing very long-chain fatty acid synthesis, it was obviously of interest to consider a different surface covering layer, suberin. The comparative structures of suberin and cutin have been elucidated by Kollatukudy and coworkers (cf. Kollatukudy, 1975). Suberin was shown to contain relatively high amounts of dicarboxylic acids, phenolics, very long-chain fatty acids and very long-chain alcohols. A convenient experimental system, in which to test for thiocarbamate effects, is the aged potato slice which makes suberin and very long-chain fatty acids very rapidly (Bolton & Harwood, 1976*a*). Three thiocarbamates were

Table 3. *Comparison of the inhibition of Δ12- and Δ15–desaturase by San 9785 in different plants.*

			% total substrate desaturated	
	Fraction	Treatment	Δ12-desaturation	Δ15-desaturation
Cucumber	Total	–	57.7	24.5
	"	+	55.3	10.3
Cucumber	MGDG	–	55.7	77.3
	"	+	54.2	46.3
Cucumber	PC	–	53.9	–*
	"	+	51.5	–
Barley	MGDG	–	76.1	43.9
	"	+	68.8	19.0
Barley	PC	–	58.9	–*
	"	+	54.8	–

Abbreviations: MGDG = monogalactosyldiacylglycerol; PC = phosphatidylcholine
* no significant radiolabelled linolenate associated with phosphatidylcholine
San 9785 used at 10^{-4}M
Data from Murphy *et al.* (1985).

Table 4. *Inhibition of very long-chain fatty acid synthesis in germinating pea by three thiocarbamate herbicides.*

Treatment	Total labelling (% control)	Fatty acid labelling (% total)				
		16:0	18:0	20:0	22:0	24:0
None	100	24	65	5	5	1
EPTC (10^{-5} M)	118	20	79	1	0	0
CDEC (10^{-5} M)	118	25	70	3	2	0
Diallate (10^{-5} M)	134	26	70	4	tr	0

Data from Harwood & Stumpf (1971) and unpublished results

tested – EPTC, diallate and triallate (S-(2,3,3-trichloroallyl) diisopropylthiocarbamate). All three compounds were found to reduce severely the proportion of C_{20}, C_{22} and C_{24} acids labelled (Table 7; Bolton & Harwood, 1976*b*). In contrast, palmitate and stearate labelling was virtually unchanged. Thus, the relative sensitivity of very long-chain fatty acid synthesis was difficult to explain in terms of a postulated interaction of the herbicides with thiol groups (Lay, Hubbell & Casida, 1975) since fatty acid

Table 5. *Relative labelling of surface and internal lipids of pea leaves in the presence of thiocarbamates.*

	Control (dpm × 10^{-6})	Percentage of control labelling (Herbicide concentration (μM)			
		1	3	6	9
Internal lipids					
Diallate	4.27	102	85	112	113
EPTC	5.62	96	93	88	84
CDEC	6.16	90	83	64	80
Surface lipids					
Diallate	2.44	82	70	67	63
EPTC	2.84	117	81	62	63
CDEC	3.54	88	88	60	48

Data from Kollatukudy & Brown (1974).

synthetase and palmitate elongase both rely on sulphydryl residues. However, the similar inhibition by thiocarbamates on very long-chain acid synthesis in both the pea and potatoes strongly suggests that these herbicides have a fundamental effect at this level of lipid metabolism.

Recently we have confirmed that the thiocarbamates selectively inhibit very long-chain fatty acid synthesis in other plant tissues, such as oat and barley leaves (K. Abulnaja & J.L. Harwood, unpublished observations). Because it has been suggested that metabolic products of thiocarbamates, such as sulphoxides, are the active compounds (cf. Fedtke, 1982) we have probed the effects of the herbicide in more detail using the germinating pea system. If seeds are treated with either diallate or triallate during germination then the microsomal fraction isolated from such peas cannot synthesise very long-chain fatty acids (Fig. 6, Table 8). This effect can either be explained by assuming that the herbicides or the active metabolites bind to the microsomal membranes and therefore inhibit synthesis. Alternatively, the presence of the thiocarbamates during germination may prevent the synthesis of the elongase protein which normally occurs during the first 24h. (cf. Harwood & Stumpf, 1970). When diallate and triallate were tested directly on microsomal fractions prepared from untreated seeds their inhibition of very long-chain fatty acid synthesis was seen (Table 9). However, total fatty acid formation was severely reduced and the rather selective effect found *in vivo* (Harwood & Stumpf, 1971) was lost. Moreover, high concentrations of the herbicides were needed. This difference between the *in vivo* and *in vitro* findings also has two obvious explanations. First, the selectivity *in vivo* is due to metabolism of the thiocarbamates to other compounds (such as sulphoxides) which are themselves more active and selective. Alternatively, the selectivity *in vivo* may be due to the different locations of very long-chain fatty acid and *de novo* fatty acid synthesis in the tissue. Thus, whereas the vast majority of *de novo* synthesis is plastid-located (cf. Harwood, 1987), very long-chain fatty acid synthesis occurs on membranes (cf. Harwood & Stumpf, 1971) and is concentrated in the outer leaf cells

Fig. 6. Effect of pre-treatment of pea seeds with thiocarbamates on microsomal fatty acid synthesis from [^{14}C]-malonyl-CoA.

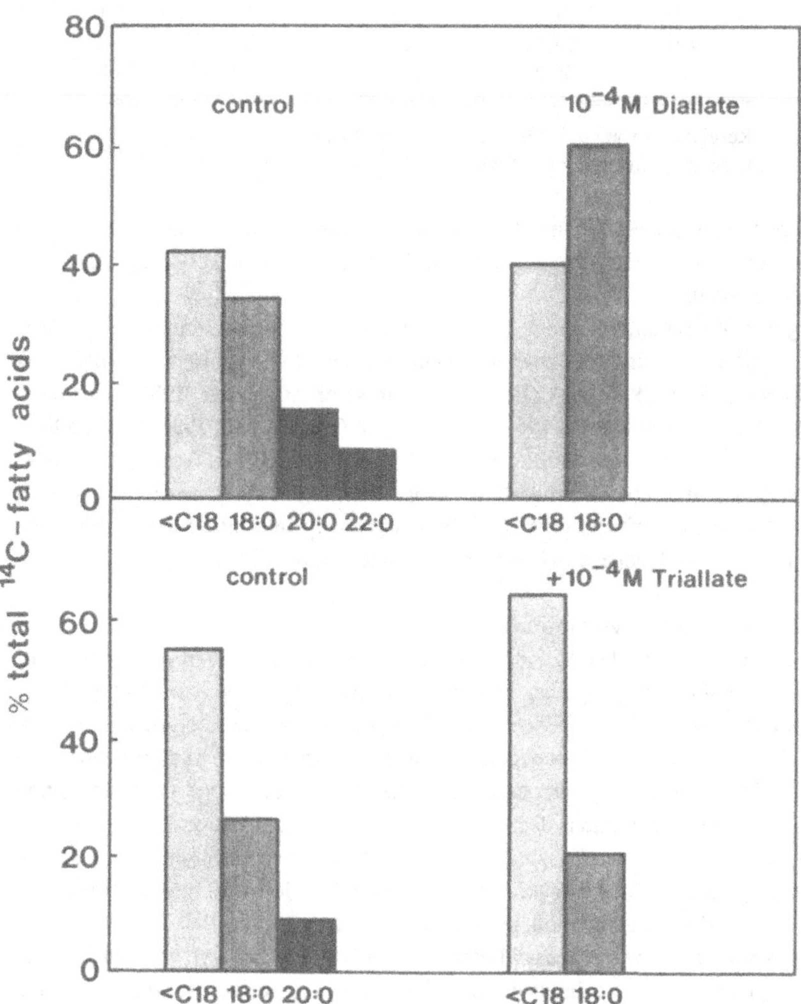

Table 6. *Changes in the distribution of radioactivity from [^{14}C]-acetate in very long-chain fatty acids caused by EPTC in pea leaves.*

[^{14}C]-Fatty acid	Concentration of EDTC (μM)			
	0	3	6	12
20:0	12.6	15.2	11.9	13.0
22:0	28.9	33.1	37.5	49.1
24:0	18.5	20.1	24.6	19.8
26:0	40.2	31.6	26.0	18.0

Results are expressed as % total [^{14}C]-very long-chain fatty acids.
Data from Kollatukudy & Brown (1984).

(Lessire & Stumpf, 1983). Thus because of poor translocation or rapid inactivation within pea, the thiocarbamates may only be able to easily inhibit very long-chain fatty acid synthesis *in vivo*.

A group of structurally similar compounds are the chloroacetamides. These show similarities in activity spectrum and symptom development to thiocarbarmates (Wilkinson, 1985, 1986) and also reduce wax synthesis (Ebert, 1982). However, whereas CDAA (Mann & Pa, 1968) and butachlor (Chang *et al.*, 1985) reduced lipid synthesis, metochlor and alachlor did not (Mellis *et al.*, 1982; Warmund, Kerr & Peters, 1985). It is possible that these compounds, and maybe the thiocarbamates, produce secondary effects on fatty acid synthesis by interfering with gibberellin metabolism (H.H. Hoppe, 1987, personal communication).

Oxyphenoxypropionic acids

In 1971 Hoechst AGF discovered that a number of phenoxypropionic acids were active selectively against graminaceous species (cf. Nestler, 1982). These compounds are post-emergence herbicides and it has been shown that most dicotyledonous and some monocotyledonous plants are tolerant to their action. The sort of effects which have been observed in sensitive species include a rapid inhibition of auxin-induced reactions, decreases in chlorophyll content and CO_2 fixation, breakdown of membranes (causing increases in membrane permeability) and necrosis of meristematic tissue (cf. Hoppe & Zacher, 1985). The destruction of the intercalary and apical meristems of the shoot precedes plant death.

The observation that the free acid of the herbicide clofop-isobutyl was also active as a hypolipidemic drug in animals prompted Hoppe and co-workers to study the action of oxyphenoxypropionic acids on lipid (and other) metabolic reactions in plants. Diclofop-methyl (Figure 7) was used for the bulk of the experiments.

Diclofop-methyl was found to affect lipid metabolism in radicles of maize. It reduced the acyl lipid content of the radicle tips (triacylglycerols and phosphoglycerides) and at the same times caused a leakage of amino acids from treated tissues (Hoppe, 1980). Incorporation studies using [^{14}C]-acetate showed that labelling of lipids in the radicle tips was strongly inhibited at herbicide concentrations which did not affect the biosynthesis of proteins or nucleic acids (Hoppe, 1981). In our own laboratory, we have shown that fluazifop-butyl has no effect on amino acid uptake, protein synthesis, carbon dioxide fixation or the formation of water-soluble organic compounds (such as sugars) in plant species where similar concentrations of herbicide cause virtually total inhibition of acyl lipid production (unpublished data).

The inhibition of lipid synthesis which was observed in maize roots treated with diclofop-methyl was examined further. Uptake of [^{14}C]-acetate was unaffected and, of the various lipid classes, only those containing acyl residues appeared to be affected. Within the acyl lipid classes all major compounds were equally affected

Fig. 7. Oxyphenoxy propionic acids (fluazifop-butyl, diclofop-methyl, haloxyfop-butyl) and sethoxydim.

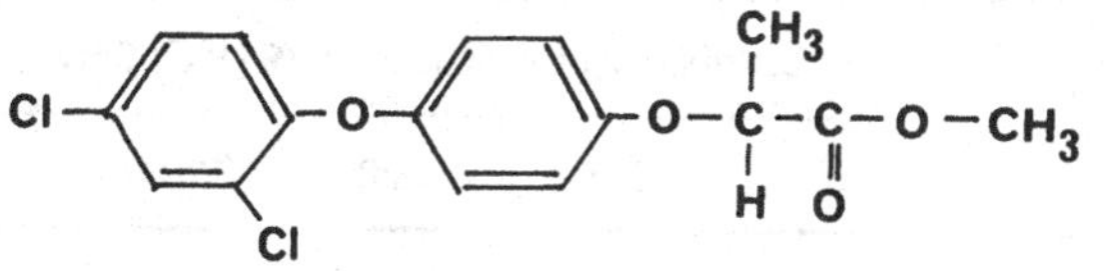

Diclofop-methyl

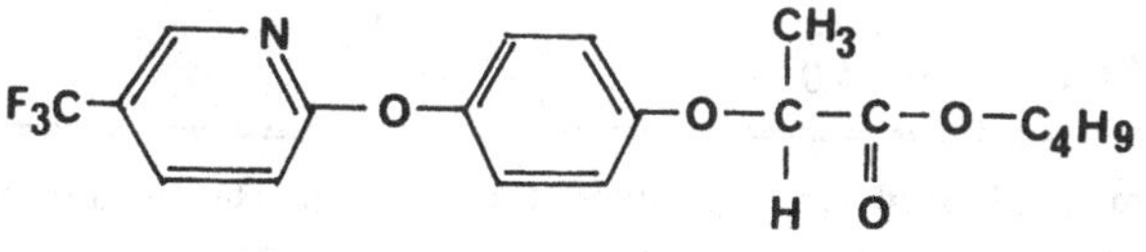

Fluazifop-butyl

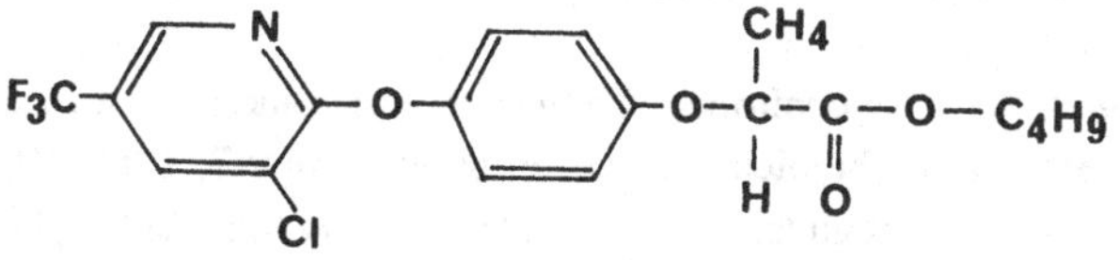

Haloxyfop-butyl

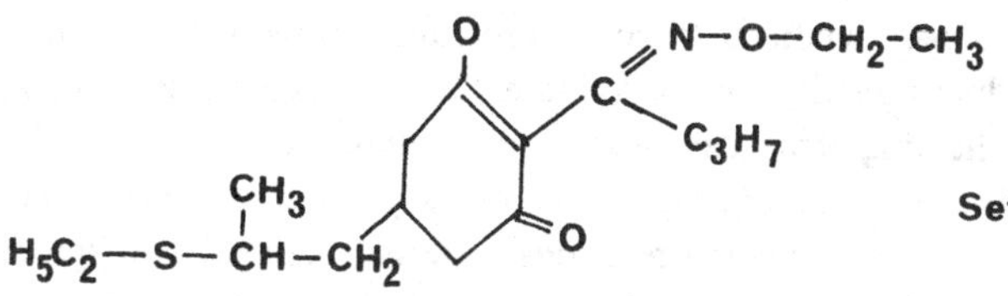

Sethoxydim

Table 7. *Effect of thiocarbamates on very long-chain fatty acid synthesis by aged potato slices.*

Inhibitor	Fatty acid labelling (% total)						
	16:0	18:0	18:1	20:0	22:0	24:0	other
None	31 ± 2	7 ± 1	36 ± 4	9 ± 1	7 ± 1	3 ± tr	7 ± 3
10^{-5} M EPTC	33 ± 3	7 ± 1	42 ± 4	11 ± 2	1 ± 1	tr	6 ± 3
None	27 ± 3	7 ± 1	49 ± 3	10 ± 1	4 ± 2	–	3 ± 2
10^{-5} M Diallate	43 ± 1	6 ± 1	43 ± 3	2 ± 1	1 ± tr	–	5 ± 3
None	31 ± 1	6 ± 1	40 ± 4	10 ± 2	6 ± 1	–	7 ± 3
10^{-5} M Triallate	40 ± 2	5 ± 1	26 ± 1	9 ± 1	t ± tr	–	9 ± 3

Labelling from [^{14}C]-acetate

Table 8. *Effect of pre-treatment of pea seeds with thiocarbamates on microsomal fatty acid synthesis from [^{14}C]-malonyl-CoA.*

	Distribution of radioactivity (% [^{14}C]-fatty acids)				
	16:0	17:0	18:0	20:0	22:0
No treatment	30.1	12.4	34.2	15.0	7.7
Diallate (10^{-4} M)	40.2	0	59.8	0	0
No treatment	55.5	0	25.9	8.8	0
Triallate (10^{-4} M)	64.0	0	19.1	0	0

Peas were germinated for 24 h in the presence of the herbicide. The microsomal fraction was prepared and incubations carried out as described by Bolton & Harwood (1977).
Data from K. Abulnaja & J.L. Harwood (unpublished).

(Hoppe & Zacher, 1982). We have confirmed these observations and also carried out similar experiments with another herbicide, fluazifop-butyl (Figure 7, Table 10). In Table 10 it will be seen that, although the two herbicides severely inhibited acyl lipid labelling from [^{14}C]-acetate in sensitive barley plants, they were without effect on resistant pea species. Pigments were labelled at near normal rates and Hoppe & Zacher (1982) found that sterols were unaffected also. Incorporation of radioactivity from [^{14}C]-malonate was reduced by diclofop-methyl in maize to an equal extent to that from [^{14}C]-acetate indicating that acetyl-CoA carboxylase was not a site for inhibition. In addition, incorporation of [^{14}C]-oleate into acyl lipids was unaffected, showing that acylation or trans-acylation reactions were not inhibited (Hoppe & Zacher, 1982). Thus, the site of action of diclofop-methyl appeared to be on fatty acid

Table 9. *Direct action of diallate and triallate on pea microsomal fatty acid synthesis from [^{14}C]-malonyl-CoA.*

	Total Synthesis (dpm)	Distribution of radioactivity (% [^{14}C]-fatty acids)				
		12:0	14:0	16:0	18:0	20:0
Control	7390	2.6	11.0	48.7	21.0	16.8
+ Triallate (10^{-5} M)	4750	2.3	7.2	44.3	26.1	20.2
+ Diallate (10^{-5} M)	5480	2.5	7.9	49.3	26.1	20.2
Control	14170	13.7	20.7	49.1	14.8	1.7
+ Triallate (10^{-4} M)	5850	9.9	14.0	47.8	28.4	0
+ Diallate (10^{-4} M)	5800	8.5	16.8	54.4	20.4	0

Microsomes prepared and incubated as described by Bolton & Harwood (1977).
K. Abulnaja & J.L. Harwood (unpublished results).

synthesis. Because endogenous fatty acid patterns of diclofop-methyl- and fenoxaprop-ethyl-treated tissues still showed appreciable polyunsaturated fatty acids, then desaturation reactions still seemed able to occur (Hoppe, 1980; Köcher, Lotzsch & Gantz, 1984). Accordingly, the fatty acid synthesis seemed to be the sensitive site for oxyphenoxypropionic acid action.

The agreement between relative toxicity and inhibition of fatty acid synthesis for different plant species has been studied in detail by Hoppe (1985) for diclofop-methyl and by ourselves for fluazifop-butyl. Incorporation of radioactivity from [^{14}C]-acetate into leaf lipids was reduced severely for sensitive species such as maize, wild oat and barnyard grass with EC_{50} values of about 10^{-7} M. Fatty acid biosynthesis in tolerant species such as bean (*Phaseolus vulgaris*), sugar beet and soybean was not affected (Hoppe, 1985). Clofop-butyl, fenthiaprop-ethyl, fenoxaprop-ethyl and the corresponding free acids show similar effects on fatty acid labelling (H. H. Hoppe, personal communication). Wheat (which is tolerant of diclofop-methyl) showed a reduction of fatty acid synthesis but, unlike sensitive plants, this was able to recover (Hoppe, 1985).

The selective inhibition by fluazifop-butyl of fatty acid synthesis in leaf pieces also occurs in chloroplasts isolated from sensitive and tolerant plants. We used barley as a sensitive and pea as a resistant species (Walker *et al.* 1988*a*). The chloroplast is actually more sensitive to the free acid and similar observations have been made for various oxyphenoxypropionic acid herbicides and fatty acid synthesis in maize leaf chloroplasts (Hoppe & Zacher, 1985). Similarly, in the latters' experiments it was found that fatty acid synthesis by chloroplasts from sensitive maize, but not from resistant beans, was inhibited. Diclofop-methyl, fenthiaprop-ethyl, fenoxaprop-ethyl and their free acids were all active (Hoppe & Zacher, 1985). Moreover, the selectivity towards the two species did not appear to be due to a different metabolism of the herbicide in maize versus beans. In addition, it is clear that the inhibition of fatty acid synthesis shows the same isomeric specificity as the herbicidal activity. The active form of diclofop-methyl is the D-enantiomer (related to the asymmetric carbon at the 2-substituted position of the propionic acid moletyl) (Nestler, 1982) and this isomer was much more active at inhibiting fatty acid synthesis in leaf pieces,

Table 10. *Effect of two oxyphenoxypropionic acid herbicides on lipid labelling from [^{14}C]acetate in sensitive and resistant species.*

			Labelling (% untreated control)	
			Total lipids	Total fatty acids
Diclofop-methyl	Barley	stems	87	64
		leaves	34	16
	Wheat	stems	127	88
		leaves	95	124
	Pea	stems	138	185
		leaves	90	74
Fluazifop-butyl	Barley	stems	38	53
		leaves	6	7
	Pea	stems	97	nm
		leaves	108	97

nm = not measured.
Plants were sprayed with 10^{-3} M solutions 4 days before [^{14}C]-acetate incorporation experiments. Pea and wheat are both resistant species.

Fig. 8. Pattern of fatty acids made by barley chloroplasts in the presence of low inhibitory concentrations of fluazifop.

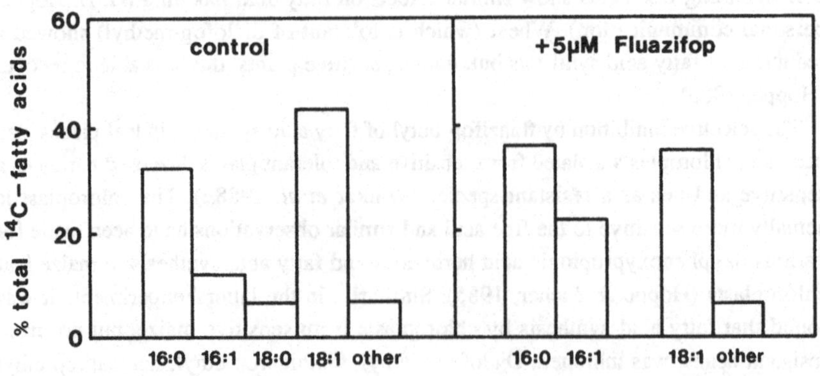

(Hoppe, H. H., personal communication) or in isolated chloroplasts (Hoppe & Zacher, 1985). Similarly, the R-isomer of fluazifop was found to be much more active than the S-isomer and the former is the more potent herbicidal form of the compound. (Walker *et al.* 1988*a,b*).

Fatty acid synthetase in higher plants is a Type II enzyme complex containing proteins which can be isolated individually and which catalyse the various partial reactions (cf. Harwood & Russell, 1984). Accordingly, it is fair to ask which of these partial reactions is sensitive to the action of oxyphenoxypropionic acid herbicides. We do not have a definite answer at present but some circumstantial evidence points to the β-ketoacyl ACP synthetase as being susceptible to inhibition. Thus, in systems where partial inhibition of fatty acid synthesis by fluazifop is measured then there is a significant decrease in the ratio of C_{18}/C_{16} products (Fig. 8). This would indicate β-ketoacyl 2 (cf. Harwood, 1987) as a target site and, by analogy, β-ketoacyl ACP synthetase 1 would be inhibited at higher herbicide concentrations. Experimental results with [^{14}C]-malonate exclude the elongation systems involved in chain lengthening stearate and very long-chain fatty acids as being inhibited by fluazifop (Fig. 9). Thus, the action of fluazifop (and, presumably, other oxyphenoxypropionic acids) appears to be on fatty acid synthetase and probably at the level of the condensing enzymes (but see Addendum).

Work in other laboratories has confirmed the inhibition of fatty acid synthesis caused by oxyphenoxypropionic acids. Fenoxaprop-ethyl inhibits lipid synthesis in susceptible maize and wild oat (Köcher *et al.*, 1982) while fenthiaprop-ethyl was active in maize but not in tolerant soybeans (Köcher *et al.*, 1984). Haloxyfop also inhibits incorporation of radiolabel from [^{14}C]-acetate into the lipids of maize suspension cells (Cho, Widholm & Slife, 1986).

Although they are not oxyphenoxypropionic acids, two cyclohexanedione derivatives, alloxydim and sethoxydim, show herbicidal activity which is very similar to the oxyphenoxy compounds. The symptoms include necrosis of meristematic

Fig. 9. Synthesis of fatty acids from [^{14}C]-malonate by barley in the presence of fluazifop.

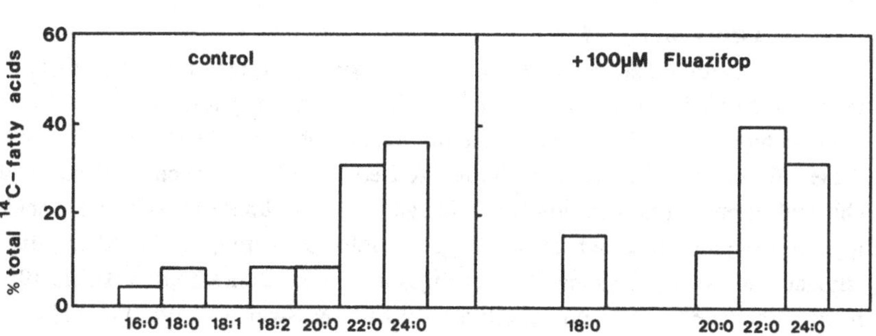

zones, leaf chlorosis, and a selectivity against grasses (Duke & Kenyon, 1987; Veerasekaran & Catchpole, 1982; Lichtenthaler & Meir, 1983). Sethoxydim has been found to inhibit fatty acid synthesis in excised maise root tips (Ishihara *et al.*, 1986), maize leaves (e.g. Burgstahler & Lichtenthaler, 1984) and isolated maize chloroplasts (Burgstahler, Retzlaff & Lichtenthaler, 1986). It is also highly inhibitory towards fatty acid synthesis in barley (K.A. Walker, S.M. Ridley & J.L. Harwood, unpublished observations). However, fatty acid synthesis in tolerant soybean plants was also inhibited (Hatzios, 1982) which may indicate that the herbicide's selectivity lies in its differential metabolism or transport in various plant species.

The physiological effects of oxypenoxypropionic acids, alloxydim and sethoxydim have been reviewed (Duke & Kenyon, 1987). The inhibition of fatty acid synthesis can explain most of the detailed findings such as membrane disruption (Hoppe, 1980), chloroplast damage (Brezeanu, Davis & Shimabukuro, 1976) and associated decreases in chlorophyll, and CO_2 fixation (Köcher *et al.*, 1982) and reduced ATP content (Gronwald, 1986) and mitochondrial dysfunction. Indeed, the time-courses of the rather slow mitochondrial and chloroplast changes are fully consistent with them being secondary to reduced acyl lipid formation (Duke & Kenyon, 1987).

The reason for the effect of diclofop and diclofop-methyl in inhibiting auxin-stimulated reactions (Shimabukuro, Shimabukuro & Walsh, 1982) could also be explained from their reduction of fatty acid synthesis. However, it was also noted that diclofop-methyl can act as a proton ionophore in some experimental systems (Lucas, Wilson, & Wright, 1984). Moreover, there seems to be some antagonism between diclofop-methyl and auxins, including 2,4-D (Todd & Stobbe, 1980). Such antagonism appears to require a sensitive (membrane?) site and is not due to direct chemical interaction (Shimabukuro, Walsh & Hoerauf, 1986).

To summarise, the oxyphenoxypropionic acids appear to be phytotoxic because they inhibit *de novo* fatty acid synthesis rather effectively. Most available evidence is consistent with a mechanism of selectivity which is based on the target site rather than on differential metabolism. We are currently attempting to obtain unequivocal evidence for this.

Other herbicides

Several other herbicides have been reported to affect lipid metabolism and, in addition, certain mutants which are herbicide-resistant may also show changed acyl lipid contents or metabolism. For example, various photosynthetic inhibitors such as those affecting the Hill reaction (Sumida & Ueda, 1973) have been shown to inhibit chloroplast but not non-chloroplast lipid synthesis and dinoseb was found to inhibit glycolipid synthesis in wheat seedlings (St John & Hilton, 1973). Alternatively, atrazine-resistant pea mutants show changes in fatty acid unsaturation (Blein, 1980), *trans* -3-hexadecenoate content (Burke, Wilson, & Swafford, 1982) or lipid class composition (Pillai & St John, 1981). Similarly, the acquisition of diuron-resistance in *E. gracilis* is associated with changes in thylakoid membrane lipids (Troton *et al.*, 1986).

Nevertheless, it is agreed generally that such affects on lipids are secondary (cf. Rivera & Penner, 1979; Duke, 1985; Mann & Pu, 1968). To date only the three classes of herbicides covered in this review can be said to inhibit lipid synthesis as a primary event in their toxicity. However, several lipids are virtually exclusive to the plant kingdom (e.g. the plant sulpholipid cf. Harwood, 1980*b*) and offer attractive targets for the development of new types of herbicides. Moreover, the species selectivity displayed by oxyphenoxypropionic acids, apparently at the target site, provides a challenge to the biochemist researching the detailed mechanism of action of such compounds.

Addendum

Since the original manuscript was submitted (April 1987) considerable progress has been made towards understanding the mode of action of the aryloxyphenoxypropionate and cyclohexanedione herbicides. These compounds inhibit *de novo* fatty acid synthesis (see Walker *et al.*, 1988*a*) and several individual herbicides have been shown to inhibit acetyl-CoA carboxylase in susceptible plants. Direct action of sethoxydim and haloxyfop (Burton *et al.*, 1987) and of haloxyfop and tralkoxydim (Secor and Cseke, 1988) on maize acetyl-CoA carboxylase has been reported. Indirect experiments with different radiolabelled precursors have also confirmed the action of different aryloxyphenoxypropionates on the acetyl-CoA carboxylase from various monocotyledons (Kobek *et al.*, 1988). Moreover, we have shown by three different indirect experiments as well as inhibitory measurements with acetyl-CoA carboxylase that fluazifop acts specifically on the enzyme in sensitive but not resistant plants (Walker *et al.*, 1988*b*). These results not only delineate the mechanism of action of these classes of herbicides but also raise intriguing questions about why the inhibition of the carboxylase is specific to grass species.

References to Addendum

Burton, J.D., Gronwald, J.W., Somers, D.A., Connelly, J.A., Gengenbach, B.G. & Wyse, D.L. (1987). Inhibition of plant acetyl-CoA carboxylase by the herbicides sethoxydim and haloxyfop. *Biochemical and Biophysical Research Communications* **148**, 1039–1044.

Kobek, K., Focke, H. & Lichtenthaler, H.K. (1988). Fatty acid biosynthesis and acetyl-CoA carboxylase as a target of diclofop, fenoxaprop and other aryloxy-phenoxy-propionic acid herbicides. *Zeitschrift für Naturforschung* **43c**, 47-54.

Secor, J. & Cseke, C. (1988). Inhibition of acetyl-CoA carboxylase activity by haloxyfop and tralkoxydim. *Plant Physiology* **86**, 10–12.

Walker, K.A., Ridley, S.M. & Harwood, J.L. (1988*a*). Effects of the selective herbicide, fluazifop, on fatty acid synthesis in pea and barley. *Biochemical Journal*, **254**, 811–17.

Walker, K.A., Ridley, S.M., Lewis, T. & Harwood, J.L. (1988*b*). Fluazifop, a grass-selective herbicide, which inhibits acetyl-CoA carboxylase in sensitive grass species. *Biochemical Journal*, **254**, 307–10.

References

Blair, J.P. (1980) *Chenopodium album* cells in batch suspension culture from atrazine susceptible or tolerant seedlings. *Physiologie Vegetale* **18**, 703–710.

Bolton, P. & Harwood, J.L. (1976*a*). Fatty acid synthesis in aged potato slices. *Phtyochemistry* **15**, 1501–6.

Bolton, P. & Harwood, J.L. (1976*b*). Effect of thiocarbamate herbicides on fatty acid synthesis by potato. *Phytochemistry* **15**, 1507–9.

Brezeanu, A.G., Davis, D.G. & Shimabukuro, R.H. (1976). Ultrastructural effects and translocation of methyl-2-(4-(2,4-dichlorophenoxy)-phenoxy) propanoate in wheat (*Triticum aestivum*) and wild oat (*Avena fatua*). *Canadian Journal of Botany* **54**, 2038–48.

Burgstahler, R.J. & Lichtenthaler, H.K. (1984). Inhibition by sethoxydim of phospho- and galactolipid accumulation in maize seedlings. In *Structure, Function and Metabolism of Plant Lipids*, eds. P-A. Siegenthaler and W. Eichenberger, pp. 619–22. Amsterdam: Elsevier.

Burgstahler, R.J., Retzlaff, G. & Lichtenthaler, H.K. (1986). Mode of action of sethoxydim: effects on the plant's lipid metabolism. *6th IUPAC Pesticide Chemistry Congress, Abstr.* 3B-11.

Burke, J.J., Wilson, R.F. & Swafford, J.R. (1982). Characterisation of chloroplasts isolated from Triazine-susceptible and Triazine-resistant biotypes of *Brassica campestris. Plant Physiology* **70**, 24–9.

Chang, S.S., Ashton, F.M. & Bayer, D.E. (1985). Butachlor influence on selected meatabolic processes of plant cells and tissues. *Journal of Plant Growth Regulation* **4**, 1–9.

Cho, H. Y., Widholm, J.M. & Slife, F.W. (1986). Effects of haloxyfop on corn (*Zea mays*) and soybean (*Glycine max*) suspension cultures. *Weed Science* **34**, 496–501.

Davies, A.O. & Harwood, J.L. (1983). Effect of substituted pyridazinones on chloroplast structure and lipid metabolism in greening barley leaves. *Journal of Experimental Botany* **34**, 1089–100.

Davis, D.G. & Dusbabek, K.E. (1973). Effect of diallate on foliar uptake and translocation of herbicides in pea. *Weed Science* **21**, 16–8.

Duke, S.O. (1985). Effects of herbicides on nonphotosynthetic biosynthetic processes. In *Weed Physiology*, Vol II, Ed. S.O. Duke, pp. 91–112. Boca Raton, Florida: CRC Press Inc.

Duke, S.O. & Kenyon, W.H. (1987). Polycyclic alkanoic acids. In *Herbicide Chemistry, Degradation and Mode of Action*, Vol. III, Eds P.C. Kearny and D. Kaufman, in press. New York: Marcel Dekker.

Ebert, E. (1982). The role of waxes in the uptake of metolachlor into sorghum in relation to the protectant CGA 43089. *Weed Research* **22**, 305–11.

Eder, F.A. (1979). Pyridazinones, their influence on the biosynthesis of carotenoids and the metabolism of lipids in plants. *Zeitschrift für Naturforschung*, Part C, **34**, 1052–4.

Ezra, G., Gressel, J.E. & Flowers, H.M. (1983). Effects of the herbicide EPTC and the protectant DDCA on incorporation and distribution of 2-[^{14}C]-acetate into major lipid fractions of maize cell suspension cultures. *Pesticide Biochemistry and Physiology* **19**, 225–34.

Fedtke, C. (1982). *Biochemistry and Physiology of Herbicide Action.* New York: Springer-Verlag.

Frosch, S., Jabben, M., Begeld, R., Kleinig, H. & Mohr, H. (1979). Inhibition of carotenoid biosynthesis by the herbicide SAN 9789 and its consequences for the action of phytochrome on plastogenesis. *Planta* **145**, 497–505.

Gronwald, J.W. (1986). Effect of haloxyfop and haloxyfop-methyl on elongation and respiration of corn (*Zea mays*) and soybean (*Glycine max*) roots. *Weed Science* **34**, 196–202.

Harwood, J.L. (1980*a*). Fatty acid synthesis. In *Biogenesis and Function of Plant Lipids*, ed. P. Mazliak, P. Benveniste, C. Costes & R. Douce, pp. 143–52. Amsterdam: Elsevier.

Harwood, J. L. (1980*b*). Plant acyl lipids: structure, distribution and analysis. In *The Biochemistry of Plants*, Vol. IV, ed. P.K. Stumpf and E.E. Conn, pp. 1–55. New York: Academic Press.

Harwood, J.L. (1987). Medium and long-chain fatty acid synthesis. In *The Structure and Function of Plant Lipids*, ed. P.K. Stumpf, J.B. Mudd and W.D. Nes, pp. 465–72. New York: Plenum Press.

Harwood, J.L. & Russell, N.J. (1984). *Lipids in Plants and Microbes*. London: George Allen and Unwin.

Harwood, J.L. & Stumpf, P.K. (1970). Synthesis of fatty acids in the intial stages of seed germination. *Plant Physiology* **46**, 500–8.

Harwood, J.L. & Stumpf, P.K. (1971). Control of fatty acids synthesis in germinating seeds. *Archives of Biochemistry and Biophysics* **142**, 281–91.

Hatzios, K.K. (1982). Effects of sethoxydim on the metabolism of isolated leaf cells of soybean (*Glycine max* L. Merr) *Plant Cell Reports* **1**, 87–90.

Hoppe, H.H. (1980). Veranderungen der membranpermenbilitat, des kohlenhydratgehaltes, des lipidgehaltes and der lipidzusammensetzung in keimwurzelspitzen von *Zea mays* L. sach behandlung mit diclofop-methyl. *Zeitschrift für Natursforschung* **100**, 415–26.

Hoppe, H.H. (1981). Einfluss von diclofop-methyl and die protein-nukleinsaure- und lipid biosynthese der keimwurzelspitzen von *Zea mays* L. *Zeitschrift für Planzenphysiologie* **102**, 189–97.

Hoppe, H.H. (1985). Differential effect of diclofop-methyl on fatty acid synthesis in leaves of sensitive and tolerant plant species. *Pesticide Biochemistry and Physiology* **23**, 297–308.

Hoppe, H.H. & Zacher, H. (1982). Hemmung der fettsaurebiosynthese durch diclofop-methyl in keimwurzelspitzen von *Zea mays*. *Zeitschrift fur Pflanzenphysiologie* **106**, 287–98.

Hoppe, H.H. & Zacher, H. (1985). Inhibition of fatty acid biosynthesis in isolated bean and maize chloroplasts by herbicidal phenoxy-phenoxypropionic acid derivatives and structurally related compounds. *Pesticide Biochemistry and Physiology* **24**, 298–305

Ishihara, K., Hosaka, H., Kubota, M., Kamimura, H. & Yasuda, Y. (1986). Effects of sethoxydim on the metabolism of excised root tips of corn. *6th IUPAC Pesticide Chemistry Congress, Abstract* 3B-1D.

Jones, V.H.M. & Harwood, J.L. (1980). Desaturation of linoleic acid from exogenous lipids by isolated chloroplasts. *Biochemical Journal* **190**, 851–4

Khan, M-U, Lem, N.W., Chandorkar, K.R. & Williams, J.P. (1979). Effects of substituted pyridazinones (San 6706, San 9774, San 9785) on glycerolipids and their associated fatty acids in leaves of *Vicia faba* and *Hordeum vulgare*. *Plant Physiology* **64**, 300–5.

Köcher, H., Kellner, H.M. Lötzsch, K. & Dorn, E. (1982). Mode of action and metabolic fate of the herbicide fenoxaprop-ethyl, Hoe 33171. *Proceedings of the British Crop Protection Conference on Weeds* 341–7.

Köcher, H., Lotzsch, K.& Gantz, D. (1984). Mode of action of HOE 35609, a new herbicide for the control of grass weeds in broad-leaved crops. *Proceedings EWRS 3rd Symposium on Weed Problems in the Mediterranean Area* 559–66.

Kollatukudy, P.E. (1975). Biochemistry of cutin, suberin and waxes, the lipid barriers of plants. In *Recent Advances in the Chemistry and Biochemistry of Plant Lipids*, ed. T. Galliard and E.I. Mercer, pp. 203–46. London: Academic Press.

Kollatukudy, P.E. (1980). Cutin, suberin and waxes. In *The Biochemistry of Plants*, Vol.. IV, ed. P.K. Stumpf and E.E. Conn, pp. 571–645. New York: Academic Press.

Kollatukudy, P.E. & Brown, L. (1974). Inhibition of cuticular lipid biosynthesis in *Pisum sativum* by thiocarbamates. *Plant Physiology* **53**, 903–6.

Laskay, G. & Lehoczki, E. (1986). Correlation between linolenic acid deficiency in chloroplast membrane lipids and decreasing photosynthetic activity in barley. *Biochimica et Biophysica Acta* **849**,77–84

Laskay, G., Lehoczki, E., Dobi, A.L. & Szalay, L. (1986). Photosynthetic characteristics of detached barley leaves during greening in the presence of SAN 9785. *Planta* **169**, 123–9

Lay, M-M., Hubbell, J.P. & Casida, J. E. (1975). Dichloroacetamide antidotes for thiocarbamate herbicides: mode of action. *Science* **189**, 287–9

Leech, R.M., Walton, C.A. & Baker, N.R. (1985). Some effects of 4-chloro-5-(dimethylamino)-2-phenyl-3(2H)-pyridazinone (San 9785) on the development of chloroplast thylakoid membranes in *Hordeum vulgare* L. *Planta* **165**, 277–83.

Lem, N.W.& Williams, J.P. (1983). Changes in the [^{14}C]-labelling of molecular species of 3-monogalactosyl-1,2-diacylglycerol in leaves of *Vicia faba* treated with compound San 9785. *Biochemical Journal* **209**, 513–18.

Lessire, R. & Stumpf, P.K. (1982). Nature of fatty acid synthetase systems in parenchymal and epidermal cells of *Allium porrum* leaves. *Plant Physiology* **73**, 614–18.

Lichtenthaler, H.K. & Meier, D. (1983). Inhibition by sethoxydim of chloroplast biogenesis, development and replication in barley seedlings. *Zeitschrift fur Natursforschung*, Part C, **39**, 115–22.

Lucas, W.J., Wilson, C. & Wright, J.P. (1984). Perturbation of *Chara* plasmalemma transport function by 2(4(2',4'-dichloro-phenoxy)phenoxy) propionic acid. *Plant Physiology* **74**, 61–6.

Mann, J.B. & Pu, M. (1968). Inhibition of lipid synthesis by certain herbicides. *Weed Science* **16**, 197–8.

Mellis, P., Pillai, P., Davies, P.E. & Truelove, B. (1982). Metolachlor and alachlor effects on membrane permeability and lipid synthesis. *Weed Science* **30**, 399–404.

Murphy, D.J., Harwood, J.L., St John, J.B. & Stumpf, P.K. (1980). Effect of a substituted pyridazinone, compound BASF 13-338, on membrane lipid synthesis in photosynthetic tissues. *Biochemical Society Transactions* **8**, 119–20.

Murphy, D.J., Harwood, J.L., Lee, K.A., Roberto, F., Stumpf, P.K. & St John, J.B. (1985). Differential responses of a range of photosynthetic tissues to a substituted pyridazinone, Sandoz 9785, Specific effects on fatty acid desaturation. *Phytochemistry* **24**, 1923–9.

Nestler, H.J. (1982). Phenoxy-phenoxypropionic acid derivatives and related compounds. In *Chemie der Pflanzenschutz-und Schadlingsbekamp Fungsmittel*, ed. R. Wegler, pp. 1–25. Berlin: Springer-Verlag.

Pillai, P. & St John, J.B. (1981). Lipid composition of chloroplast membranes from weed biotypes differentially sensitive to triazine herbicides. *Plant Physiology* **68**, 585–7.

Rivera, C.M. & Penner D. (1979). Effect of herbicides on plant cell membrane lipids. *Residue Reviews* **70**, 45–76.

St John, J.B. (1976). Manipulation of galactolipid fatty acid composition with substituted pyridazinone. *Plant Physiology* **57**,38–40

St John, J.B. (1982). Effects of herbicides on the lipid composition of plant membranes. In *Biochemical Responses Induced by Herbicides* ed. D.E. Moreland, J.B. St John and F.D. Hess, pp. 97–109. Washington: American Chemical Society

St John, J.B. & Christiansen, M.N. (1976). Inhibition of linolenic acid synthesis and modification of chilling resistance in cotton seedlings. *Plant Physiology* **57**, 257–9.

St John, J.B. & Hilton, J.L. (1973). Lipid metabolism as a site of herbicide action. *Weed Science* **21**, 477–80.

St John, J.B. & Hilton, J.L. (1976). Structure versus activity of substituted pyridazinones as related to mechanism of action. *Weed Science* **24**, 579–82.

St John, J.B., Christiansen, M.N., Ashworth, E.N. & Gentner, W.A. (1979). Effect of BASF 13-338, a substituted pyridazinone, on linolenic acid levels and winter hardiness of cereals. *Crop Science* **19**, 65–9.

St John, J.B. Schirmer, U., Rittig, F.R. & Bleiholder, H. (1984). Substituted pyridazinone herbicides: Structural requirements for action on membrane lipids. In *Pesticide Synthesis through Rational Approaches*, eds. M. Kohn, P.S. Magee and J.J. Menn, pp. 145–62. Washington: American Chemical Society.

Shimabukuro, M.A., Shimabukuro, R.H. & Walsh, W.C. (1982). The antagonism of IAA-induced hydrogen ion extrusion and coleoptile growth by diclofop-methyl. *Physiologia Plantarum* **56**, 444–452.

Shimabukuro, R.H., Walsh, W.C. & Hoerauf, R.A. (1986). Reciprocal antagonism between the herbicides, diclofop-methyl and 2,4-D in corn and soybean tissue culture. *Plant Physiology* **80**, 612–17.

Still, G.G., Davis, D.G., & Zander, G.L. (1970). Plant epicuticular lipids: alteration by herbicidal carbamates. *Plant Physiology* **46**, 307–14.

Sumida, S. & Ueda, M. (1973). Studies of pesticide effects on *Chlorella* metabolism. 1. Effect of herbicides on complex lipid biosynthesis. *Plant and Cell Physiology* **14**, 781–5.

Todd, B.G. & Stobbe, E.H. (1980). The basis of the antagonistic effect of 2,4-D on diclofop-methyl toxicity to wild oat. *Weed Science* **28**, 371–7.

Troton, D., Calvayrac, R., Pham Thi, A.T. & Laval-Martin, D. (1986). Modifications of thylakoid lipids in *Euglena gracilis* during diuron-adaptation. *Phytochemistry* **25**, 393–9.

Veerasekaran, P. & Catchpole, A.H. (1982). Studies on the selectivity of alloxydim-sodium in plants. *Pesticide Science* **13**, 452–62.

Warmund, M.R. Kerr, H.D. & Peters, E.J. (1985). Lipid metabolism in grain sorghum (*Sorghum bicolor*) treated with alachlor plus flurazole. *Weed Science* **33**, 25–8.

Wharfe, J. &Harwood, J. L. (1978). Fatty acid biosynthesis in the leaves of barley, wheat and pea. *Biochemical Journal* **174**, 163–9.

Wilkinson, R.E. (1985). CDAA inhibition of kaurene oxidation in etiolated sorghum coleoptiles. *Pesticide Biochemistry and Physiology* **23**, 19–23.

Wilkinson, R.E. (1986). Diallate inhibition of gibberellin biosynthesis in sorghum coleoptiles. *Pesticide BIochemistry and Physiology* **25**, 93–7.

Wilkinson, R.E. & Hardcastle, W.S. (1969). EPTC effects on sicklepod petiolar fatty acids. *Weed Science* **17**, 335–8.

Wilkinson, R.E. & Smith, A.E. (1975). Thiocarbamate inhibition of fatty acid biosynthesis in isolated spinach chloroplasts. *Weed Science* **23**, 100–4.

Wilkinson, R. E. & Smith, A.E. (1976). Butylate, pebulate and vernolate inhibition of plant fatty acid biosynthesis. *Phytochemistry* **15**, 841–2

Willemot, C. (1977). Simultaneous inhibition of linleic acid synthesis in winter wheat roots and frost hardening by BASF 13-338, a derivative of pyridzinone. *Plant Physiology* **60**, 1–4

Willemot, C., Slack, C.R., Browse, J. & Roughan, P.G. (1982). Effect of BASF 13-338, a substituted pyridazinone, on lipid metabolism in leaf tissue of spinach, pea, linseed and wheat. *Plant Physiology* **70**, 78–81.

JOHN R. COGGINS

The shikimate pathway as a target for herbicides

Introduction

The observation reported in 1980 by Amrhein and his colleagues (Amrhein, Schab & Steinrücken, 1980; Steinrücken & Amrhein, 1980) that the herbicide glyphosate was a highly specific inhibitor of the shikimate pathway enzyme 5-enolpyruvylshikimate 3-phosphate (EPSP) synthase served to focus the attention of many agricultural scientists on the possibility of finding other herbicides which acted by inhibiting the biosynthesis of macromolecular precursors and especially amino acids. Although knowledge of most pathways of amino acid biosynthesis in higher plants was still fairly primitive the idea that amino acid biosynthetic enzymes were good targets for herbicides was soon reinforced by the recognition that the sulfonyl urea and imidazolinone based herbicides (La Rossa & Schloss, 1984; La Rossa & Falco, 1984; Shaner, Anderson & Stidham, 1984; see Chapter 7) acted by inhibiting acetolactate synthase, a key enzyme in branched chain amino acid biosynthesis, and phosphinothricin (Leason *et al.*, 1982; see Chapter 8) acted by inhibiting glutamine synthetase. All these compounds had resulted from random screening procedures but it is now widely believed in the agrochemical industry that it should be possible to accelerate the development of new herbicides by the judicious application of knowledge of the biochemistry and molecular biology of amino acid biosynthesis.

The principal information required for rational herbicide design is the structure and mechanism of the target enzymes. This immediately highlights a difficulty. There was, and continues to be, a general lack of mechanistic and structural information for plant amino acid biosynthetic enzymes. Until recently this was also true for the corresponding microbial enzymes. In this Chapter I shall illustrate, for the shikimate pathway, how this situation is rapidly changing because of the application of both molecular biological techniques and the new fast, high performance chromatographic methods for enzyme purification.

The shikimate pathway

In plants, as in micro-organisms, the biosynthesis of all the aromatic compounds involved in primary metabolism proceeds by way of the shikimate pathway (Haslam, 1974; Weiss & Edwards, 1980; Conn, 1986) (see Fig. 1). The seven steps of this pathway lead from erythrose 4-phosphate via shikimate to chorismate, which is the common precursor of all the aromatic amino acids and many other important compounds such as *p*-aminobenzoic acid, ubiquinone and vitamin K (see Fig. 2). The pathway is of particular importance in plants since it provides the precursors for lignin, the flavonoids and a host of secondary metabolites. These aromatic compounds account for up to 35% of the dry weight of higher plants. This indicates that a substantial proportion of carbon flux in plants must be through the shikimate pathway (Boudet, Graziana & Ranjeva, 1985).

Fig. 1. The reactions of the shikimate pathway. The numbers refer to the enzymes of the pathway: (1) 3-deoxy-D-*arabino*-heptulosonic acid 7-phosphate (DAHP) synthase (EC 4.1.2.15), (2) 3-dehydroquinate (DHQ) synthase (EC 4.6.1.3), (3) 3-dehydroquinase (EC 4.2.1.10), alternative name 3-dehydroquinate dehydratase, (4) shikimate dehydrogenase (EC 1.1.1.25), (5) shikimate kinase (EC 2.7.1.71), (6) 5-enolpyruvylshikimate 3-phosphate (EPSP) synthase (EC 2.5.1.19), alternative name 3-phosphoshikimate 1-carboxyvinyltransferase, (7) chorismate synthase (EC 4.6.1.4).

The *arom* multifunctional enzyme of *N. crassa* catalyses the reactions numbered 2 to 6 in the above scheme.

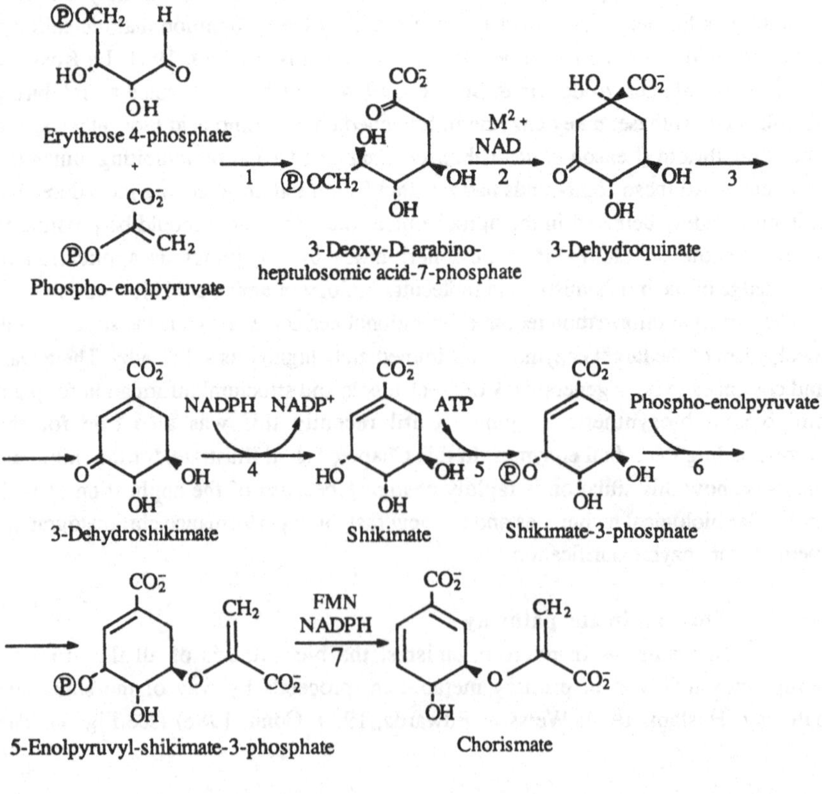

Erythrose 4-phosphate + Phospho-enolpyruvate —1→ 3-Deoxy-D-arabino-heptulosomic acid-7-phosphate —2 (M^{2+}, NAD)→ 3-Dehydroquinate —3→

3-Dehydroshikimate —4 (NADPH, $NADP^+$)→ Shikimate —5 (ATP)→ Shikimate-3-phosphate —6 (Phospho-enolpyruvate)→

5-Enolpyruvyl-shikimate-3-phosphate —7 (FMN, NADPH)→ Chorismate

The chemical intermediates between erythrose 4-phosphate and chorismate were first isolated and characterized from micro-organisms by Davis, Weiss, Sprinson & Gibson and their collaborators more than 20 years ago (Davis, 1955; Levin & Sprinson, 1964; Gibson & Pittard, 1968) and this work has recently been reviewed (Weiss, 1986). It is generally believed that the same intermediates occur in plants (Gilchrist & Kosuge, 1980) although the evidence is much less complete than for micro-organisms. There is also some data which suggests that in plants a parallel pathway for aromatic biosynthesis occurs which involves quinate as an intermediate (Boudet, Graziana & Ranjeva, 1985).

Although the chemistry of the shikimate pathway has been an active research area ever since the pathway was elucidated it is only very recently that the detailed study of the individual enzymes has been undertaken. This was principally because the enzymes are present at very low levels especially in plants and their purification was difficult. Also the assay of the individual enzymes by simple continuous spectrophotometric methods was generally not possible because the substrates and coupling enzymes were not generally available. Thus, in 1980, Steinrücken & Amrhein, who first demonstrated that EPSP synthase was the site of action of glyphosate, were obliged to study the enzyme activity in crude extracts of *Aerobacter aerogenes*. At that

Fig. 2. Some of the aromatic compounds derived from chorismic acid in plants.

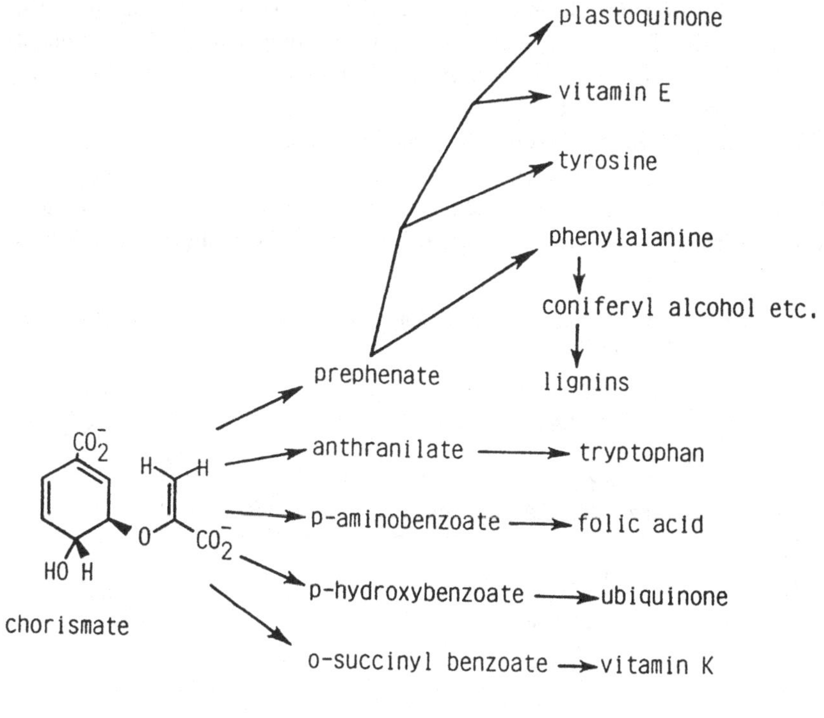

plastoquinone
vitamin E
tyrosine
phenylalanine
coniferyl alcohol etc.
lignins
prephenate
CO_2^-
H H
O CO_2^-
HO H
chorismate
anthranilate → tryptophan
p-aminobenzoate → folic acid
p-hydroxybenzoate → ubiquinone
o-succinyl benzoate → vitamin K

time no monofunctional, higher plant EPSP synthase had been purified and the only homogenous preparations of EPSP synthase available were the multifunctional *arom* enzymes of *Neurospora crassa* (Lumsden & Coggins, 1977; 1978; Gaertner & Cole, 1977) and *Euglena gracilis* (Patel & Giles, 1979).

The enzymes of the shikimate pathway from *Neurospora crassa*

In 1981 following the work from Amrhein's laboratory on the effect of glyphosate on EPSP synthase there was only one organism, *N. crassa*, from which all the shikimate pathway enzymes were available in pure form (see Coggins, 1986). Five of these enzyme activities (including EPSP synthase) occurred as the *arom* complex which is a multifunctional enzyme consisting of two identical pentafunctional polypeptide chains (Lumsden & Coggins, 1977; 1978; Gaertner & Cole, 1977; for reviews see Coggins & Boocock, 1986; Coggins *et al.*, 1987*a*, 1987*b*). The remaining two enzymes DAHP synthase (Nimmo & Coggins, 1981) and chorismate synthase (Welch *et al.*, 1974; Boocock, 1983; White, Millar & Coggins, 1988) were separate entities. Although these pure enzymes were only available in small quantities (1 to 5 mg) they were to prove invaluable as model enzymes for kinetic studies and to screen potential herbicidal compounds. It was at this stage that my laboratory in Glasgow entered the herbicide field.

Our first project was a detailed study of the steady state kinetics of *N. crassa* EPSP synthase and its kinetics of inhibition by glyphosate (Boocock & Coggins, 1983; Boocock, 1983). The kinetic patterns indicated a compulsory order sequential mechanism in which either phosphoenolpyruvate or glyphosate bound to an enzyme: shikimate 3-phosphate complex (Fig. 3).

This type of inhibition, involving a ternary dead-end complex, explained the biological effectiveness of glyphosate. The build-up of shikimate 3-phosphate or shikimate in poisoned plant tissue could not relieve the inhibition; only a build-up of phosphoenolpyruvate which is highly unlikely in view of the multiple metabolic uses

Fig. 3. The proposed kinetic mechanism of the *N. crassa arom* EPSP synthase (Boocock, 1983).

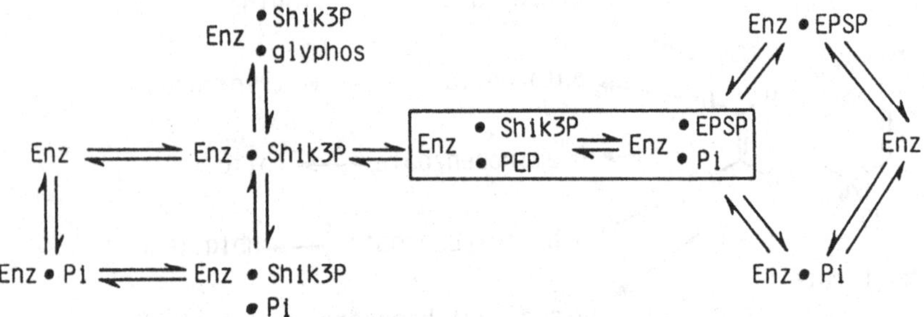

Table 1. *Insensitivity of* arom *EPSP synthase to inhibition by structural analogues of glyphosate (Boocock, 1983). (Most of the analogues were supplied by Dr. T. Lewis, ICI Agrochemicals.)*

Structure	Trivial name	Estimated K_I' / K_M' (PEP)
Ⓟ $CH_2\ NH\ CH_2\ CO_2H$	glyphosate	0.32
Ⓟ $CH_2\ NH\ CH_3$		> 200
Ⓟ $CH_2\ NH_2$		> 30
$H_2N\ CH_2\ CO_2H$	glycine	> 100
$Me\ CO\ CO_2H$	pyruvate	> 100
$Me\ CH(OH)CO_2H$	L - lactate	> 100
Ⓟ $CH_2\ O\ CH_2\ CO_2H$		> 60
Ⓟ $CH_2\ S\ CH_2\ CO_2H$		> 180
Ⓟ $CH_2\ NH\ CO\ CCl_3$		> 50
$Me—PO(OH)—CH_2\ NH\ CH_2\ CO_2H$		> 50
Ⓟ $CH_2\ N(Me)CH_2\ CO_2H$		> 90
Ⓟ $CH_2\ NH\ CH(Me)\ CO_2H$		> 150
Ⓟ $CH_2\ NH\ CH_2\ CH_2\ CO_2H$		> 50
Ⓟ $CH_2\ CH_2\ NH\ CH_2\ CO_2H$		> 120
(Ⓟ $CH_2)_2\ N\ CH_2\ CO_2H$	glyphosine	> 4.0
Ⓟ $CH_2\ CH(NH_2)\ CO_2H$		> 200
Ⓟ $O\ CH_2\ CH(NH_2)\ CO_2H$	phosphoserine	> 20
Ⓟ $CH_2\ CH_2\ CH(NH_2)\ CO_2H$		> 50
Ⓟ epoxide (H, H, Me, O)	fosfomycin	> 100

of this intermediate, would be expected to compete out the glyphosate inhibition. It is interesting to note that Cornish-Bowden (Cornish-Bowden, 1986) has suggested that drugs and herbicides should be specifically designed to achieve this kind of effect.

The next stage of our work involved an extensive study of the effect of structural analogues of glyphosate on EPSP synthase. The results which are shown in Table 1 showed very clearly that glyphosate was an isolated toxophore. Even very small changes in the structure of glyphosate resulted in the virtual abolition of its ability to inhibit EPSP synthase. The only compound which shared significant inhibitory activity was glyphosine and it was later suggested that this was probably due to glyphosate contamination in our sample (J. Franz, personal communication).

Our principal difficulty at this time was that we did not know whether the *N. crassa* EPSP synthase was a good model for the plant enzyme or whether the monofunctional bacterial enzyme might be a better model. The same question applied to the other *N. crassa* enzymes which we were interested in using as model enzymes to aid in herbicide design. The only solution was to purify and characterize some of the bacterial and plant enzymes. Initially we chose to concentrate our attention on the shikimate pathway enzymes of *E. coli* since there was a substantial amount of background genetic information and because the cloning of the genes encoding these enzymes was clearly feasible using the rapidly developing techniques for genetic engineering.

The shikimate pathway enzymes of *Escherichia coli*

The genetic studies of Pittard and his collaborators (Pittard & Wallace, 1966) showed that the genes encoding the seven shikimate pathway enzymes of *E. coli* were widely scattered about the bacterial genome and indicated that each enzyme is carried on a separate polypeptide chain. It was also known that the five central actives (those which in *N. crassa* are carried on the *arom* multifunctional enzyme) were separable on sucrose gradients (Berlyn & Giles, 1969) and did not form a multienzyme complex.

The purification of the *E. coli* enzymes proved to be very demanding since all are present at very low levels. Purification factors in the range 2000 to 20,000 are typically required (Lewendon & Coggins, 1983; Chaudhuri & Coggins, 1985). However, despite considerable experimental difficulties all the *E. coli* shikimate pathway enzymes have now been purified to homogeneity: the DAHP synthases in the laboratory of Herrmann (Herrmann, 1983), DHQ synthase in Knowles laboratory (Frost *et al.*, 1984; Mehdi, Frost & Knowles, 1987) and the remaining five enzymes in Glasgow (Chaudhuri *et al.*, 1986, 1987*b* ; Chaudhuri & Coggins, 1985; Chaudhuri *et al.*, 1987*a* ; Millar *et al.*, 1986*b* ; Lewendon & Coggins, 1983, 1987; White, Millar & Coggins, 1988). Details of the quaternary structure of the five central enzymes of the pathway are given in Table 2. These data, together with extensive limited proteolysis and chemical modification data, lead us to propose that the fungal *arom* multifunctional enzymes have arisen by the fusion of five structural domains each one of which was homologous to the subunit of the corresponding monofunctional *E. coli* enzyme (Coggins *et al.*, 1985). This hypothesis has now been confirmed by extensive sequencing work (Duncan, Edwards & Coggins, 1987).

Cloning the genes for the *E. coli* shikimate pathway enzymes

We now had access to the pure shikimate pathway enzymes from two species. However, the relatively small amounts of enzyme available (generally from 10 to 100 μg) still made large-scale screening experiments and very detailed structural and mechanistic studies impossible. The obvious way forward was to use molecular biological techniques to clone and over-express the appropriate genes. For *E. coli* the

Table 2. *The polypeptide chain organization of the five central shikimate pathway enzymes in micro-organisms. The* N. crassa arom *polypeptide chain is approximately the same length as the yeast* arom *polypeptide chain*

Enzyme	Polypeptide chain length (amino acids)	Calculated M_r	Quaternary structure
3-Dehydroquinate synthase	362	38,880	monomer
3-Dehydroquinase	240	26,377	dimer
Shikimate dehydrogenase	272	29,380	monomer
Shikimate kinase	173	18,937	monomer
EPSP synthase	427	46,112	monomer
Total	1474	159,689	
Yeast *arom*	1588	174,555	dimer

locations of the genes encoding the shikimate pathway enzymes were known and it proved to be a straightforward matter to clone (or in most cases to sub-clone) the genes using relief of auxotrophy for selection. The genes for the three DAHP synthases (*aro* F, *aro* G and *aro* H, see Herrmann, 1983, and Pittard, 1987) and for 3-dehydroquinase (*aro* D, Kinghorn *et al.*, 1981) were cloned in other laboratories while in Glasgow we cloned the genes for DHQ synthase (*aro* B, Duncan & Coggins, 1984), shikimate dehydrogenase (*aro* E, Anton & Coggins, 1984), shikimate kinase (*aro* L, Millar *et al.*, 1986*b*) EPSP synthase (*aro* A, Duncan & Coggins, 1984) and chorismate synthase (*aro* C, Millar *et al.*, 1986*a*). The genes have now all been sequenced and the enzyme sequences deduced and confirmed by N-terminal protein sequencing (the DAHP synthases by Herrmann's group, see Herrmann, 1983; the remaining enzymes in Glasgow: Millar & Coggins, 1986; Duncan *et al.*, 1986; Anton & Coggins, 1988; Millar *et al.*, 1986*b* ; Duncan, Lewendon & Coggins, 1984*b* ; White, Millar & Coggins, 1988). The availability of the cloned genes has in every case allowed the construction of overproducing strains. Even the 'first generation' overproducing strains which simply utilized multicopy plasmids greatly simplified the enzyme isolation and gave milligram rather than microgram quantities of enzyme (Duncan, Lewendon & Coggins, 1984*a*). More recently the use of special expression vectors has made the enzymes readily available on the 100 milligram scale (Anton & Coggins, 1978).

The shikimate pathway enzymes in plants

When our work on herbicides began the only shikimate pathway enzyme which had been purified from plants was the bifunctional enzyme shikimate dehydrogenase/3-dehydroquinase (Polley, 1978; Koshiba, 1978). This was hardly

Table 3. *Comparison of the kinetic parameters of three purified EPSP synthases*

Kinetic parameter	*E. coli* enzyme	*N. crassa* enzyme	*P. sativum* enzyme
K_m for PEP	16 μM	2.7 μM	5.2 μM
K_m for shik 3-P	2.5 μM	0.36 μM	7.7 μM
K_m for EPSP	3.0 μM	0.25 μM	5.2 μM
K_m for phosphate	2.5 mM	1.8 mM	4.0 mM
K_i for glyphosate	0.9μM	1.1 μM	0.08 μM

surprising since in plants the shikimate pathway enzymes are present at even lower levels than in micro-organisms, for example, the amount of EPSP synthase found in pea seedling shoot tissue (on a fresh weight basis) is approximately one-tenth that found in *E. coli*. Our work with the plant enzymes has been greatly facilitated by our experience with purifying (and assaying) the microbial enzymes and by the advent of excellent high performance columns, particularly for ion exchange chromatography of enzymes (Mousdale & Coggins, 1984, 1986*b*, 1987; Mousdale *et al.*, 1987). Thus we found that the procedure developed for the purification of *E. coli* EPSP synthase worked very well for the *Pisum sativum* enzyme (Mousdale & Coggins, 1984). This enzyme, like the *E.coli* enzyme, is monomeric and has a subunit M_r of 50,000 (Mousdale & Coggins, 1984). Although the amount of enzyme obtained was very small, it has allowed the kinetic characterization of the plant enzyme and its comparison with the microbial enzymes (see Table 3).

To assist comparative kinetic studies on EPSP synthase, and in particular to allow the sensitivity to different inhibitors to be assessed, a simple, rapid procedure has been developed which enables the enzyme to be purified to near homogeneity from a variety of plant species within one working day (Mousdale & Coggins, 1986, 1987). The method takes advantage of the known sub-cellular localization of the enzyme (Mousdale & Coggins, 1985*a*), the use of cellulose phosphate as an affinity column (Lewendon & Coggins, 1983; Mousdale & Coggins, 1984) and the excellent resolution of the soluble proteins of the chloroplast that can be obtained on a Mono-Q anion exchange column (Mousdale & Coggins, 1985*a*, 1986*a*). Washed chloroplasts are first isolated, then lysed and after centrifugation the soluble proteins are separated on an anion exchange column (Fig. 4) which separates the central shikimate pathway enzymes. Further chromatography of the EPSP synthase peak on a cellulose phosphate column gives protein that is essentially homogeneous as judged by polyacrylamide gel electrophoresis in the presence of sodium dodecyl sulphate. The method works for species as diverse as maize, spinach and pea (Mousdale &

Coggins, 1986*a*; 1987). By using similar methods it should soon be possible to isolate microgram quantities of all the shikimate pathway enzymes from plants.

Three general points seem to be emerging from our work on the plant enzymes. First, the enzymes are predominantly, although in the case of shikimate dehydrogenase/3-dehydroquinase not exclusively, chloroplastic (Mousdale & Coggins, 1985*a* ; Mousdale, Campbell & Coggins, 1987). Secondly, the quaternary

Table 4. *Shikimate pathway enzyme activities*

Enzyme	*P. sativum* chloroplast stroma (nkat/mg protein)	*N. crassa* crude extract (nkat/mg protein)
DAHP synthase	0.026	0.05
Dehydroquinate synthase	0.166	0.13
Dehydroquinase	0.169	0.16
Shikimate dehydrogenase	0.972	0.42
Shikimate kinase	0.101	0.13
EPSP synthase	0.073	0.12
Chorismate synthase	0.033	0.08

Fig. 4. Elution profile of the central shikimate pathway enzymes from a spinach chloroplast lysate. The lower line shows absorbance at 280 nm; the upper lines show enzyme activity: (▲) EPSP synthase; (●) shikimate dehydrogenase; (○) 3-dehydroquinase; (Δ) shikimate kinase; (❑) 3-dehydroquinate synthase. A Pharmacia Mono-Q column was used; full experimental details are given in Coggins (1986).

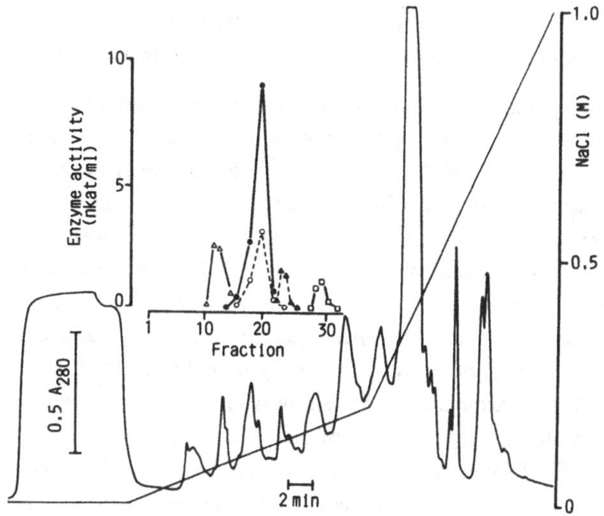

structure of the plant enzymes is very similar to that of the corresponding *E. coli* enzymes and thirdly, although the quaternary structure of the fungal enzymes is very different from that of the plant and microbial enzymes, the functional domains involved in catalysis are conserved between micro-organisms and plants.

Although the work with the plant enzymes is still rather fragmented expertise gained with the preparation (Coggins *et al.*, 1987*a, b*) and analysis (Mousdale & Coggins, 1985*b*) of the substrates and the assay (Coggins *et al.*, 1987*a*) of the microbial enzymes has allowed us to make much more progress with the plant enzymology than would have otherwise been the case. For the first time all the plant enzymes have been detected (Mousdale & Coggins, 1986*b*) and their relative activities determined (Table 4). The microbial enzymes are available for screening potential herbicidal compounds and for mechanistic and structural studies and the plant enzymes are available in either purified or partially purified form for kinetic characterization.

Strategies for obtaining herbicide resistance

The new knowledge of the enzymology and the molecular biology of the shikimate pathway is already assisting the search for novel herbicidal compounds that act by inhibiting the enzymes of the pathway. This knowledge can also be used to address the problem of selectivity since it can be used to aid the rational development of herbicide-resistant varieties of crop plants.

There are several mechanisms by which herbicide resistance might arise in plants, for example:

1. The uptake of the herbicide may become impaired.
2. Plants may acquire the ability to detoxify the herbicide by conjugation or degradation.
3. The target enzyme may become insensitive to the herbicide by mutation.
4. The plants may become resistant by adapting to overproduce the target enzyme.

The latter two approaches have attracted most attention and several groups have now obtained glyphosate-resistant plants. It was first noted that the overproduction of EPSP synthase in *E. coli* results in glyphosate resistance (Rogers *et al.*, 1983; Duncan, Lewendon & Coggins, 1984*a*). Plant cell cultures and bacterial cultures grown in the presence of moderate concentrations of glyphosate were found to adapt to grow at nearly normal rates by over-expressing EPSP synthase (Amrhein *et al.*, 1983; Nafziger *et al.*, 1984; Smart *et al.*, 1985). Also two groups have showed that glyphosate resistance in micro-organisms could also arise by the acquisition of a glyphosate-resistant form of EPSP synthase by mutation (Comai, Sen & Stalker, 1983; Schulz, Sost & Armhein, 1984). These results have now been exploited to construct glyphosate-resistant varieties of crop plants. The gene for a glyphosate-resistant form of EPSP synthase was cloned from *Salmonella typhimurium* (Stalker, Hiatt & Comai, 1985) and it was then transferred to and expressed in tobacco plants.

The resulting plants were glyphosate-tolerant (Comai *et al.*, 1985) but the introduced enzyme was not targeted as it should have been to the chloroplasts (Mousdale & Coggins, 1985*a*; Shields, 1985). Recently the Monsanto group have achieved glyphosate tolerance by proper targeting of a mutant *E. coli* EPSP synthase gene in a plant (Della-Cioppa *et al.*, 1987). They have also engineered a glyphosate-resistant petunia plant which over-expresses EPSP synthase (Shah *et al.*, 1986).

The other possible ways of achieving glyphosate resistance using mechanisms (1) and (2) above have not yet succeeded. Too little is yet known about the uptake and translocation of glyphosate for there to be any hope of rationally interfering with this process at present. It is known that soil bacteria readily degrade glyphosate and therefore must have the necessary degrading enzymes but, to date, all attempts to purify these enzymes and to characterize their genes have failed. However, the possibility of introducing into plants the ability to degrade or detoxify glyphosate remains an attractive option particularly since an analogous approach has recently been used successfully to produce phosphinothricin-resistant plants (De Block *et al.*, 1987).

Current work

The major current effort is to learn more about the structures and mechanisms of the individual enzymes of the pathway. Not surprisingly, because of the commercial interest, most progress has been made with EPSP synthase. Recent work from the Monsanto group, on the now readily available *E. coli* enzyme, has suggested that the active site contains a lysine residue, Lys-22 (Huynh *et al.*, 1988) and a histidine residue (Huynh, 1987) and the enzyme has also been crystallized (Abdel-Mequid *et al.*, 1985). Once its three-dimensional structure has been determined, it will be possible to employ protein engineering techniques to try and construct improved glyphosate-resistant forms of the enzyme. Plant genes for EPSP synthase have also now been cloned (Shah *et al.*, 1986; Klee, Muskopf & Gasser, 1987) and the over produced plant enzymes will soon be available. It is interesting to note that the sequences of the plant enzymes indicate that they are more homologous to the corresponding *E. coli* and *S. typhimurium* monofunctional enzymes than to the corresponding domain of the fungal multifunctional enzymes (Klee, Muskopf & Gasser, 1987; Duncan *et al.*, 1987). More precise work on the subcellular localization (Smart & Amrhein, 1987) and on the targeting (Della-Cioppa *et al.*, 1986) of EPSP synthase has recently been reported. In addition, the ready availability of the other shikimate pathway enzymes from bacteria is promoting their study and new herbicides may well arise from this work.

Conclusion

In this chapter I have reviewed recent work on the enzymology and molecular biology of the shikimate pathway and have attempted to show how strategies for herbicide invention have changed as more knowledge has become available. The materials and techniques necessary to facilitate the rational invention and improvement of herbicides directed against this pathway, and by analogy other similar pathways, are clearly now available.

Acknowledgement

I would like to thank all the members of my Research Group in Glasgow and especially Dr David Mousdale for all their support over many years. It is a pleasure also to acknowledge the financial support of the Science and Engineering Research Council, ICI Plant Protection Division and The Royal Society of Edinburgh.

References

Abdel-Mequid, S., Smith, W.W. & Bild, G.S. (1985) Crystallization of 5-enolpyruvylshikimate 3-phosphate from *Escherichia coli*. *Journal of Molecular Biology*, **186**, 673.

Amrhein, N. (1986). Specific inhibitors as probes into the biosynthesis and metabolism of aromatic amino acids. In *The Shikimic Acid Pathway* (Conn, E.E., Ed.) pp. 83–117, Plenum Press, New York.

Amrhein, N., Schab, J. & Steinrücken, H.C. (1980) The mode of action of the herbicide glyphosate. *Naturwissenschaften* **67**, 356–7.

Amrhein, N., Johanning, D., Schab, J. & Schulz, A. (1983). Biochemical basis for glyphosate-tolerance in a bacterium and a plant tissue culture. *Federation of European Biochemical Societies Letters* **157**, 191–6.

Anton, I. & Coggins, J.R. (1984). Subcloning of the *Escherichia coli* shikimate dehydrogenase gene (*aro* E) from transducing bacteriophage *spc1*. *Biochemical Society Transactions* **12**, 275–6.

Anton, I.A., Duncan, K.& Coggins, J.R. (1987). A eukaryotic repressor protein, the qa-1S gene product of *Neurospora crassa*, is homologous to part of the *arom* multifunctional enzyme. *Journal of Molecular Biology*, **197**, 367–71.

Anton, I.A. & Coggins, J.R. (1988). Sequencing and overexpression of the *Escherichia coli* gene encoding shikimate dehydrogenase. *Biochemical Journal* **249**, 319–26.

Baillie, A.C., Corbett, J.R., Dowsett, J.R. & McCloskey, P. (1972). Inhibitors of shikimate dehydrogenase as potential herbicides. *Pesticide Science* **3**, 113–20.

Berlyn, M.B. & Giles, N.H. (1969). Organization of the enzymes in the polyaromatic synthetic pathway: separability in bacteria. *Journal of Bacteriology* **99**, 222–30.

Boocock, M.R. (1983). Aspects of the structure and function of the *arom* multifunctional enzyme from *Neurospora crassa*. PhD Thesis, Glasgow University.

Boocock, M.R. & Coggins, J.R. (1983). Kinetics of 5-enolpyruvylshikimate 3-phosphate synthase inhibition by glyphosate. *Federation of European Biochemical Societies Letters* **154**, 127–33.

Boudet, A.M., Graziana, A. & Ranjeva, R. (1985). Recent advances in the regulation of the prearomatic pathway. In *The Biochemistry of Plant Phenolics* (Van Sumere, C.F. & Lea, P.J., eds.) *Annual Proceedings of the Phytochemical Society of Europe* 25, Clarendon Press, Oxford.

Chaudhuri, S., & Coggins, J.R. (1985). The purification of shikimate dehydrogenase from *Escherichia coli*. *Biochemical Journal*, **226**, 217–23.

Chaudhuri, S., Lambert, J.M., McColl, L.A. & Coggins, J.R. (1986). Purification and characterization of 3-dehydroquinase from *Escherichia coli. Biochemical Journal* **239**, 699–704.

Chaudhuri, S., Anton, I.A., & Coggins, J.R. (1987*a*). Shikimate dehydrogenase from *Escherichia coli. Methods in Enzymology* **142**, 315–20.

Chaudhuri, S., Duncan, K., & Coggins, J.R. (1987*b*). 3-Dehydroquinate dehydratase from *Escherichia coli. Methods in Enzymology* **142**, 320–4.

Coggins, J.R. (1986). Enzymology and molecular biology as aids for the invention and improvement of herbicides. In *Biotechnology and Crop Improvement and Protection* (Day, P.R., ed.) pp. 101–10, British Crop Protection Council, Croydon.

Coggins, J.R. & Boocock, M.R. (1986). The *arom* multifunctional enzyme. In *Multidomain Proteins–Structure and Evolution* (Hardie, D.G. & Coggins, J.R., eds.) pp. 259–81, Elsevier, Amsterdam.

Coggins, J.R., Boocock, M.R., Campbell, M.S., Chaudhuri, S., Lambert, J.M., Lewendon, A., Mousdale, D.M. & Smith, D.D.S. (1985). Functional domains involved in aromatic amino acid biosynthesis. *Biochemical Society Transactions* **13**, 299–303.

Coggins, J.R., Boocock, M.R., Chaudhuri, S., Lambert, J.M., Lumsden, J., Nimmo, G.A. & Smith, D.D.S. (1987*a*) The *arom* multifunctional enzyme of *Neurospora crassa. Methods in Enzymology* **142**, 325–41.

Coggins, J.R., Duncan, K., Anton, I.A., Boocock, M.R., Chaudhuri, S., Lambert, J.M., Lewendon, A., Millar, G., Mousdale, D.M. & Smith, D.D.S. (1987*b*). The anatomy of a multifunctional enzyme. *Biochemical Society Transactions* **15**, 754–9.

Comai, L., Facciotti, D., Hiatt, W.R., Thompson, G., Rose, R.E. & Stalker, D.M. (1985). Expression in plants of a mutant *aro A* gene from *Salmonella typhimurium* confers tolerance to glyphosate. *Nature (London)* **317**, 741–4.

Comai, L., Sen, L.C. & Stalker, D.M. (1983). An altered *aro A* gene product confers resistance to the herbicide glyphosate. *Science* **221**, 370–1.

Conn, E.E. (1986). *The Shikimic Acid Pathway*. Plenum Press, New York.

Cornish-Bowden, A. (1986). Why is uncompetitive inhibition so rare? A possible explanation, with implications for the design of drugs and pesticides. *Federation of European Biochemical Societies Letters* **203**, 3–6.

Davis, B.D. (1955). Intermediates in amino acid biosynthesis. *Advances in Enzymology* **16**, 247–312.

De Block, M., Botterman, J., Vandewiele, M., Dockx, Thoen, C., Gossele, V., Rao Movva, N., Thompson, C., Van Montagu, M. & Leemans, J. (1987). Engineering herbicide resistance in plants by expression of a detoxifying enzyme. *EMBO J.* **6**, 2513–18.

Della-Cioppa, G., Bauer, S.C., Klein, B.K., Shah, D.M., Frayley, R.T. & Kishore, G.M. (1986). Translocation of the precursor of 5-endopyruvylshikimate-3-phosphate synthase into chloroplasts of higher plants *in vitro. Proc. Natl. Acad. Sci. USA* **83**, 6873–7.

Della-Cioppa, G., Bauer, S.C., Taylor, M.L., Rochester, D.E., Klein, B.K., Shah, D.M., Frayley, R.T., Kishore, G.M. (1987). Targeting a herbicide resistant enzyme from *Escherichia coli* to chloroplasts of higher plants. *Biotechnology* **5**, 579–84.

Duncan, K. & Coggins, J.R. (1984). Subcloning of the *Escherichia coli* genes *aro A* (5-enolpyruvylshikimate 3-phosphate synthase) and *aro B* (3-dehydroquinate synthase). *Biochemical Society Transactions* **12**, 274–5.

Duncan, K., Lewendon, A., & Coggins, J.R. (1984*a*). The purification of 5-enolpyruvylshikimate 3-phosphate synthase from an overproducing strain of

Escherichia coli. Federation of European Biochemical Societies Letters **165**, 121–7.

Duncan, K., Lewendon, A. & Coggins, J.R. (1984*b*). The complete amino acid sequence of *Escherichia coli* 5-enolpyruvylshikimate 3-phosphate synthase. *Federation of European Biochemical Societies Letters* **170**, 59–63.

Duncan, K., Chaudhuri, S., Campbell, M.S. & Coggins, J.R. (1986). The overexpression and complete amino acid sequence of *Escherichia coli* 3-dehydroquinase. *Biochemical Journal* **238**, 475–83.

Duncan, K., Edwards, R.M. & Coggins, J.R. (1987). The pentafunctional *arom* enzyme of *Saccharomyces cerevisiae* is a mosaic of monofunctional domains. *Biochemical Journal* **246**, 375–86.

Frost, J.W., Bender, J.L., Kadonaga, J.T. & Knowles, J.R. (1984). Dehydroquinate synthase from *Escherichia coli*: purification, cloning, and construction of overproducers of the enzyme. *Biochemistry* **23**, 4470–5.

Gaertner, F.H. & Cole, K.W. (1977). A cluster gene: evidence for one gene, one polypeptide, five enzymes. *Biochemical and Biophysical Research Communications* **75**, 259–64.

Gibson, F. & Pittard, J. (1968). Pathways of biosynthesis of aromatic amino acids and vitamins and their control in micro-organisms. *Bacteriological Reviews* **32**, 465–92.

Gilchrist, D.G. & Kosuge, T. (1980). Aromatic amino acid biosynthesis and its regulation. In *The Biochemistry of Plants* (Miflin, B.J., ed.), Volume 5, pp. 507–31, Academic Press, New York.

Haslam, E. (1974). *The Shikimate Pathway*. Butterworths, London.

Herrmann, K.M. (1983). The common aromatic biosynthetic pathway. In *Amino Acids: Biosynthesis and Genetic Regulation* (Herrmann, K.M. & Sommerville, R.L., eds.) pp. 301–22, Addison-Wesley, Reading, USA.

Huynh, Q.K. (1987). Reaction of 5-enolpyruvyl-shikimate-3-phosphate synthase with diethylpyrocarbonate: evidence for an essential histidine residue. *Archives of Biochemistry and Biophysics* **258**, 233–9.

Huynh, Q.K., Kishore, G.M. & Bild, G.S. (1988). 5-enolpyruvyl-shikimate-3-phosphate synthase from *Escherichia coli:* identification of Lys-22 as a potential active site residue. *Journal of Biological Chemistry* **263**, 735–9.

Kinghorn, J.R., Schweizer, M., Giles, N.H. & Kushner, S.R. (1981). The cloning and analysis of the *aro D* gene of *Escherichia coli:* K-12. *Gene* **14**, 73–80.

Klee, H.J., Muskopf, Y.M. & Glasser, C.S. (1987). Cloning of the *Arabidopsis thaliana* gene encoding 5-enolpyruvyl-shikimate-3-phosphate synthase: sequence analysis and manipulation to obtain glyphosate-tolerant plants. *Molecular and General Genetics* **210**, 437–42.

Koshiba, T. (1978). Purification of two forms of the associated 3-dehydroquinate hydro-lyase and shikimate: NADP oxidoreductase in *Phaseolous mungo* seedlings. *Biochimica et Biophysica Acta* **522**, 10–18.

LaRossa, R.A. & Falco, S.C. (1984). Amino acid biosynthetic enzymes as targets of herbicide action. *Trends in Biotechnology* **2**, 158–61.

LaRossa, R.A. & Schloss, J.V. (1984). The sulfonylurea herbicide sulfometran methyl is an extremely potent and selective inhibitor of acetolactate synthase in *Salmonella typhimurium. Journal of Biological Chemistry* **259**, 8753–7.

Leason, M., Cunliffe, D., Parkin, D., Lea, P.J. & Miflin, B. J. (1982). Inhibition of *Pisum sativum* leaf glutamine synthetase by methionine sulfoximine, phosphinothricin, and other glutamate analogs. *Phytochemistry* **21**, 855–7.

Levin, J.G. & Sprinson, D.B. (1964). The enzymatic formation and isolation of 3-enolpyruvyl-shikimate-5-phosphate. *Journal of Biological Chemistry* **239**, 1142–50.

Lewendon, A. & Coggins, J.R. (1983). Purifications of 5-enolpyruvylshikimate-5-phosphate synthase from *Escherichia coli. Biochemical Journal* **213**, 187–91.
Lewendon, A. & Coggins, J.R. (1987). 3-Phosphoshikimate 1-carboxyvinyltransferase from *Escherichia coli. Methods in Enzymology* **142**, 342–8.
Lumsden, J. & Coggins, J.R. (1977). The subunit structure of the *arom* multienzyme complex of *Neurospora crassa* : a possible pentafunctional polypeptide chain. *Biochemical Journal* **161**, 599–607.
Lumsden, J. & Coggins, J.R.(1978). The subunit structure of the *arom* multienzyme complex of *Neurospora crassa:* evidence from peptide 'maps' for the identity of the subunits. *Biochemical Journal* **169**, 441–4.
Mehdi, S., Frost, J.W. & Knowles, J.R. (1987). Dehydroquinate synthase from *Escherichia coli* and its substrate 3-deoxy-D-arabino-heptulosonic acid 7-phosphate. *Methods in Enzymology* **142**, 306–14.
Millar, G., & Coggins, J.R. (1986). The complete amino acid sequence of 3-dehydroquinate synthase of *Escherichia coli* K12. *Letters* **200**, 11–17.
Millar, G., Anton, I., Mousdale, D.M., White, P.J. & Coggins, J.R. (1986*a*). Cloning and overexpression of the *Escherichia coli aro C* gene encoding the enzyme chorismate synthase. *Biochemical Society Transactions* **14**, 262–3.
Millar, G., Hunter, M.G., Lewendon, A. & Coggins, J.R. (1986*b*). The cloning and overexpression of the *aro L* gene from *Escherichia coli* K12: purification and complete amino acid sequence of shikimate kinase II the *aro L* gene product. *Biochemical Journal* **237**, 427–37.
Mousdale, D.M. & Coggins, J.R. (1984). Purification and properties of 5-enolpyruvylshikimate-3-phosphate synthase from seedlings of *Pisum sativum L. Planta* **160**, 78–83.
Mousdale, D.M. & Coggins, J.R. (1985*a*). Subcellular localisation of the common shikimate pathway enzymes in *Pisum sativum L. Planta* **163**, 241–9.
Mousdale, D.M. & Coggins, J.R. (1985*b*). High performance liquid chromatography of shikimate pathway intermediates. *Journal of Chromatography* **329**, 268–72.
Mousdale, D.M. & Coggins, J.R. (1986*a*). Rapid chromatographic purification of glyphosate sensitive 5-enolpyruvylshikimate 3-phosphate synthase from higher plant chloroplasts. *Journal of Chromatography* **367**, 217–22.
Mousdale, D.M. & Coggins, J.R. (1986*b*). Detection and subcellular localisation of higher plant chorismate synthase. *Federation of European Biochemical Societies Letters* **205**, 328–32.
Mousdale, D.M. & Coggins, J.R. (1987). 3-Phosphoshikimate 1-carboxyvinyl transferase synthase from *Pisum sativum. Methods in Enzymology* **142**, 348–54.
Mousdale, D.M., Campbell, M.S. & Coggins, J.R. (1987). Purification and characterisation of bifunctional dehydroquinase-shikimate: NADP oxidoreductase from pea seedlings. *Phytochemistry* **26**, 2665–7.
Nafziger, E.D., Widholm, J.M., Steinrücken, H.C. & Killmer, J.L. (1984) Selection and characterisation of a carrot cell line tolerant to glyphosate. *Plant Physiology* **76**, 571–4.
Nakatani, H.Y. & Barber, J. (1977). An improved method for isolating chloroplasts retaining their outer membranes. *Biochimica et Biophysica Acta* **461**, 510–2.
Netzer, W.J. (1984). Engineering herbicide tolerance: when is it worthwhile? *Biotechnology* **2**, 940–4.
Nimmo, G.A. & Coggins, J.R. (1981). Some kinetic properties of the tryptophan-sensitive 3-deoxy-D-*arabino*-heptulosonate 7-phosphate synthase from *Neurospora crassa. Biochemical Journal* **199**, 657–65.

Padgette, S.R., Huynh, Q.K., Borgmeyer, J., Shah, D.M., Brand, L.A., Biest Re, D., Bishop, R.F., Rogers, S.G., Frayley, R.T. & Kishore, G.M. (1987). Bacterial expression and isolation of *Petunia hybrida* 5-enolpyruvyl shikimate-3-phosphate synthase. *Archives of Biochemistry and Biophysics* **258**, 564–73.

Patel, V.B. & Giles, N.H. (1979). Purification of the *arom* multienzyme aggregate from *Euglena gracilis*. *Biochimica et Biophysica Acta* **567**, 24–34.

Pittard, A.J. (1987). Biosynthesis of the aromatic amino acids. In *Escherichia coli and Salmonella typhimurium: Cellular and Molecular Biology* (Neidhardt, F.C., ed.), pp. 368–94, American Society for Microbiology, Washington, DC.

Pittard, A.J. & Wallace, B.J. (1966). Distribution and function of genes concerned with aromatic amino acid biosynthesis in *Escherichia coli* .*Journal of Bacteriology* **91**, 1494–508.

Polley, L.D. (1978). Purification and characterisation of 3-dehydroquinate hydrolyase and shikimate oxidoreductase: evidence for a bifunctional enzyme. *Biochimica et Biophysica Acta* **526**, 259–66.

Rogers, S.G., Brand, L.A., Holder, S.B., Sharps, E.S. & Brackin, M.J. (1983). Amplification of the *aro A* gene from *Escherichia coli* results in tolerance to the herbicide glyphosate. *Applied and Environmental Microbiology* **46**, 37–43.

Schulz, A., Sost, D. & Amrhein, N. (1984). Insensitivity of 5-enolpyruvylshikimate 3-phosphate synthase confers resistance to this herbicide in a strain of *Aerobacter aerogenes*. *Archives of Microbiology* **137**, 121–3.

Shah, D.M., Horsch, R.B., Klee, H.J., Kishore, G.M., Winter, J.A., Tumer, N.E., Hironaka, C.M., Sanders, P.R., Gasser, C.S., Aykent, S., Siegel, N.R, Rogers, S.G. & Frayley, R.T., (1986). Engineering herbicide tolerance in transgenic plants. *Science* **233**, 478–81.

Shaner, D.L., Anderson, P.C. & Stidham, M.A. (1984). Imidazolinones: potent inhibitors of acetohydroxyacid Synthase. *Plant Physiology* **76**, 545–6.

Shields, R. (1985). Engineering Herbicide resistance. *Nature (London)* **317**, 668.

Smart, C.C., Johanning, D., Muller, G. & Amrhein, N. (1985). Selective overproduction of 5-enolpyruvylshikimic acid 3-phosphate in a plant cell culture which tolerates high doses of the herbicide glyphosate. *Journal of Biological Chemistry* **260**, 16338–46.

Smart, C.C. & Amrhein, N. (1987). Ultrastructural localisation by protein A-gold immunocytochemistry of 5-enolpyruvyl-shikimic acid 3-phosphate synthase in a plant cell culture which overproduces the enzyme. *Planta* **170**, 1–6.

Stalker, D.M., Hiatt, W.R. & Comai, L. (1985). A single amino acid substitution in the enzyme 5-enolpyruvylshikimate-3-phosphate synthase confers resistance to the herbicide glyphosate. *Journal of Biological Chemistry* **260**, 4724–8.

Steinrücken,H.D. & Amrhein, N. (1980). The herbicide glyphosate is a potent inhibitor of 5-enolpyruvylshikimate acid 3-phosphate synthase. *Biochemical and Biophysical Research Communications* **94**, 1207–12.

Weiss, U. & Edwards, J.M. (1980). *The Biosynthesis of Aromatic Compounds*. J. Wiley, New York.

Weiss, U. (1986). Early research on the shikimate pathway. In *The Shikimic Acid Pathway* (Conn. E.E., ed.). pp. 1–12, Plenum Press, New York.

Welch, G.R., Cole, K.W. & Gaertner, G. H. (1974). Chorismate synthase of *Neurospora crassa:* a flavoprotein. *Archives of Biochemistry and Biophysics* **165**, 505–18.

White, P.J., Millar, G. & Coggins, J.R. (1988). The overexpression, purification and complete amino acid sequence of chorismate synthase from *Escherichia coli* K12 and its comparison with the enzyme from *Neurospora crassa*. *Biochemical Journal* **251**, 313–22.

TIMOTHY R. HAWKES, JOY L. HOWARD AND SARAH E. PONTIN

Herbicides that inhibit the biosynthesis of branched chain amino acids

Abbreviations

Standard biochemical nomenclature has been used where appropriate. Trivial names and company numbers have been used to refer to some chemicals. Their structures are given in the chemical glossary, page 261.

chlorsulfuron (page 175)
sulfometuron methyl
imazapyr
imazaquin
AC 222164
N-phthalyl-L-valine anilide
2-nitro-6-methyl sulphonanilide

Introduction

Sulphonylureas (E.I. du Pont de Nemours and Co) and imidazolinones (American Cyanamid Co) are new classes of herbicides characterised by low use rates (as little as 4 g/ha for some sulphonylureas), pre- and post-emergent activity and low toxicity. Variants have been developed to control weeds in a wide range of different crops.

Sulphonylureas, imidazolinones and sulphonanilides (a type of herbicide described in recent patent from Dow Co) kill plants in an identical and distinctive fashion. The symptoms of plant death first appear in the meristematic tissues where growth ceases soon after treatment. Chlorosis and the necrosis of the tissue soon follows with die back to the more mature parts of the plant taking a further 3–4 weeks. In this chapter a review of the current ideas of the mechanism of action of these herbicides will be presented. Some of this work has not been reported elsewhere.

Physiological studies of herbicide mode of action

Studies of the physiological effects of the sulphonylureas on plants provided the first clues to their mode of action. Corn seedlings stopped growing within two hours of a foliar application of chlorsulfuron (Ray, 1982*a*, *b*). Trace amounts (28 nM) of this herbicide were sufficient to prevent the growth of excised pea roots (Ray, 1984). An early effect on plant tissue was a decrease in mitotic frequency accompanied by inhibition of [^{3}H] thymidine incorporation into DNA. Protein and RNA synthesis were only slightly affected and there was no direct effect on either photosynthesis (the Hill reaction) or respiration. However, although inhibition of DNA synthesis was the major effect, chlorsulfuron did not appear to inhibit DNA polymerase directly and neither could its effects be reversed by addition of the four nucleotide precursors.

The effects of imidazolinones were similar to sulphonylureas (Shaner & Reider, 1986). Twenty-four hours after treatment of intact corn seedlings with imazapyr, the rate of respiration in excised root cultures decreased slightly. No major changes in the rates of lipid synthesis ([^{14}C] acetate incorporation), protein synthesis ([^{14}C] leucine and [^{14}C] cystine incorporation) or RNA synthesis ([^{14}C] uridine incorporation) were observed (after allowing for herbicide-induced decreases in the uptake of radiolabelled precursors). Again, a fall (*c.* 65%) in the rate of DNA synthesis ([^{14}C] thymidine incorporation) was the major effect. However, it seemed unlikely that this was the primary event. Corn seedlings sprayed with imazapyr began to stop growing after only 3 hours whereas the rate of synthesis of DNA remained unabated for a further 5 hours (Shaner *et al.*, 1985*a*). Foliar treatment with imazapyr also caused a slight increase in the neutral sugar content of corn leaves (probably due to a non-specific reduction in sink demand) and, after 24 hours, a 40% decrease in the level of extractable soluble proteins. Fifty hours after treatment, leaves contained increased (1.3–20 fold) levels of most amino acids but markedly reduced levels of valine and leucine. Imazapyr had similar effects on amino acid levels in cultured corn cells (Anderson & Hibberd, 1985*a*).

The primary action of the herbicides was difficult to discern on the basis of the physiological responses of plant tissues. The synthesis of DNA was always severely inhibited but not, apparently, as a direct effect of either type of herbicide. The important breakthrough came with the use of an inhibition/reversal approach towards identifying biosynthetic processes that the herbicides might block. Thus, the discovery that, in a minimal medium containing valine, sulfometuron methyl-induced bacteriostasis of *Salmonella typhimurium* was specifically reversed by the addition of isoleucine to the medium (LaRossa & Schloss, 1984) provided the first evidence that sulphonylureas inhibit the biosynthesis of branched chain amino acids. Excised pea root culture (Ray, 1984) and soya bean culture (Scheel & Casida, 1985) provided sensitive and quantifiable systems in which to extend this approach to plant tissues. Only trace levels of chlorsulfuron (*c.* 4 nM and 1 nM, respectively) were required to inhibit growth by 50%. In both systems, mixtures of isoleucine and valine (but neither amino acid alone) were effective in counteracting the herbicide. Similar data were obtained for the imidazolinones (Anderson & Hibberd, 1985*a*; Shaner & Reider, 1986). In pea root culture, mixtures of 2-ketoisovaleric acid and 2-keto-3-

methylvaleric acid (the penultimate intermediates in the branched chain amino acid pathway) also reversed the effect of chlorsulfuron (Ray, 1984). This further localised the probable site of inhibition to an early step in the pathway.

The genetics of herbicide resistance in plants and microbes

The possibility of developing herbicide-resistant crop plants has stimulated a great deal of work on the genetics of herbicide resistance. This work has also furthered our understanding of herbicide action.

Chlorsulfuron was shown to be an extremely potent inhibitor (I_{50} *c.* 20 nM) of acetohydroxyacid synthase, the first enzyme in the pathway for the biosynthesis of branched chain amino acids (Ray, 1984). Genetic data have provided additional evidence that acetohydroxyacid synthase is the target of the sulphonylurea herbicides.

Tobacco mutants that were resistant to sulphonylureas were isolated from cell cultures by selection for growth in the presence of sulfometuron methyl or chlorsulfuron (Chaleff & Mauvais, 1984). Plants regenerated from these lines retained the resistant phenotype and a particular mutation (S4) conferred a 100-fold level of resistance to chlorsulfuron (i.e. the plants tolerated 100 fold more herbicide than did the wild type). Significantly, acetohydroxyacid synthase extracted from resistant plants was far less sensitive ($I_{50} > 8\ \mu M$) to inhibition by chlorsulfuron than normal (I_{50} 14 nM). Heterozygotes for the S4 mutation were constructed by a series of backcrosses of the resistant plants with the parental variety. Self-fertilisation of the heterozygotes yielded homozygous mutants, heterozygotes and homozygous wild-type progeny in the ratio 1 to 2 to 1 which is the pattern expected for the segregation of a single semi-dominant nuclear allele. The chlorsulfuron-resistant phenotype of the acetohydroxyacid synthase extracted from these plants segregated in exactly the same way. Thus sensitive homozygous wild types contained fully sensitive enzyme, the most highly resistant S4 mutants contained the most resistant enzyme and the heterozygotes contained enzyme with an intermediate degree of resistance. Since the S4 mutation had been 'purified' by a series of backcrosses there can be little doubt that the alteration in acetohydroxyacid synthase was the sole determinant of sulphonylurea resistance. The S4 mutant that was selected on chlorsulfuron was also resistant to sulfometuron methyl. At both the plant and enzyme level, the resistance was less than for chlorsulfuron. This discrimination suggests that the S4 mutation is likely to lie within the structural gene for acetohydroxyacid synthase (rather than, for example, at a control site that regulates the expression of acetohydroxyacid synthase).

Inhibition of acetohydroxyacid synthase by the sulphonylureas was so potent as to render the genetic pointers to the mode of action almost superfluous. Imidazolinones, on the other hand, are much weaker inhibitors of the enzyme *in vitro* (at least three orders of magnitude less inhibitory than sulphonylureas: cf. Shaner *et al.*, 1984).

The concentration of imazapyr required to inhibit the growth of corn seedlings from excised mature embryos by 50% was similar (*c.* 20 nM) to that required to inhibit the growth of corn cells in culture (Anderson & Hibberd, 1985*a*). It therefore seemed reasonable to assume that the basis of herbicide action in cultured cells was the same as in whole plants. A line of imidazolinone-resistant corn cells (XA17) was selected by a process of gradually increasing the concentration of imazaquin in the subculturing media to achieve about a 30-fold level of resistance (Shaner *et al.*, 1985*b*). The resistant callus contained acetohydroxyacid synthase with a markedly reduced sensitivity to inhibition by all imidazolinones (e.g. greater than 500 μM). Herbicide resistance was maintained in plant lines that were regenerated from the callus and, similar to the S4 mutation in tobacco, inherited in a semi-dominant Mendelian fashion (Anderson & Hibberd, 1985*b*). No data were presented with respect to the extraction of herbicide-resistant acetohydroxyacid synthase from the regenerated plants. The resistant phenotype of acetohydroxyacid synthase from the culture line XA17 constitutes very strong evidence that acetohydroxyacid synthase is the site of action of the imidazolinones (particularly since it was also substantially cross-resistant to the sulphonylurea, chlorsulfuron) but does not entirely exclude the possibility that additional changes were co-determinants of herbicide resistance.

Changes in DNA sequences that determine sulphonylurea-resistance have been identified in micro-organisms. Single amino acid changes in the sequence of isozyme II of acetohydroxyacid synthase from *E. coli* (valine for alanine) and acetohydroxyacid synthase from yeast (serine for proline) confer resistance to sulfometuron methyl (Falco & Dumas, 1985; Yadav *et al.*, 1986).

Acetohydroxyacid synthase

Acetohydroxyacid synthase (EC 4.1.3.18) is the first step in a combined pathway responsible for the biosynthesis of valine, leucine and isoleucine (see Fig. 1). Two alternative reactions are catalysed, one, for the the synthesis of valine and leucine, produces 2-acetolactate and CO_2 from the condensation of two molecules of pyruvate and the other, for isoleucine, produces 2-acetohydroxybutyric acid and CO_2 from pyruvate and 2-ketobutyrate. The enzyme has normally been assayed on the basis of the homologous condensation of pyruvate. The product, 2-acetolactate, is conveniently determined as a coloured complex following acid-catalysed decarboxylation to acetoin (Westerfield, 1945).

The only detailed information about acetohydroxyacid synthase has come from work on bacterial enzymes. Three distinct isozymes have been identified in *Escherichia coli* K12 and *Salmonella typhimurium* LT2 (De Felice, Squires & Levinthal, 1978; De Felice *et al.*., 1982). These differ in their feedback regulation, levels of expression and substrate specificities. Isozymes I and III are subject to inhibition by valine whereas isozyme II is insensitive. Logically enough, isozyme I is probably biased towards the production of acetolactate (Shaw, Berg & Sobol, 1980) and isozyme II (Abell, O'Leary & Schloss, 1985) toward 2-acetohydroxybutyric acid (isoleucine synthesis is regulated through feedback inhibition of threonine deaminase).

All three isozymes are not normally present and their functions are interchangeable. Thus *Escherichia coli* K12 expresses mainly isozyme I, low levels of isozyme III and does not express isozyme II since the wild type structural gene (*ilv* G) is cryptic (Lawther *et al.*, 1981). *Salmonella typhimurium* LT2, on the other hand, expresses isozymes I and II but not III. Only isozyme I (Grimminger & Umbarger, 1979; Eoyang & Silverman, 1984) and isozyme II (Schloss *et al.*, 1985) have been purified to homogeneity. In both cases the enzyme was an $\alpha_2\beta_2$ tetramer of subunits with molecular weights of *c*. 60 and 9.7 kDa.

It is intriguing that acetohydroxyacid synthase requires FAD although the catalytic reaction involves no overall change in redox. Other 'non-redox' flavoenzymes include mandelonitrile lyase (Jorns, 1979), chorismate synthase (Hasan & Nester, 1978) and glyoxylate carboligase (Cromartie & Walsh, 1976). Possibly the mechanism of catalysis involves cyclic internal changes in the oxidation state of the flavin. However,

Fig. 1. The pathway for the biosynthesis of branched chain amino acids.

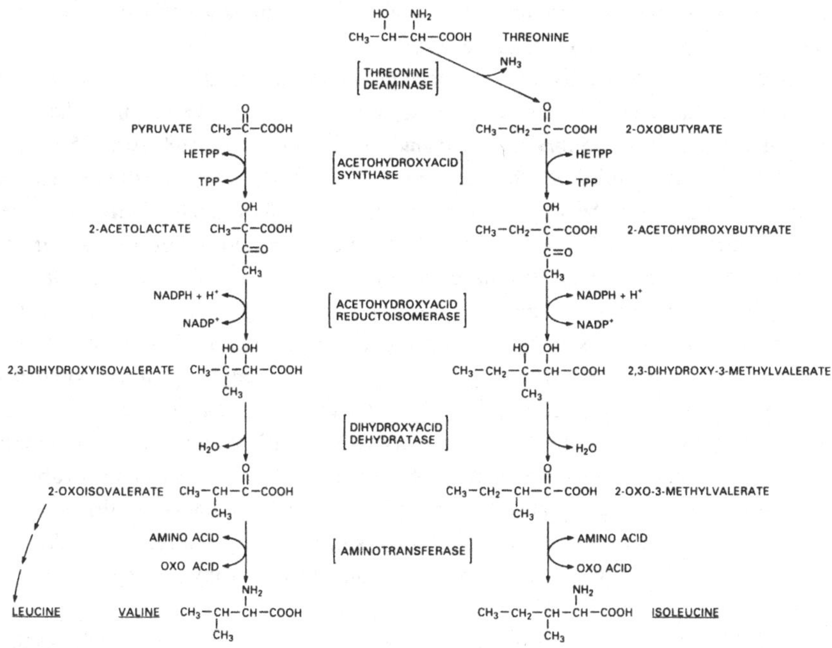

it appears that a non-biosynthetic form of the acetolactate synthase (involved in the fermentation of glucose to acetoin by *Aerobacter aerogenes*) neither contains nor requires FAD (Stormer, 1968). Furthermore it has been shown recently that FAD can be removed from isozyme II of *Salmonella typhimurium* and and replaced with 5-deaza-FAD (which can only participate in two electron redox processes) or reduced *in situ* to (dihydro) $FADH_2$ without significant loss of enzyme activity (J.V. Schloss, personal communication). Remarkably, the deduced amino acid sequence of pyruvate oxidase (encoded by *poxB*) shares a high degree of homology with the large subunits of isozymes I, II and III of bacterial acetohydroxyacid synthase (Grabau & Cronan, 1986). It seems likely that the FAD in acetohydroxyacid synthase is an evolutionary fossil which originates from a pyruvate oxidase-like ancestral flavoenzyme and which now fulfils a purely structural role.

In the presence of oxygen, isozyme II of acetohydroxyacid synthase slowly loses activity. This is probably due to the gradual autooxidation of enzyme-bound thiamine pyrophosphate to the tight-binding inhibitor, thiamine thiazolone pyrophosphate (Schloss *et al*., 1985). Consistent with this hypothesis, inactivation was both slowly reversible and dependent on the presence of divalent cations which are required for thiamine pyrophosphate to bind to the enzyme.

Acetohydroxyacid synthase has not been purified from plants. Part of the difficulty has been the fact that plant tissues appear to contain so little of the enzyme. The V_{max} of pure isozyme II from *Salmonella typhimurium* (Schloss *et al*., 1985) was 25.3 µmol acetolactate formed/min/mg whereas the specific activity in crude extracts of plant material has generally been some 5000 fold less than this value (cf. Miflin & Cave, 1972; Relton *et al*., 1986 and Shaner *et al*., 1984). The procedure that we use to prepare acetohydroxyacid synthase (specific activity *c*. 20 nmol acetolactate formed/min/mg) from pea shoots is outlined in the legend to Table 1. Further purification by anion exchange chromatography (Pharmacia FPLC mono Q) and hydrophobic interaction chromatography (Pharmacia phenyl sepharose) yielded a twenty-fold increase in specific activity but at the expense of massive (80%) losses in the total activity (data not shown). For successful purification it will be necessary to define conditions under which the enzyme is stable. Pilot experiments (Table 1) indicated that FAD, pyruvate, dithiothreitol and glycerol stabilised the activity. The specific effect of FAD is interesting since it is indicative that, like the better characterised bacterial enzymes, plant acetohydroxyacid synthase is a flavoenzyme.

Unlike bacteria, plants probably contain a single form of acetohydroxyacid synthase. Resistance to sulphonylurea herbicides was inherited as a single Mendelian trait (see above) and leucine and valine were cooperative inhibitors of acetohydroxyacid synthase activity in extracts of barley (Miflin, 1971).

Table 1. *Factors affecting the stability of AHAS from* Pisum sativum.

Additions to dialysis buffer	% of initial activity remaining after 36 hours
None	49
0.1 mM FAD	77
1.0 mM TPP	38
5.0 mM pyruvate	88
30% v/v glycerol	95
0.5 M NaCl	9
10 mM $MgCl_2$	35

Assay conditions and methods were as described by LaRossa & Schloss (1984). Units were nmol acetolactate formed/min and the final volume of the assay solution was *c*. 0.65 ml. AHAS was prepared as follows. Young pea shoots were ground in a liquid N_2-cooled mortar and pestle. The powder was resuspended in an equal volume of ice cold 40 mM tricine/NaOH buffer at pH 8.0 containing 10 mM EDTA, 0.08 mM FAD, 5.0 mM pyruvate, 1.0 mM dithiothreitol and 0.15% v/v Tween 80. The mixture was blended, allowed to stand and centrifuged for one hour at 26,000 g av. The supernatant was further clarified by precipitation with protamine sulphate and low speed centrifugation and then mixed with an equal volume of 50% polyethylene glycol (M_r 3600). The precipitate was collected by low speed centrifugation and resuspended in buffer (as above) containing 25% v/v glycerol. The protein was then exchanged into 40 mM tricine/NaOH buffer at pH 8.0 containing 10 mM EDTA, 1.0 mM dithiothreitol and 25% v/v glycerol. The final preparation had a specific activity of *c*. 20 nmol acetolactate formed/min/mg and was stored as frozen beads at liquid N_2 temperature.

Stability experiments were carried out by dialysing small samples of the protein against 40 mM tricine/NaOH buffer (pH 8.0) containing 10 mM EDTA, 1mM dithiothreitol and various different additives (as indicated in the table) for 36 hours at 4 °C. Prior to assay all the dialysis bags were transferred to the above buffer (containing no additives) for 3 hours to avoid the problem of different additives being carried over into the assays.

The mechanism of acetohydroxyacid synthase

Acetohydroxyacid synthase is a non-oxidative thiamine pyrophosphate-containing decarboxylase and its probable mechanism (Scheme 1) has been described by Walsh (1979). The initial decarboxylation of pyruvate yields a hydroxyethyl-thiamine pyrophosphate intermediate. Unlike, for example, pyruvate decarboxylase this intermediate does not break down to release acetaldehyde and thiamine pyrophosphate but survives to act as a nucleophile on the 2-keto group of a second molecule of pyruvate of 2-ketobutyrate. The relative preferences of 2-ketobutyrate and pyruvate for the first substrate position (carboxyl loss) versus the second position (carboxyl retention) of bacterial isozyme II have been investigated (Abell *et al.*, 1985). Isotope mass ratio spectroscopy of the release of enzyme-derived CO_2 from an equimolar mixture of 2-ketobutyrate and ^{13}C carboxyl-labelled pyruvate indicated that pyruvate was preferred over 2-ketobutyrate in the first site (19 to 1) and that the converse was true at the second site. This mechanism would minimise the non productive decarboxylation of 2-ketobutyrate (although the enzyme was shown to be able to catalyse the homologous condensation of 2-ketobutyrate).

Scheme 1: The chemical and kinetic mechanism of acetohydroxyacid synthase.

$$\underset{1}{E} \underset{k_2}{\overset{k_1S}{\rightleftharpoons}} \underset{2}{ES} \xrightarrow[\searrow CO_2]{k_3} \underset{3}{E'S} \underset{k_6}{\overset{k_5S}{\rightleftharpoons}} \underset{4}{E'SS} \underset{k_8}{\overset{k_7}{\rightleftharpoons}} \underset{5}{Ep} \xrightarrow{k_9} E+p$$

E = Enzyme/thiamine pyrophosphate S = Pyruvate

E'S = Enzyme/hydroxyethyl–thiamine pyrophosphate p = α acetolactate

$$\frac{e}{v} = \frac{1}{\phi_0} + (\phi_{s_1} + \phi_{s_2})\frac{1}{S} \qquad \phi_{s_1} = \frac{(k_2+k_3)}{k_1 k_3}$$

$$\phi_0 = \frac{1}{k_9} + \frac{1}{k_3} + \frac{(k_8+k_9)}{k_7 k_9} \qquad \phi_{s_2} = \frac{(k_8+k_9)(k_6+k_7)}{k_5 k_7 k_9} - \frac{k_8}{k_5 k_9}$$

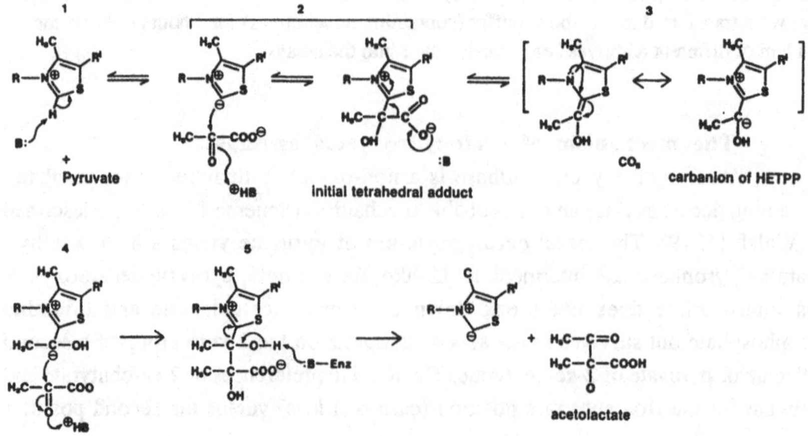

The steady state kinetics of acetohydroxyacid synthase have been described (Schloss *et al.*, 1985). Except for the enzyme from barley (Miflin, 1971) saturation by pyruvate appears hyperbolic and the plot of 1/S (pyruvate) v.1/V (initial rate) linear (cf Schloss, 1984; Figs. 6–9). The simplest mechanism that corresponds with the likely chemistry is shown in Scheme 1. The important assumption is that product release is irreversible under conditions of low product accumulation (i.e. rates governed by k4 and k10 can be ignored). Terms in $1/S^2$ that might have introduced curvature into the double reciprocal plots are dependent on the product of k2, k4 and the concentration of CO_2. Certainly, at pH 8.0, CO_2 would be removed by hydration. The kinetic carbon isotope (^{13}C) effect on the release of CO_2 from carboxyl-labelled pyruvate was surprisingly small (Abell *et al.*, 1985). This was interpreted in terms of a low value for k2 and 'sticky' binding of the first pyruvate. The K_m for pyruvate is *c.* 8 mM for isozyme II of acetohydroxyacid synthase from *Salmonella typhimurium* (Schloss, 1984; Schloss *et al.*, 1985) and about 1.6 mM for the enzyme from peas (see Figs. 2 and 6–9). Using the pea enzyme a slight complication in the interpretation of our data was substrate inhibition at high concentrations of pyruvate. We did not fully characterise this effect (possibly due to non-specific reaction of the reactive carbonyl with lysine residues etc.) and, for the purposes of our steady state kinetic work, have effectively ignored it by using a range of pyruvate concentrations below 15 mM. Inhibition was very slight below 15 mM although, it must be said, it still creates something of a problem in the weighting of the reciprocal plots.

Inhibition by sulfometuron methyl

Inhibition of isozyme II of acetohydroxyacid synthase by sulfometuron methyl developed slowly with time and could best be modelled (LaRossa & Schloss, 1984; Schloss, 1984; see Scheme 2) as a biphasic process involving the rapid formation of an initial, relatively weak, enzyme/inhibitor (EI) complex that slowly isomerises to a more tightly bound form (EI*). Both the initial (K_i 0.35 μM) and final (K_i 0.04 μM) stages of inhibition by sulfometuron methyl were reversible and competitive with respect to pyruvate. The apparent first order rate constant for isomerisation tended, with increasing concentrations of inhibitor, to a limiting value (k_1) of 0.15–0.25/min. On this basis, the expected 'off rate' (k_2 in Scheme 2) can be calculated as approximately 0.02/min. Inhibition was independent of variation (2–200 μM) in the concentrations of thiamine pyrophosphate and FAD in the asay. By contrast isozyme I of acetohydroxyacid synthase from *Salmonella typhimurium* is relatively insensitive to inhibition by sulfometuron methyl (LaRossa & Smulski, 1984).

There is some evidence to suggest that the inhibitor binds more tightly to the form of the enzyme with one pyruvate bound (form 3 in Scheme 1) than with no pyruvate bound (form 1). Slow isomerisation to the tight complex only occurred in the presence of pyruvate (LaRossa & Schloss, 1984). Enzyme, preincubated with pyruvate and sulfometuron methyl, recovered activity slowly (corresponding to k_2 in Scheme 2) after the protein was reseparated from the inhibitor solution. Without pyruvate present at the outset, the full activity returned with no lag (J.V. Schloss, personal communication). Furthermore, on the basis of preliminary stopped flow experiments, it seems likely that the steady state concentration of the enzyme-bound hydroxyethyl thiamine pyrophosphate intermediate increases in response to inhibition by sulfometuron methyl (Ciskanik & Schloss, 1985).

Sulphonylureas inhibit acetohydroxyacid synthase from plants (Ray, 1984) in a similar manner to the enzyme from bacteria. Our data indicate that sulfometuron methyl initially binds as an extremely potent non-competitive inhibitor (K_i slope 5 nM, K_i intercept 20 nM; Fig. 2) of the enzyme from peas. Inhibition developed further with time (Fig. 3), and slowly (k_{app} *c.* 0.025/min) became at least five-fold more potent. For slow and tight binding competitive inhibitors it is notoriously easy to generate false intercept effects. Error can arise through failing to allow for the more rapid onset of tight inhibition that occurs at higher inhibitor concentrations or through adding so much enzyme that the concentration of inhibitor free in solution is

Scheme 2: The biphasic model of inhibition.
Computer fitting of progress curves. Progress curves for assays which included slow acting inhibitors (see e.g. Fig. 4) were computer fitted to the equation below (LaRossa & Schloss, 1984). The values of P_o, v, v_f and the apparent first order rate constant (k_{app}) were allowed to float. With very low concentration of inhibitor the final steady rate, v_f, was not attained even after 500 minutes. In these cases v_f could not be allowed to vary and its value was fixed on the basis of the ratio v/v_f determined at higher inhibitor concentrations. The observed rate constant (k_{app}) was related to the true rate constant (k_1) by the equation '$k_{app} = f * k_1$' where f is the fraction of the enzyme in the initial complex EI (this ignores the relatively small reverse rate constant k_2). Experimentally f was obtained from $1-(v/v_o)$ (where v_o is the initial rate without inhibitor) or $1-(I_{50}/(I_{50} + I))$ (where I_{50} is the concentration of inhibitor that (initially) gives 50% inhibition and I is the concentration of inhibitor).

$$\mathbf{E} + \mathbf{I} \rightleftharpoons \mathbf{EI} \underset{k_2}{\overset{k_1}{\rightleftharpoons}} \mathbf{EI^*}$$

Enzyme — Inhibitor — Initial complex — Tight complex

$$p = p_0 + ((v_f - v)/k_1)\, e^{-k_1 T} + ((v - v_f)/k_1) + v_f T$$

p_0 is product present initially
p is product present at time T
T is time
v is initial steady rate
v_f is final steady rate
k_1 is the rate constant for transition from the initial to the final steady rate

Fig. 2. Initial inhibition of acetohydroxyacid synthase from peas by sulfometuron methyl.

The enzyme was isolated and assayed as described in the legend to Table 1 except that the assays were run at 30°C rather than 37 °C. Calibration of the vertical axis is 98 nmol acetolactate/unit absorbance at 530 nm.

Parameters were as follows: assay time 25 min; total protein 0.08 mg; V_{max} 14.7 +/- 1.2 units/mg; K_m 1.3 +/- 0.2 mM pyruvate; K_i slope 5.2 +/- 2 nM; K_i intercept 18.2 +/- 6 nM.

To investigate the initial binding of slow acting inhibitors (Scheme 2) it was necessary to run the assays for the shortest time that still produced enough acetolactate for us to measure. At low concentration of inhibitor, the slope of the progress curve declines gradually (cf. Fig 4) and, after 25 minutes, still approximates to the tangent at time zero. At the highest concentration of sulfometuron methyl (40 nM), the apparent first order rate constant (k_{app}) for the onset of 5-fold tighter inhibition (see Fig. 3 and Fig. 5) can be calculated as *c.* 0.035/min. Thus, after 25 minutes, the curvature in the progress curve would, at worst, lead to a *c.* 25% underestimate of the rate at time zero.

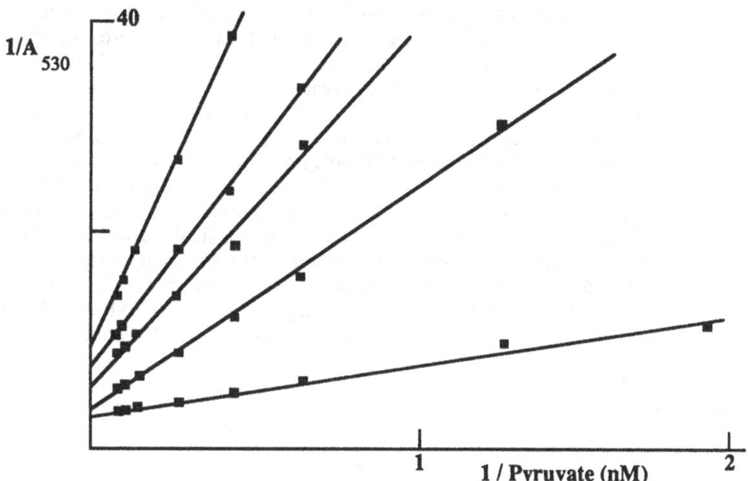

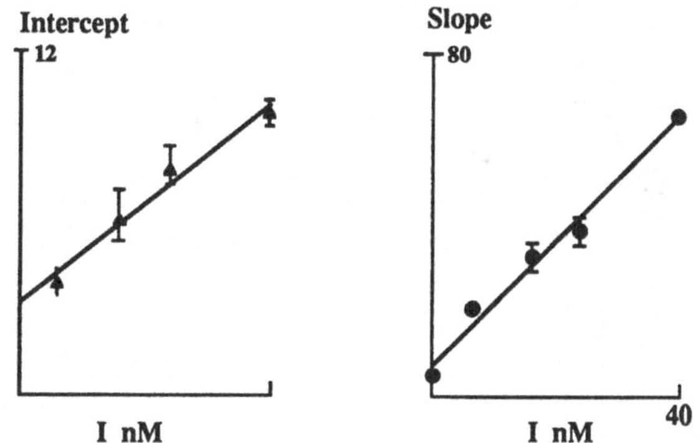

significantly diminished. Nevertheless we have effectively allowed for the first effect (Fig. 2, legend) and, based on the V_{max} of pure bacterial isozyme II (25 μmol/min/mg), calculate that the maximum concentration of acetolactate synthase in our assays (*c.* 0.4 nM?) was probably no more than 5% of our minimum inhibitor concentration (basing the calculation on the tenuous assumption that the plant and bacterial enzymes have about the same k_{cat}). Overall, it seems most likely that sulfometuron methyl is a non-competitive rather than a competitive inhibitor of the plant enzyme.

Inhibition by imidazolines

Imidazolinones were originally reported to be uncompetitive inhibitors with respect to pyruvate of acetohydroxyacid synthase from corn with K_i values of 12 μM

Fig. 3. Progress curves for inhibition of acetohydroxyacid synthase from peas by different inhibitors. Protein was isolated and assayed as described in Table 1. To improve linearity over long time periods assays used degassed buffer and were run at 30 °C. The concentration of pyruvate was 50 mM and assays were started with the addition of protein. The four graphs had similar intercepts on the vertical axis but have been offset for clarity. The calibration of the vertical scale is 115 nmol acetolactate per unit change in absorbance at 530 nm. The data were computer fitted to the equation described in Scheme 2. The parameters were as follows: ◆ Control: protein 0.0043 mg; rate 18.5 units/mg; ▲ Sulfometuron methyl: protein 0.03 mg; inhibitor 11.20 nM; initial rate (v) 10.1 units/mg; final rate (v_f) 2.1 units/mg; value of k_{app} 0.021/min; ■ Imazapyr: protein 0.067 mg; inhibitor 52.5 μM; initial rate (v) 9.75 units/mg; final rate (v_f) 0.44 units/mg; value of k_{app} 0.0245/min; ● 2-NO_2 6-Me sulphonanilide: protein 0.056 mg; inhibitor 37.5 nM; initial rate (v) 8.53 units/mg; final rate (v_f) 0.753 units/mg; value of k_{app} 0.028/min.

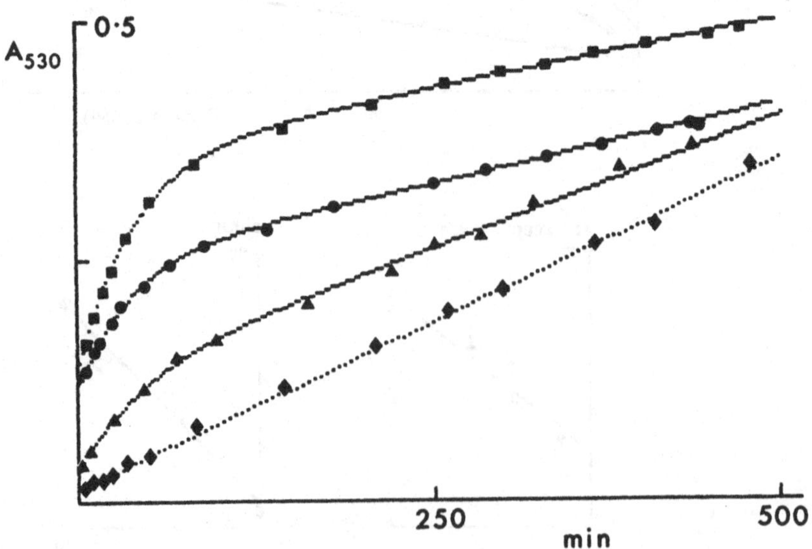

for imazapyr, 3.4 μM for imazaquin, and 1.7 μM for the benzoic acid-containing analogue, AC 222164 (Shaner *et al.*, 1984). In our hands, imazapyr was a simple non-competitive inhibitor (K_i *c.* 60 μM) of the enzyme from peas (Fig. 6). Again, we observed a slow and, in this case, dramatic (22-fold) increase in inhibition with time (Fig. 4). The biphasic model of inhibition (Scheme 2) provided a good fit to the data and with increasing concentrations of imazapyr the apparent first order rate constant for the development of the second stage of inhibition tended to a limiting value (k_1) of 0.05/min (Fig. 5). Thus, even when fully developed, inhibition by imazapyr was some three orders of magnitude less potent than for sulfometuron methyl.

Inhibition by L-valine

L-valine was (approximately) a competitive inhibitor but unlike the sulphonylureas was markedly non-linear in its effect (see Fig. 7). Depending on the concentration of pyruvate, inhibition tended to an upper limit that was considerably less than 100%. For example, at 1.0 mM pyruvate, concentrations of 0.35 mM and 3.5 mM L-valine inhibited the enzyme by 45% and 65% respectively. Glucose-6-phosphate is another example of a regulatory molecule that competes with a structurally dissimilar substrate (competitive inhibition of hexokinase with respect to

Fig. 4. Progress curves with different concentrations of imazapyr. The conditions were exactly as described in the legend to Fig. 3. Again the graphs have been offset for clarity. ● Imazypyr (3.5 μM): protein 0.012 mg; initial rate (v) 17.3 units/mg; final rate (v_f) 1.2 units/mg; value of k_{app} 0.0042/min; ■ Imazypyr (17.5 μM): protein 0.028 mg; initial rate (v) 14.5 units/mg; final rate (v_f) 0.5 units/mg; value of k_{app} 0.011/min; ▲ Imazapyr (52.5 μM) protein 0.067 mg; initial rate (v) 9.75 units/mg; final rate (v_f) 0.44 units/mg; value of k_{app} 0.0245/min.

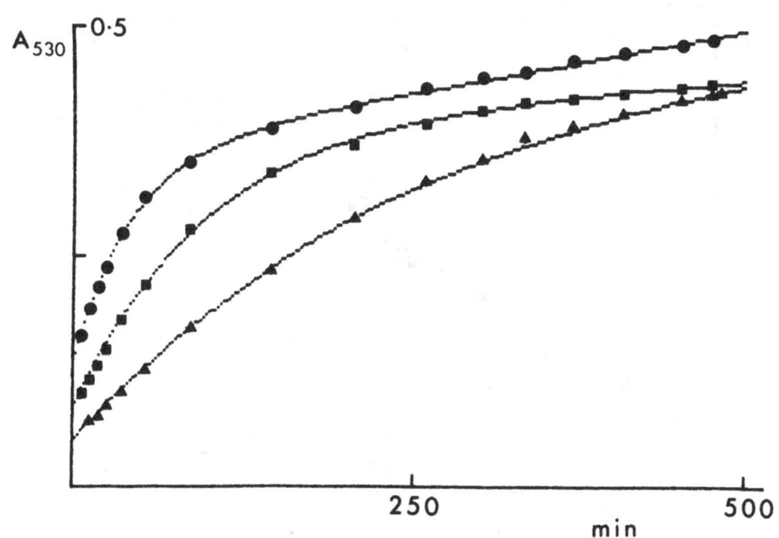

ATP). Again, like the sulphonylureas and imidazolinones, we found that inhibition by valine increased slowly with time although we did not, in this case, attempt to obtain quantitative data.

Inhibition by N-phthalyl-L-valine anilide

N-phthalyl-L-valine anilide is a very weakly herbicidal derivative of L-valine which was reported to be a much better inhibitor of acetohydroxyacid synthase from corn (I_{50} 2 μM) than L-valine itself (Huppatz & Casida, 1985). Since the D-analogue was relatively poor (as D-valine is poor relative to L-valine) it was proposed that the phthalimido anilide derivatives might bind at the same site as valine. In contrast, we found that, unlike valine, this compound was a linear non-competitive inhibitor (K_i *c.* 35 μM; Fig. 8). Again we observed a slow increase in inhibition with time. The data (not shown) were very reminiscent of imazapyr and were fitted to an initial K_i value of 40 μM, final K_i value of *c.* 2.0 μM and a probable first order rate constant for the transition of *c.* 0.05/min (Fig. 5).

Fig. 5. Extrapolation to determine the rate constant k_1 for the onset of slow inhibition (Scheme 2). Data were collected from a number of progress curves similar to those described in Figs. 3 and 4. The values of v and k_{app} obtained from fits at different inhibitor concentrations were used to calculate the true rate constant (k_1) as described in the legend to Scheme 2. Thus the graph of k_{app} (vertical axis) versus 1 -v/v_o (horizontal axis) yields a straight line where the extrapolated value of k_1 corresponds to 1.0 on the horizontal axis. The line is a linear regression fit to the data set for imazapyr and extrapolates to a value of 0.048/min for k_1. Single data points from various other types of inhibitor lie close to the line for imazapyr. ▲ Imazapyr; ■ Sulfometuron methyl; ● 2-NO_2 6-Me sulphonanilide; ◆ N-phthalyl-L-valine anilide.

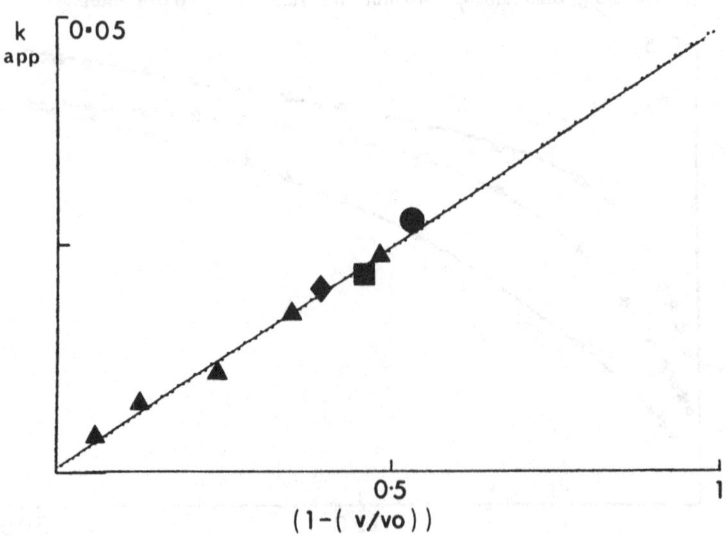

Fig. 6. Initial inhibition by imazapyr. Conditions were as described in the legend to Fig. 2. Again it was necessary to use a short assay time and relatively small range of inhibitor concentrations to achieve the best approximation to inhibition at time zero. At worst, after 25 minutes, the curvature in the progress curve at the highest concentration of imazapyr (53 μM) should cause no more than a 22% underestimate of the initial rate. The net effect will be a slight overestimate of K_i slope and underestimate of K_i intercept: assay time 25 min; total protein 0.08 mg; V_{max} 12.5 +/- 0.4 units/mg; K_m 1.8 mM pyruvate; K_i slope 0.085 +/- 0.03 mM; K_i intercept 0.072 +/- 0.01 mM. Combined with other data (not shown) our average values for K_i slope and K_i intercept were 0.07 and 0.055 mM.

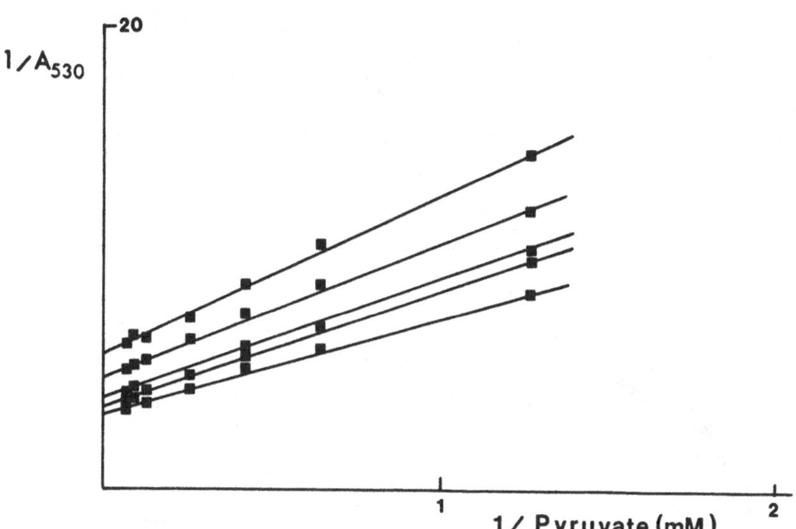

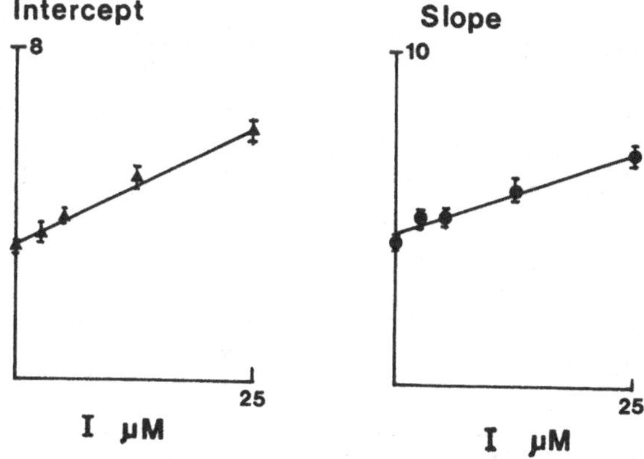

Inhibition by 2-NO_2 6-Me sulphonanilide

1,2,4-triazolo[1,5-a]-1,3,5-triazine-2-sulphonamides are herbicides described in a recent patent from the Dow Co.(Kleschick *et al.*, 1984). The similarity of the symptoms of plant death induced by these compounds to the sulphonylureas and imidazolinones led us to investigate whether they also inhibit acetohydroxyacid

Fig. 7. Initial inhibition by L-valine. Conditions were the same as described in the legend to Fig. 6: assay time 30 min; total protein 0.08 mg; V_{max} 15.4 +/- 0.2 units/mg; K_m 1.7 +/- 0.2 mM pyruvate.

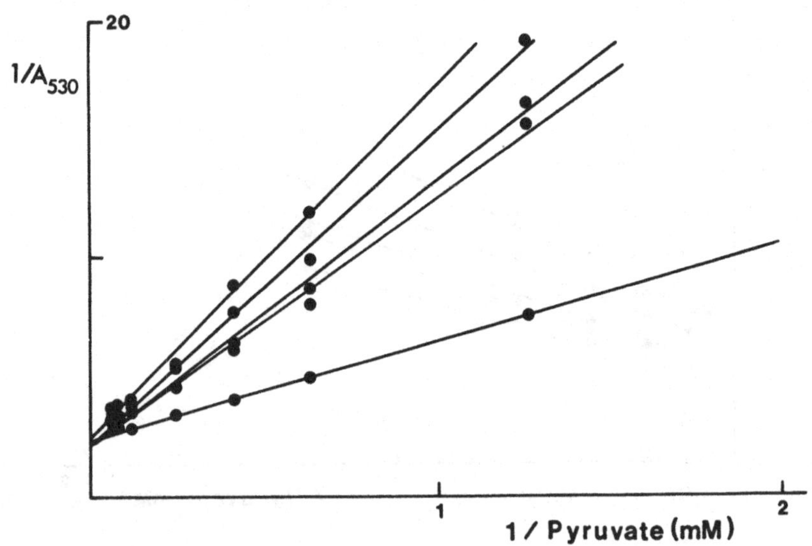

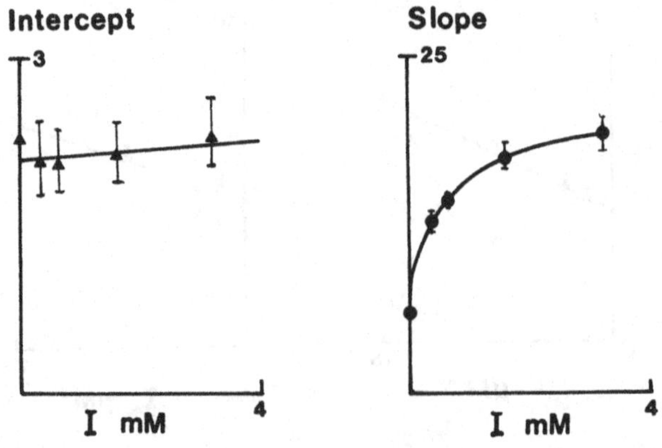

synthase. One example from the patent (2-NO_2 6-Me sulphonanilide) was a potent (K_i 28 nM) simple non-competitive inhibitor with respect to pyruvate (Fig. 9). Again, time dependent inhibition was similar to the sulphonylureas and imidazolinones (Fig. 3); a first order rate constant of 0.025/min (extrapolating to a probable value for k_1 of

Fig. 8. Initial inhibition by N-phthalyl-L-valine anilide. Conditions were the same as in Fig. 6 and the same comments apply with regard to the accuracy: assay time 30 min; total protein 0.08 mg; V_{max} 15.2 +/- 0.3 units/mg; K_m 1.3 +/- 0.2 mM; K_i slope 0.047 +/- 0.02 mM; K_i intercept 0.030 +/- 0.006 mM.

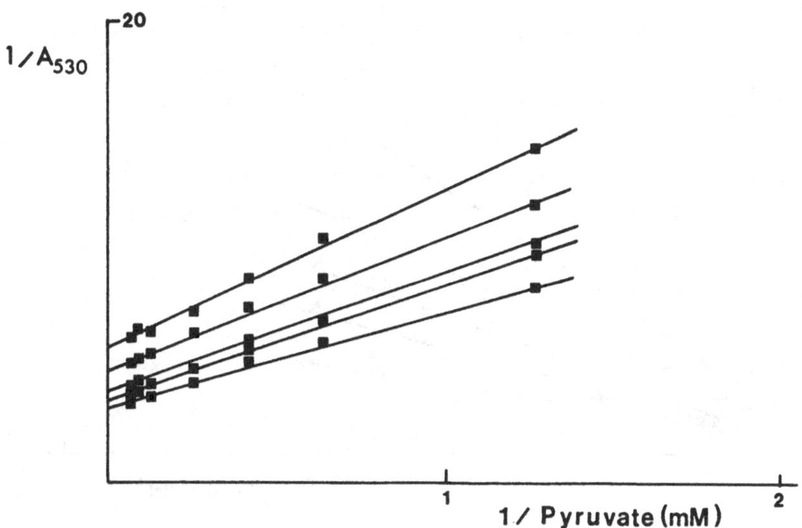

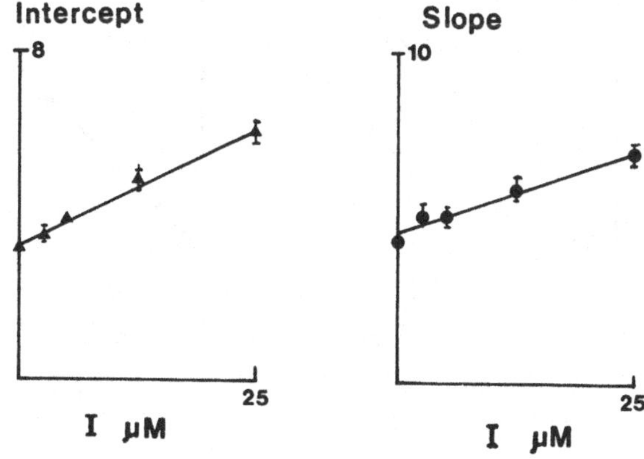

c. 0.05/min; Fig. 5) governed a slow change to full inhibition (K_i *c.* 2.2 nM). It is therefore almost certain that this herbicide also acts through inhibition of acetohydroxyacid synthase.

Fig. 9. Initial inhibition by 2-NO_2 6-Me sulphonanilide. Conditions were the same as in Fig. 6 and the same comments apply with regard to accuracy: assay time 25 min; total protein 0.08 mg; V_{max} 12.9 +/- 0.4 units/mg; K_m 2.2 +/- 0.4 mM pyruvate; K_i slope 29 +/- 9 nM; K_i intercept 28 +/- 8 nM.

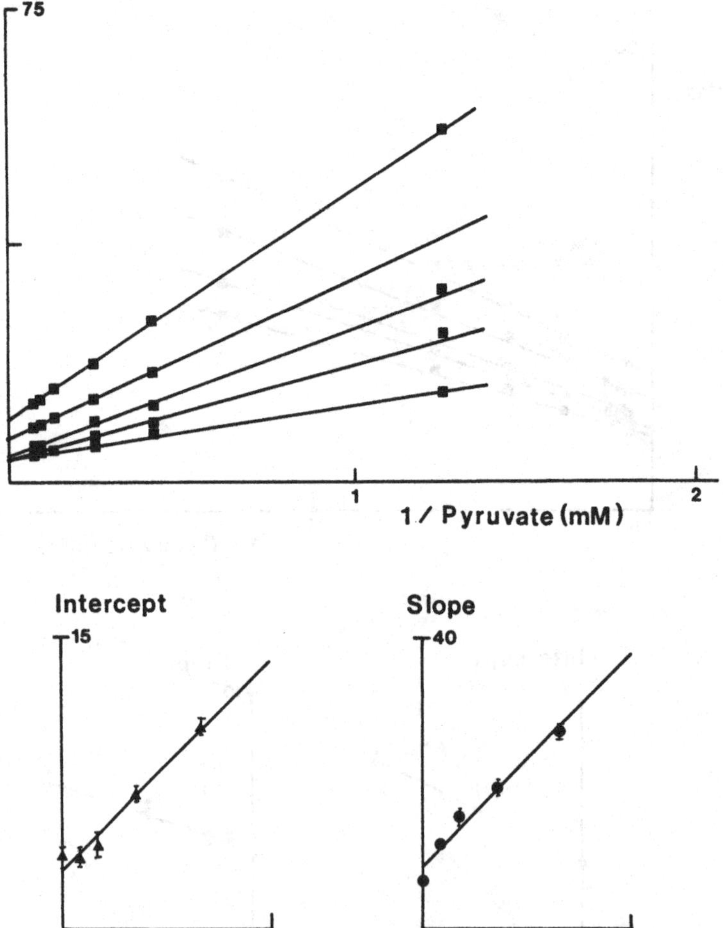

Relationships between inhibitors and substrates

Inhibition could always be modelled (Scheme 2) in terms of the rapid formation of an initial, relatively weak, EI complex followed by a slow transition to a more tightly bound form EI*. Using acetohydroxyacid synthase from peas, the value (*c.* 0.05/min) of the first order rate constant (k_1) governing the transition appeared (Fig. 5) to be independent of the particular inhibitor although the degree to which inhibition increased varied (ranging from perhaps 5 fold for sulphonylureas to *c.* 20 fold for imidazolinones). The onset of full inhibition seems to occur 4–5 fold more rapidly in bacterial isozyme II (Schloss, 1984) and presumably corresponds to some conformational shift in the protein.

The imidazolinones, phthalimido anilides and sulphonanilides were (within experimental error) simple non-competitive inhibitors. Presumably they all bind at sites which do not overlap with pyruvate.

Inhibition by sulfometuron methyl was 'mainly competitive' (i.e. the effect on the intercept of the double reciprocal plots was much less marked than the effect on the slope). Presumably binding is the tightest (but not exclusive) to species 1 and/or 3 in Scheme 1. Binding appears to be tighter following decarboxylation of pyruvate (species 3) in the bacterial enzyme.

Apart from competition with pyruvate there is further indirect evidence that sulfometuron methyl binds near the 'active site' of acetohydroxyacid synthase. Inhibition by sulfometuron methyl caused a slight hypochromic shift in the visible spectrum of the flavin in purified isozyme II (Schloss, 1984). Furthermore reduction of the flavin *in situ* lessened the degree of inhibition by sulfometuron methyl (J.V. Schloss, personal communication). Obviously the interaction with flavin might be long range and indirect. Alternatively, the inhibitor might bind close to or even directly contact some part of the isoalloxazine ring. It is also notable that the visible spectrum of FAD became gradually bleached during the course of the enzyme reaction. This process was too slow to be consistent with the formation of a catalytically important intermediate. Schloss has suggested that partial bleaching occurs due to a side reaction where the hydroxyethyl thiamine pyrophosphate anion (see Scheme 1) occasionally forms an adduct at the N_5 position of flavin (*cf.* George *et al.*, 1984). There is indirect evidence to substantiate this hypothesis based on the sequence homology between acetohydroxyacid synthase and pyruvate oxidase (Grabau & Cronan, 1986). In pyruvate oxidase thiamine pyrophosphate is thought to be close to FAD and, as the reduced hydroxyethyl thiamine pyrophosphate anion, transfer electrons to FAD via the N_5 position (Mather *et al.*, 1982). Taking all of these data into account, it would seem most likely that sulfometuron methyl binds adjacent to thiamine pyrophosphate and FAD in acetolactate synthase and partly overlaps the binding site of the second (non-decarboxylated) substrate. The insensitivity of isozyme I to sulfometuron methyl might be related to its predilection to catalyse the homologous condensation of pyruvate. The pocket for binding the second substrate might be smaller than in isozyme II and less able to accommodate sulphonylureas (LaRossa & Smulski, 1984).

Inhibition by valine appears to be competitive with pyruvate but non-linear. Probably, as a feedback regulator, it does not bind directly to the pyruvate binding site but mediates its effect by some more long-range mechanism. This would be consistent with the observation that some (valr-7) but not all valine resistant mutants of tobacco contain acetohydroxyacid synthase with a higher than normal K_m value for pyruvate (Relton *et al.*, 1986).

The simplest hypothesis is that all of the various inhibitors (with the probable exceptions of valine and leucine) bind at overlapping sites and only in the case of the sulphonylureas does part of the molecule directly hinder the binding of pyruvate. The cross resistance of the imidazolinone-resistant corn line (XA17) to chlorsulfuron lends some support to this view. There is no evidence (and indeed evidence to the contrary) that any of the inhibitors (apart from thiamine thiazolone pyrophosphate) act by displacing the FAD or thiamine pyrophosphate cofactors. We are only likely to understand the chemical basis of inhibition through X-ray crystallography of the various enzyme/inhibitor complexes. Certainly none of the herbicides are of a type that could have been designed or predicted on the basis of the enzyme mechanism and, in all cases, they have been discovered through the traditional random screening approach.

Relationship of enzyme inhibition to herbicidal activity

The useful biological selectivity of the sulphonylureas and imidazolinones is a consequence of the different rates and mechanisms of metabolic detoxification that occur in crops and weeds (Sweetser *et al.*, 1982; Orwick *et al.*, 1982; Shaner & Robson, 1985). The intrinsic activity of different sulphonylureas did not seem to vary a great deal between enzymes from different plants and there was a good correlation between herbicidal efficacy and enzyme inhibition (Ray, 1985).

It is difficult to understand on the basis of *in vitro* data why the imidazolinones should be herbicidal. Imazapyr is perhaps only 5 fold less herbicidal than sulfometuron methyl but more than three orders of magnitude less potent an inhibitor of acetohydroxyacid synthase. The uptake and transport properties of imidazolinones may be favourable compared to sulphonylureas. Apart from the whole plant data, it is also necessary to account for the high sensitivity of corn cell culture to inhibition by imazapyr (I_{50} *c.* 20 nM) relative to extracted enzyme. Perhaps, as a weak acid, imazapyr is taken up and concentrated more than 100 fold in the corn cell plastids (the probable location of acetohydroxyacid synthase). Possibly it will be necessary to look more closely at the evidence that acetohydroxyacid synthase is the sole site of action of the imidazolinones. Does the imidazolinone-resistant corn line XA17 contain important unrecognised mutations in addition to the change in acetohydroxyacid synthase? Recent work from the group at American Cyanamid (M. Stidham, personal communication) indicates that imazapyr has effects *in vivo* that might not have been anticipated on the basis of the *in vitro* data. Following foliar treatment, the concentration of radiolabelled imazapyr in the leaf mesophyll rose to a few hundred nM. After a few hours the specific activity of acetohydroxyacid synthase that could subsequently be extracted from the leaf (by conventional purification) fell dramatically. Presumably either existing acetohydroxyacid synthase had become irreversibly inactivated or, if the enzyme normally turns over rapidly, there had been some specific block in its biosynthesis.

Using the sulphonylureas as a benchmark, the relative herbicidal activities of the other types of inhibitor can be understood in terms of the inhibition of acetohydroxyacid synthase *in vitro*; i.e. N-phthalyl-L-valine anilide is virtually non-herbicidal and 2-NO_2 6-Me sulphonanilide is somewhat less herbicidal than sulfometuron methyl.

Why is inhibition of acetohydroxyacid synthase herbicidal?

Inhibition of the synthesis of DNA and of mitotic frequency appear to be important secondary events that follow from the inhibition of branched chain amino acid biosynthesis. How these events are linked and go on to cause plant death (rather than a reduction in growth rate for example) is not clear. It is interesting to compare the effects of imidazolinones and sulphonylureas with glyphosate which also inhibits the production of amino acids (aromatic amino acids and other products of chorismate). One might suppose that a block in the synthesis of any amino acid might trigger the same series of events starting with a reduced ability to make vital additions to the total pool of protein. However glyphosate has been reported to exert an early effect on a broad range of plant processes (notably cellular absorption of ions, biosynthesis of chlorophyll, photosynthetic uptake of CO_2, protein synthesis and nucleic acid biosynthesis; Cole *et al.*, 1983; Brecke & Duke, 1980; Foley *et al.*, 1983) amongst which inhibition of DNA synthesis is not especially significant. Possibly it is the lack of chorismate-derived products other than amino acids (e.g. quinones and folate) that determine the effects of glyphosate. The imidazolinones and sulphonylureas appear to cause a reduction in the level of total soluble protein without affecting the rate of protein synthesis. Presumably the plant must respond to amino acid starvation by increasing the turnover rate of the protein pool in the meristems. Amino acid degradation would then cause the pool of protein to decline gradually. It has also been suggested that toxic accumulation of 2-ketobutyrate is an important factor in the phytotoxicity of acetolactate synthase (Van Dyk & LaRossa, 1986).

Acknowledgements

The authors wish to thank the synthetic chemists and biologists at Jealott's Hill for their help.

References

Abell L.M., O'Leary M.H. & Schloss J.V. (1985). Determination of carbon isotope effects and substrate preference for acetolactate synthase by isotope ratio mass spectroscopy. *Biochemistry* **24**, 3357 (*Am. Chem. Soc.* abstracts).

Anderson P.C. & Hibberd K.A. (1985*a*). Evidence for the interaction of an imidazolinone herbicide with leucine, valine and isoleucine metabolism. *Weed Science* **33**, 479–83.

Anderson P.C. & Hibberd K.A. (1985*b*). Herbicide resistance in plants. European Patent Application 154204.

Brecke B.J. & Duke W.B. (1980). Effect of glyphosate on intact bean plants (*Phaseolus vulgaris* L.) and isolated cells. *Plant Physiol.* **66**, 656–9.

Chaleff R.S. & Mauvais C.J. (1984). Acetolactate synthase is the site of action of two sulfonylurea herbicides in higher plants. *Science* **224**, 1443–5.

Ciskanik L.M. & Schloss J.V. (1985). Reaction intermediates of the acetolactate synthase reaction: effect on sulfometuron methyl. *Biochemistry* **24**, 3357 (*Am. Chem. Soc.* abstracts).

Cole D.J., Caseley J.C. & Dodge A.D. (1983). Influence of glyphosate on selected plant processes. *Weed Research* **23**, 173–83.

Cromartie T.H. & Walsh C.T. (1976). *Escherichia coli* glyoxylate carboligase: properties and reconstitution with 5-deaza FAD and 1,5-dihydrodeaza $FADH_2$. *J. Biol. Chem.* **251**, 329–33.

De Felice M., Lago C.T., Squires, C.H. & Calvo J.M. (1982). Acetohydroxy acid synthase isoenzymes of *Escherichia coli* K12 and *Salmonella typhimurium*. *Ann. Microbiol.* **133a**, 251–6.

De Felice M., Squires C. & Levinthal M. (1978). A comparative study of the acetohydroxy acid synthase isoenzymes of *Escherichia coli* K12. *Biochim. Biophys. Acta* **541**, 9–17.

Eoyang L. & Silverman P.M. (1984). Purification and subunit composition of acetohydroxyacid synthase I from *Escherichia coli* K12. *J. Bacteriol.* **157**, 184–9.

Falco S.C. & Dumas K.S. (1985). Genetic analysis of mutants of *Saccharomyces cerevisiae* resistant to the herbicide sulfometuron methyl. *Genetics* **109**, 21–35.

George G.N., Shaw L., Bray R.C. & Engel P.C. (1984). Structure of the flavin N-5 adduct free radical obtained following inhibition of short chain acyl-CoA dehydrogenase by proprionyl-CoA, in *Flavins and Flavoproteins* (Bray R.C., Engel P.C. and Mayhew S.G., eds.) pp. 421–33, de Gruyter, Berlin.

Grabau C. & Cronan Jr. J.E. (1986) Nucleotide sequence and deduced amino acid sequence of *Escherichia coli* pyruvate oxidase, a lipid-activated flavoprotein. *Nucleic Acids Research* **14**, 5449–60.

Grimminger H. & Umbarger H.E. (1979). Acetohydroxy acid synthase I of *Escherichia coli*: purification and properties. *J. Bacteriol.* **137**, 846–53.

Hasan N. & Nester E.W. (1978). Purification and properties of chorismate synthase from *Bacillus subtilis*. *J. Biol. Chem.* **253**, 4993–8.

Huppatz J.L. & Casida J.E. (1985). Acetohydroxyacid synthase inhibitors: N-phthalyl-L-valine anilide and related compounds. *Z. Naturforsch.* **40c**, 652–6.

Jorns M.S. (1979). Mechanism of catalysis by the flavoenzyme oxynitrilase. *J. Biol. Chem.* **254**, 12145–52.

Kleschik W.A., Ehr R.J., Gerwick II B.C., Monte W.T., Pearson N.R., Costales M.J. & Meikle R.W. (1984). Novel substituted 1,2,4-triazolo-1,5-a pyrimidine-2-sulfonamides and methods of controlling undesired vegetation and suppressing the nitrification of ammonium nitrogen in soil. European Patent Application 142152.

LaRossa R.A. & Smulski D.R. (1984). *ilv* B-encoded acetolactate synthase is resistant to the herbicide sulfometuron methyl. *J. Bacteriol.* **160**, 391–4.

LaRossa R.A. & Schloss J.V. (1984). The sulfonylurea herbicide sulfometuron methyl is an extremely potent and selective inhibitor of acetolactate synthase in *Salmonella typhimurium*. *J. Biol Chem.* **259**, 8753–7.

Lawther R.P., Calhoun D.H., Adams C.W., Hauser C.A., Gray J. & Hatfield G.W. (1981). Molecular basis of valine resistance in *Escherichia coli*. K-12. *Proc. Nat. Acad. Sci. (USA)* **78**, 922–5.

Mather M., Lawrence M.S., Massey V. & Gennis R.B. (1982). Studies of the flavin adenine dinucleotide binding region in *Escherichia coli* pyruvate oxidase. *J. Biol Chem.* **257**, 12887–92.

Miflin, B.J. (1971). Cooperative feedback of barley acetohydroxy acid synthetase by leucine, isoleucine and valine. *Arch. Biochem. Biophys.* **146**, 542–50.

Miflin B.J. & Cave P.R. (1972). The control of leucine, isoleucine and valine biosynthesis in a range of higher plants. *J. Exp. Bot.* **23**, 511–6.

Orwick P.L., Marc P.A., Umeda K., Shaner D.L., Los M. & Ciarlante D.R. (1982). AC 252 214, a new broad spectrum herbicide for soybeans: greenhouse studies. *Proc. South. Weed Sci. Soc.* **36**, 90.

Ray T.B. (1982*a*). The mode of action of chlorsulfuron: a new herbicide for cereals. *Pestic. Biochem. Physiol.* **17**, 10–7.

Ray T.B. (1982*b*). The mode of action of chlorsulfuron: the lack of direct inhibition of plant DNA synthesis. *Pestic. Biochem. Physiol.* **18**, 262–6.

Ray T.B. (1984). Site of action of chlorsulfuron. *Plant. Physiol.* **75**, 827–31.

Ray T.B. (1985). The site of action of the sulfonylurea herbicides, in *Proceedings of British Crop Protection Conference* **1**, pp. 131–8, BCPC publications.

Relton J.M., Wallsgrove R.M., Bourgin J.-P. & Bright S.W.J. (1986). Altered feedback sensitivity of acetohydroxyacid synthase from valine-resistant mutants of tobacco (*Nicotinia tabacum* L.). *Planta* **169**, 46–50.

Scheel D. & Casida J.E. (1985). Sulfonylurea herbicides: growth inhibition in soybean cell suspension cultures and in bacteria correlated with block in biosynthesis of valine, leucine or isoleucine. *Pestic. Biochem. Physiol.* **23**, 398–412.

Schloss J.V. (1984). Interaction of the herbicide sulfometuron methyl with acetolactase synthase: a slow binding inhibitor, in *Flavins and Flavoproteins* (Bray R.C., Engel P.C. & Mayhew S.G. eds) pp. 737–40, de Gruyter, Berlin.

Schloss J.V., Van Dyke D.E., Vasta J.F. & Kutny R.M. (1985). Purification and properties of *Salmonella typhimurium* acetolactate synthase isozyme II from *Escherichia coli* HB101/pDU9. *Biochemistry* **24**, 4952–9.

Shaner D.L., Anderson P.C. & Stidham M.A. (1984). Imidazolinones – potent inhibitors of acetohydroxyacid synthase. *Plant Physiol.* **76**, 545–6.

Shaner D.L., Stidham M., Muhitch M., Reider M., Robson P. & Anderson P. (1985*a*). Mode of action of the imidazolinones, in *Proceedings of the British Crop Protection Conference* **1**, pp. 147–54, BCPC publications.

Shaner D.L., Malefyt T. & Anderson P.C. (1985*b*). Herbicide resistant maize through cell culture selection, in Monograph No. 32 *Biotechnology and its Application to Agriculture* (Copping L.G. & Rodgers P., eds), pp. 45–9, BCPC publications.

Shaner D.L. & Robson P.A. (1985). Absorption, translocation and metabolism of AC 252 214 in soybean (*Glycine max*), common cocklebur (*Xanthium strumarium*) and velvetleaf (*Abutilon theophrasti*). *Weed Sci.* **33**, 469–71.

Shaner D.L. & Reider M.L. (1986). Physiological responses of corn (*Zea mays*) to AC 243,997 in combination with valine, leucine and isoleucine. *Pestic. Biochem. Physiol.* **25**, 248–57.

Shaw, K.J., Berg C.M.& Sobol T.J. (1980). *Salmonella typhimurium* mutants defective in acetohydroxy acid synthases I and II. *J. Bacteriol.* **141**, 1258–63.

Stormer F.C. (1968). The pH 6 acetolactate-forming enzyme from *Aerobacter aerogenes*: II. Evidence that it is not a flavoprotein. *J. Biol Chem.* **243**, 3740–1.

Sweetser P.B., Schow G.S. & Hutchison J.M. (1982). Metabolism of chlorsulfuron by plants: biological basis for selectivity of a new herbicide for cereals. *Pestic. Biochem. Physiol.* **17**, 18–23.

Van Dyk T.K. & LaRossa R.A. (1986). Contributions of alpha-ketobutyrate accumulation to sulfonylurea herbicide toxicity. *J. Cell. Biochem.* (Suppl. 10C, 56).

Walsh C. (1979), in *Enzymatic Reaction Mechanisms*, pp. 688–9, W.H. Freeman & Co., San Francisco.

Westerfield W.W. (1945). A colorimetric determination of blood acetoin. *J. Biol Chem.* **161**, 495.

Yadav N., McDevitt R., Bernard S. & Falco S.C. (1986). Single amino acid substitutions in the enzyme acetolactate synthase confer resistance to the herbicide sulfometuron methyl. *Proc. Nat. Acad. Sci. (USA)* **83**, 4418–22.

PETER J. LEA AND STUART M. RIDLEY

Glutamine synthetase and its inhibition

Introduction

Glutamine synthetase (GS) [L-glutamate: ammonia ligase (ADP-forming) EC 6.3.1.2] catalyses reaction 1 :

$$\text{L - Glutamate} + \text{ATP} + \text{NH}_3 \rightarrow \text{L - Glutamine} + \text{ADP} + \text{P}_i + \text{H}_2\text{O} \qquad (1)$$

There is now a considerable amount of evidence to indicate that GS is the first enzyme involved in the conversion of inorganic nitrogen to an organic form in most organisms (Fig. 1). Following the formation of glutamine, the amide nitrogen is transferred to the 2-amino position of glutamate by the enzyme glutamate synthase which exists as two forms, L-glutamate: NAD^+ oxidoreductase (transaminating) EC 1.4.1.14, and L-glutamate: ferredoxin oxidoreductase (transaminating), EC 1.4.7.1. These enzymes catalyse reaction 2:

$$\begin{aligned}&\text{2-Oxoglutarate} + \text{L-glutamine} + (\text{NADH}_2 \text{ or ferredoxin}_{\text{red}}) \rightarrow \\ &2\ \text{L-glutamate} + (\text{NAD}^+ \text{ or ferredoxin}_{\text{ox}})\end{aligned} \qquad (2)$$

GS and glutamate synthase are able to act in conjunction to form the glutamate synthase cycle (Fig. 2). The evidence that this cycle operates is based on labelling

Fig. 1. The metabolism of nitrogen within higher plants.

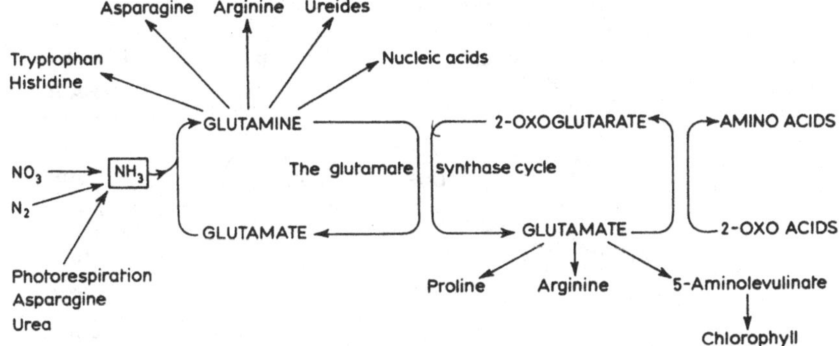

studies, enzyme kinetics and the action of inhibitors (Miflin & Lea, 1976; Wallsgrove *et al.*, 1983; Lea, Wallsgrove & Miflin, 1985). In bacteria and fungi grown at high levels of ammonia, GS is subject to complex regulation and the enzyme glutamate dehydrogenase (GDH) [EC 1.44.1.3] is considered to carry out the assimilation of ammonia. Despite the occasional claim to the contrary (Loyola-Vargas & Jimenez, 1984; Yamaya *et al.*, 1986) there is no good evidence that GDH plays an important role in ammonia assimilation in higher plants, although it may be involved in glutamate deamination.

Plants obtain their nitrogen from two major sources, either as nitrate taken up from the soil or by nitrogen fixation carried out by ***Rhizobium*** bacteria in legume root nodules. Nitrate reduction may take place either in the root or shoot depending upon a number of factors (Andrews, 1986). However, the primary assimilation of ammonia is not the major role of GS in higher plants, as ammonia may be generated internally by a number of independent reactions.

During the germination of seedlings (in particular legumes), seed storage proteins are broken down to yield keto acids and the liberated ammonia is reassimilated via GS and transferred to asparagine (Lea & Joy, 1983). Asparagine and ureides are the major transport compounds in higher plants (Lea & Miflin, 1980), upon their arrival at the sink (either the young developing leaves or a maturing fruit) they are metabolised via

Fig. 2. The glutamate synthase cycle of ammonia assimilation.

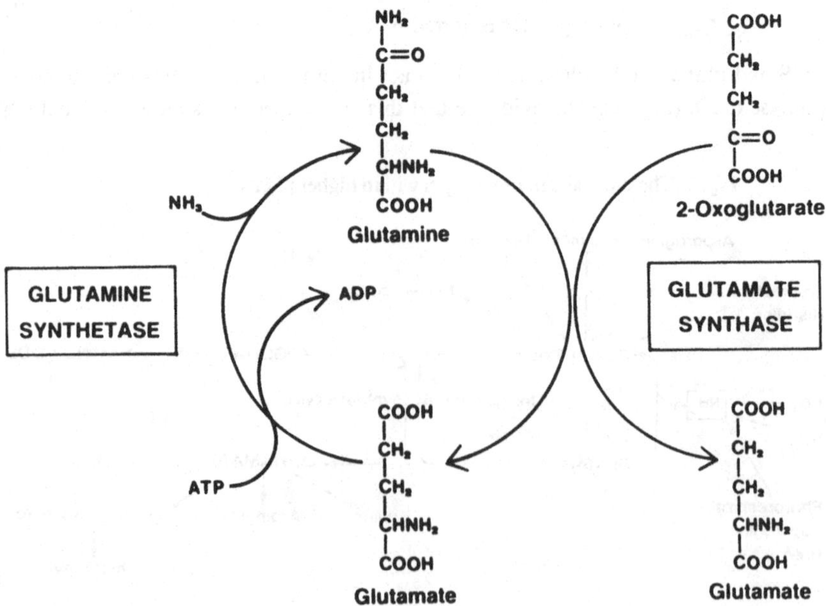

free ammonia which is again reassimilated via GS (Ta, Joy & Ireland, 1984). Quantitatively the most important role of GS is in the leaves of C_3 plants, which carry out the process of photorespiration in the light. The enzyme RuBP carboxylase oxygenase has the ability to synthesise phosphoglycollate in the presence of oxygen. Phosphoglycollate is metabolised via the photorespiratory nitrogen cycle in the chloroplasts, peroxisomes and mitochondria to phosphoglycerate (Keys *et al.*., 1978; Tolbert, 1980).

In the mitochondria, glycine is converted to serine as shown in reaction 3:

$$2\ \text{Glycine} + H_2O \rightarrow \text{Serine} + CO_2 + NH_3 + 2H^+ + 2e^- \qquad (3)$$

The CO_2 is lost to the atmosphere and the reaction is considered to contribute to the lower yields of C_3 plants. As the flux through the pathway has been calculated to be at least ten times higher than the rate of primary nitrogen assimilation, it is essential if the plant is not to die of nitrogen starvation that the ammonia is rapidly reassimilated by GS. Mitochondria are only able to reassimilate less than 1% of the ammonia released from glycine (Keys *et al.* 1978, Bergman, Gardestrom & Ericson, 1981; Yamaya *et al.*, 1986), probably due to the lack of GS in the organelle (Wallsgrove *et al.*, 1980). The ammonia diffuses through the cytoplasm, where some assimilation may take place, and into the chloroplast the major site of GS activity (Woo & Osmond, 1982). It is the rapid turnover of ammonia in leaves in the light (Wallsgrove *et al.*, 1983) that has made GS activity an important target for herbicide action.

Distribution and localisation of GS in plant tissues

In non-photosynthetic tissue (e.g. roots and tissue culture) the major proportion of GS activity is located in the cytoplasm (see Oaks & Hirel, 1985 for review) although finite amounts have recently been detected in the plastids (Vézina *et al.*, 1987). In legume roots infected with *Rhizobium* there is a dramatic increase in GS activity that is either concomitant with or slightly in advance of the development of nitrogenase activity (Robertson, Farnden & Banks, 1975; Lara *et al.*, 1983; Padilla *et al.*, 1987). The increase in GS in *Phaseolus vulgaris* was due entirely to the appearance of a novel form of the enzyme GS_{n1} in the nodule cytoplasm, the activity of a second form GS_{n2} remained constant throughout nodule development. The two forms of nodule GS have been purified to homogeneity and the enzymological and immunological data suggests that GS_{n2} and the root enzyme GS_r are identical (Cullimore *et al.*, 1983; Cullimore & Miflin, 1984).

In pea leaves, GS was originally isolated and purified from the chloroplast (O'Neal & Joy, 1973). However, it then became clear that a second distant cytoplasmic form existed in some plants (Mann, Fentem & Stewart, 1979; Wallsgrove *et al.*, 1980). McNally *et al.* (1983) have examined the GS isoenzyme content of a wide range of plants. They concluded that four groups of plants could be distinguished:

(1) Achlorophyllous higher plant parasites nutritionally dependent on their hosts (e.g. *Orobanche* species) and containing only the cytoplasmic form GS_1.
(2) A group containing a range of plants including spinach, tobacco and lupin which have only the chloroplastic form GS_2.
(3) A group of C_3 grasses and temperate legumes which have a minor GS_1 (5–30%) and a major GS_2 (70–95%) form. (Fig. 3).
(4) A group of C_4 and CAM plants and also tropical legumes that have a high proportion of cytoplasmic GS_1 which may be up to 80% of the total.
Hirel *et al.* (1982) demonstrated that during the greening of etiolated rice leaves, the total GS level increased two fold. However there was a five fold increase in the chloroplastic GS_2, whilst the cytoplasmic GS_1 decreased correspondingly.
In a detailed study of the development of wheat leaves in the light, Tobin, Ridley & Stewart (1985) showed that if GS was expressed on a cell basis the activity increased 50 fold from the base to the tip. The activity increase was mainly due to the chloroplast enzyme, the cytoplasmic GS remained constant in all segments of the leaf (Figs. 3 & 4).

In a developmental study of maturing pea cotyledons, Sodek, Lea & Miflin (1980) showed that high levels of GS activity were present in the testa until the final stages of maturation. However, GS activity in the cotyledons was initially low, but showed a rapid increase during the last 8 days of the maturation process. It has in fact been suggested that even in green pea seeds there are two distinct forms of GS located in the cytoplasm and chloroplast (Pushkin *et al.*, 1985).

In the majority of studies on higher plant GS the molecular weight of the active enzyme varies between 320–380 kD, although there have been reports by Russian workers of values of 480–520 kD (Pushkin *et al.*, 1985). The plant enzyme is made

Fig. 3. (a) Elution profile for the separation by FPLC (Pharmacia Mono Q anion exchange column) of the activity of GS isoenzymes from an extract of 7 day old wheat leaves. (b) Elution profile, as above, of GS_2 from the chloroplast stroma of 7 day old wheat leaves (Tobin *et al.*, 1985).

o—o GS activity: ---- NaCl concentration

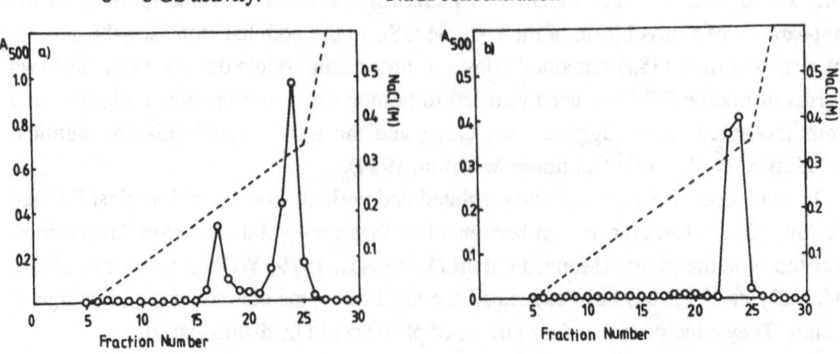

up of eight subunits in a similar manner to the mammalian enzyme (Meister, 1973) but different to the twelve subunits of the *E. coli* enzyme (Almassy *et al.*, 1986).

In the *Phaseolus vulgaris* root there are two types of polypeptide subunit with molecular weights of 39–43 kD. The α-subunit is synthesised early in the development of the root whilst the β-subunit becomes more predominant in the mature root (Ortega *et al.*, 1986). In the root nodule, GS_{n2} contains only β-subunits and is also very similar to the cytoplasmic GS found in the leaf (Lara *et al.*, 1984). GS_{n1} contains predominantly a third polypeptide termed the γ-subunit which is nodule specific (Cullimore & Bennett, 1988). In the leaf chloroplast GS, four polypeptides of molecular weight 45,000 may be detected following I.E.F., that have been termed subunits a,b,c and d (Hirel *et al.*, 1984; Lara *et al.*, 1984). It has been shown that glutamine synthetase subunits are encoded by a small multigene family showing organ-specific expression (Cullimore *et al.*, 1984; Gebhard *et al.*, 1986). In pea the chloroplast GS is encoded by a nuclear gene (Tingey, Walker & Coruzzi, 1987), and in *P. vulgaris* the enzyme protein is synthesised as a high MW precursor in the cytosol prior to transport into the chloroplast (Lightfoot *et al.*, 1988).

The mechanism of the glutamine synthetase reaction

The overall reaction (1), takes place in two steps as outlined in Fig. 5.

1. involves the formation of γ-glutamyl phosphate from ATP and glutamate.
2. involves attack by NH_3 on γ-glutamyl phosphate to form a tetrahedral adduct with the subsequent liberation of glutamine, ADP and Pi.

Fig. 4. Changes in GS activity and chloroplast and mesophyll cell numbers during development of 7 day old primary wheat leaf.

o—o GS activity.
●—● Chloroplast per cell.
×—× Mesophyll cells per 0.5 cm section (× 10^{-4}).

Error bars denote ± standard error (n = 90), for plastids per cell; (n = 20) for mesophyll cells per section. (Tobin *et al.*, 1985).

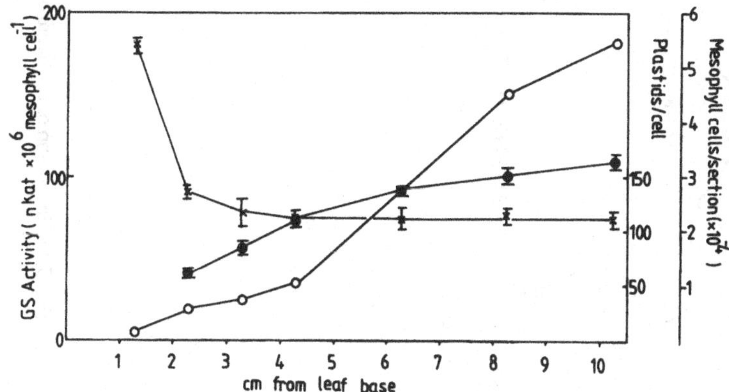

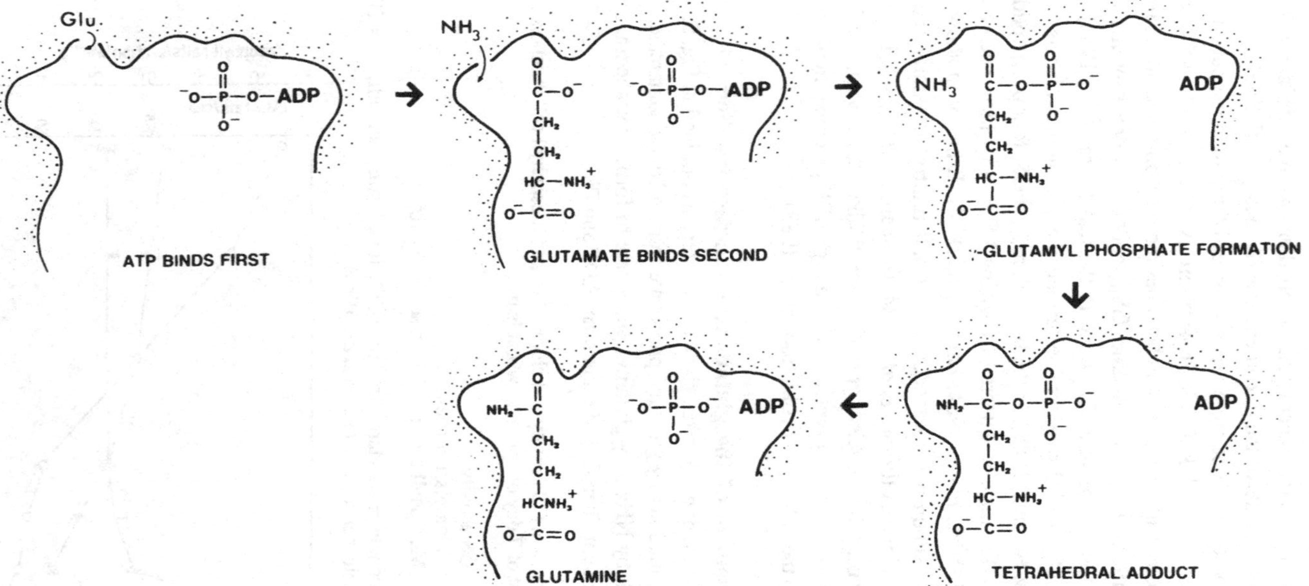

Fig. 5.An illustration of the binding of substrates in the glutamine synthetase reaction involving a tetrahedral adduct transition state.

Determination of substrate binding order using 2-amino-4-phosphonobutyric acid (APBA)

To help understand the chemical mechanism of the GS reaction, knowledge of the order in which substrates are bound and products released can be obtained from kinetic analyses. Meek & Villafranca (1980) used a dead-end reversible inhibitor, APBA, to determine a binding order of substrates on GS from *E.coli*. They established a sequential binding of ATP, followed by glutamate, and then NH_3. Some significant differences have been found in the chemical reaction profiles between GS from *E.coli*, and GS from pea seeds and ovine brain particularly in the extent to which the reaction intermediate is stabilized (Wedler, Horn & Roby, 1980).

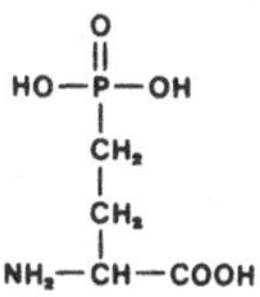

```
         O
         ||
    HO—P—OH
         |
        CH₂
         |
        CH₂
         |
  NH₂—CH—COOH
```

APBA

The binding order of the substrates to pea chloroplast GS_2 has also been examined using APBA. With K_i of 0.60 mM, APBA bound almost 20-fold more tightly to pea GS_2 than Glu (K_m = 11.4 mM) (Ridley & Howard, unpublished data); but the compounds show weak herbicidal activity (Maier & Lea, 1983).

Fig. 6 shows double reciprocal plots of GS activity versus substrate at various APBA doses. The inhibition by APBA was uncompetitive for ATP, competitive for glutamate, and non-competitive for NH_3. This is diagnostic of the same ordered binding of substrates established by Meek & Villafranca (1980) for *E.coli* GS.

Inhibitors of glutamine synthetase

1. Methionine sulphoximine

Occurrence. Methionine sulphoximine (MSO) was initially isolated by the Research Association of British Flour Millers in 1949. The compound was crystallised as the toxic constituent of nitrogen chloride (agene) treated zein, which was known to cause hysteria, convulsions and epileptic fits in a number of animals (Bentley *et al.*, 1949). MSO was shown to inhibit the growth of bacteria (an effect that could be reversed by glutamine) and inhibit the germination of seedlings. Using an acetone powder of sheep brain, Pace & McDermott (1952) were able to show that

Figure 6. Kinetic analyses for the inhibition of pea GS_2 by APBA.
(a) Uncompetitive inhibition versus APT; (b) competitive inhibition versus L-glutamate; (c) non-competitive inhibition versus NH_3. Reaction mixtures (150 µl) contained 5 mM EDTA, 20 mM $MgSO_4$, 50 mM Tris-HCl, pH 7.6, and substrate concentrations when fixed were 1 mM ATP, 0.2 mM NH_4Cl, and 70 mM L-Glu. Samples were incubated for 10 min at 30 °C, and GS activity was recorded by the production of inorganic phosphate using a method based on that of Lanzetta *et al.*, (1979); after the incubation, 750 µl of a Malachite Green colour reagent were added, followed 1 to 3 min later by 150 µl of 34% citric acid (analytical grade). [The colour reagent was made up by mixing 3 parts of 0.045% Malachite Green, HCl (Sigma) in water with 1 part of 4.2% ammonium molybdate in 4N HCl and leaving for at least 30 min before filtering (by injection through a 0.45 µ filter). Then 0.1 part of 1.5% Triton N-101 was added.]

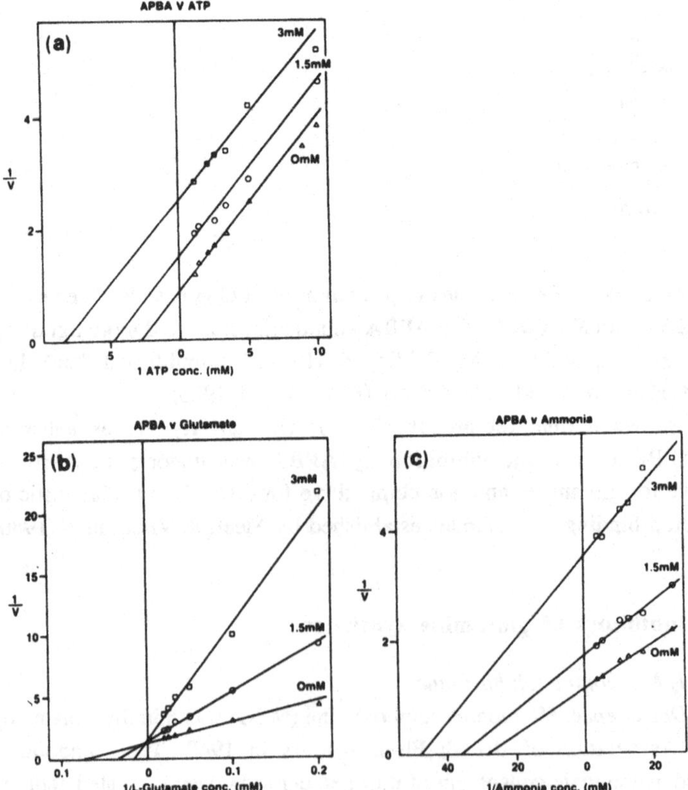

MSO inhibited a 'glutamine synthesising enzyme' assayed by a method employing glutamine and hydroxylamine.

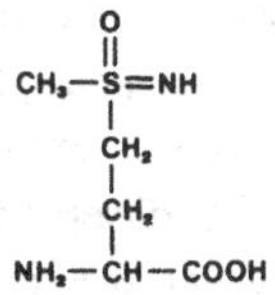

O
||
CH_3—S=NH
|
CH_2
|
CH_2
|
NH_2—CH—COOH

Methionine sulphoximine
(MSO)

It was later shown that of the four diastereoisomers of MSO, only one, i.e. L-methionine-S-sulphoximine caused convulsions and acted as an inhibitor of glutamine synthetase (Manning *et al.*, 1969). More recently a compound with antibiotic activity identified as L-(N^5-phosphono)methionine-S-sulphoximinyl-L-alanyl-L-alanine has been isolated from *Streptomyces* (*cf.* phosphinothricin). This tripeptide inhibited the growth of bacteria, an action that was reversed by glutamine. MSO could be liberated by the action of a leucine aminopeptidase and a phosphatase (Scannell *et al.*, 1972; Pruess *et al.*, 1973).

Effect on isolated glutamine synthetase. Following the pioneering work of Meister and his colleagues, it became clear that MSO was a potent inhibitor of sheep brain GS. In the presence of ATP and metal ions, MSO bound irreversibly to the active site due to the formation of phosphorylated MSO. Computer studies have indicated that the sulphoximine oxygen atom of MSO binds to the oxygen binding site for the γ-carboxyl group of glutamate and that the methyl group binds to the ammonia site. In this configuration the sulphoximine nitrogen atom is in the correct position to be phosphorylated. Thus for the sheep brain enzyme under initial rate conditions, inhibition by MSO was competitive with respect to glutamate, but preincubation with ATP showed that the inhibition became completely irreversible (Ronzio, Rowe & Meister, 1969). In a series of studies by Wedler & Horn (1976) and Wedler, Horn & Roby (1980), a comparison was made of the mechanism of inhibition of GS purified from pea seed and *E.coli*. MSO inhibited the pea seed enzyme noncompetitively, with a K_i of 200 μM, whereas with the *E.coli* enzyme there was a large change in the K_m for glutamate at 5–10 μM MSO (K_i = 1.5 μM), at which level of inhibition the pattern was competitive, but changed to noncompetitive at concentrations of MSO of 25 μM and above. It was originally thought that MSO could not be removed from the enzyme and activity restored. However, Maurizi & Ginsburg (1982) have been able to recover stoichiometric amounts of MSO from the *E.coli* enzyme at pH 4.0 in the presence of 1 M KCl and 0.4 M $(NH_4)_2SO_4$.

Ammonia was not competitive with MSO for the pea enzyme, a result which is not in agreement with the earlier suggestion that the methyl group of MSO binds to the ammonia binding site. The kinetics of the inhibition of plant GS by MSO have also been examined in relatively crude extracts of pea leaves (Leason *et al.*, 1982) and in a highly purified preparation from spinach chloroplasts (Ericson, 1985). Competitive inhibition (K_i values, pea = 160 μM; spinach = 100 μM) was observed at low MSO concentrations for GS from both sources. However, at the low glutamate concentrations used for the pea enzyme, non-linear double reciprocal plots were obtained, whilst at the high glutamate concentrations used for the spinach enzyme, such inhibition was not evident. Of particular interest is the strikingly low K_i for the *E.coli* enzyme (over 100-fold) as compared to the three plant enzymes tested.

As mentioned previously the bacterial enzyme has a different structure from that found in higher plants and mammals. The enzyme of molecular weight 620,000 consists of two planar hexagons of subunits of 50,000 daltons. Low concentrations of MSO have been shown to stabilise intersubunit binding, and active structures of 4,6,8 and 10 subunits have been isolated in the presence of 6 M guanidine hydrochloride (Maurizi & Ginsburg, 1982). It was this information, together with a detailed X-ray crystallographic examination of *Salmonella typhimurium* GS, that allowed Almassy *et al.* (1986) to determine that the active site of bacterial GS is situated at the interface of two adjacent subunits in the planar rings.

Action of MSO on plant metabolism Despite the wide use of MSO in animal studies, the inhibitor was little used in experiments on plants until the introduction of the GS/GOGAT pathway of ammonia assimilation (see Introduction). Following this, MSO was frequently used to emphasise the importance of GS in a wide range of plants. Almost without fail, the addition of even low concentrations of MSO caused dramatic increases in the levels of free ammonia and corresponding falls in most soluble amino acids (see Table 1). Notable exceptions have been those instances where glutamate dehydrogenase has been known to operate, e.g. in ectomycorrhizal fungi or green algae grown on high concentrations of ammonia.

A particular interesting use of MSO has been to establish rates of ammonia evolution in the photorespiratory nitrogen cycle (Keys *et al.*, 1978). However, in most cases, rates of ammonia evolution were not as high as had been measured by other techniques involving CO_2 release (Platt & Anthon, 1981; Platt & Rand, 1983; Martin *et al.*, 1983, Berger & Fock, 1985; Archireddy *et al.*, 1983; Lea *et al.* 1984; Johansson & Larsson, 1986). The major reason for these low rates of ammonia evolution is probably due to the effect of MSO on photosynthetic CO_2 assimilation. This effect will be discussed in the later section on herbicidal action.

Table 1. *Action of MSO on plant metabolism*

Plant Tissue	Effect	Reference
Lemna minor	Rapid accumuluation of NH_3. $^{14}CO_2$ assimilation into glutamine decreased, into 2-oxoglutarate increased.	Stewart & Rhodes (1976)
Pea chloroplasts	Inhibited GS but not glutamate synthase activity	Anderson & Done (1977)
Rice roots	Assimilation of $^{15}NH_3$ and $^{15}NO_3^-$ prevented	Arima & Kumazawa (1977)
Tobacco cell cultures	Inhibits incorporation of $^{13}NO_3^-$ and $^{13}NH_3$ into glutamine	Skokut *et al.* (1978)
Spinach leaf	Ammonia accumulated and glutamine decreased in the light	Ito *et al.* (1978)
Datura roots	$^{15}NO_3^-$ accumulated in NH_3 in MSO treated roots	Probyn & Lewis (1979)
Soybean leaves	MSO inhibited conversion of ^{14}C-aspartate to asparagine	Stewart (1979)
Lemna	Completely blocked assimilation of $^{15}NH_3$	Rhodes *et al.* (1980)
Spinach leaf cells	MSO inhibited flow of $^{14}CO_2$ into glutamine	Woo & Canvin (1980)
Chlamydomonas	Photorespiratory NH_3 derived from protein breakdown	Cullimore & Sims (1980)
Oat leaves	Comparison of MSO and tabtoxin	Frantz *et al.* (1982)
Barley plants	Blocks assimilation of $^{15}NO_3^-$ into intact plants	Fentem *et al.* (1983*a*)
Pea, barley and maize shoots	A comparison of PPT, PPO and MSO amino acid levels	Lea *et al.* (1984)
Cyanobacteria *Anacystis*	Potential use of MSO commercially for photoproduction of NH_3	Ramos *et al.* (1982)
Amaranthus & *Vigna*	MSO induced ammonia production increased by adding glycine	Kumar *et al.* (1984)
Dwarf bean *Phaseolus*	MSO relieves ammonia inhibition of nitrate uptake	Breteler & Siegerist (1985)
Alfalfa root nodules	MSO prevents incorporation of $^{14}CO_2$ into amino acids exported in xylem sap.	Snapp & Vance (1986)

The most thorough examination of the action of MSO on plant metabolism has been carried out by Walker, Keys & Givan (1984*a, b*), utilising excised wheat leaves. By feeding [^{14}C]-glutamate they were able to show MSO stimulated the deamination and decarboxylation of glutamate to succinate, malate and CO_2. The incorporation of $^{14}CO_2$ into glycine, serine, alanine, glutamate and aspartate was decreased by MSO, presumably due to the lack of available amino donors. The decline in labelling in the amino acids was accompanied by an increased percentage incorporation into phosphorylated products and into malic acid. The increased malate may either be due to an increased carboxylation of phosphoenolpyruvate or the operation of the tricarboxylic acid cycle.

In a recent detailed examination of nitrogen metabolism in MSO treated *Lemna*, Rhodes *et al.* (1986) detected the established increase in ammonia and decrease in the amino acids, glutamine, glutamate, aspartate, asparagine, alanine, glycine and serine. However, proline, valine, leucine, isoleucine, threonine, lysine, methionine, phenylalanine, tyrosine and tryptophan accumulated over a 24-hour period. The authors were able to show by [^{15}N]-feeding that the accumulated amino acids were not formed by *de novo* synthesis. It is likely that the amino acids are formed by protein breakdown (as has been previously shown by Cullimore & Sims (1980) in *Chlamydomonas*), but are not further metabolised. The amino acids showing the major decreases are those shown by Ta & Joy (1986) to be actively involved in the photorespiratory nitrogen cycle and act as donors for glycine synthesis.

2. *Phosphinothricin*

Phosphinothricin (2-amino-(methylphosphinyl)-butanoic acid (PPT) is an analogue of glutamate that occurs naturally as a tripeptide antibiotic (L-phosphinothricyl-L-alanyl-L-alanine), and was first isolated from cultures of *Streptomyces viridochromogenes* (Bayer *et al.*, 1972) (cf. the MSO tripeptide). Omura *et al.* (1984*a, b*) have isolated another tripeptide antibiotic containing PPT from a soil actinomycete *Kitasatosporia phosalacinea,* called phosalacine (L-phosphinothricyl-L-alanyl-L-leucine).

Phosphinothricin is potently herbicidal both as the tripeptide and as PPT alone; both the natural product, called bialaphos, and PPT have been developed as herbicides. PPT is between five and ten-fold more potent than MSO in whole plant tests. However, Tachibana *et al.* (1986) found that the tripeptide bialaphos was not active even at 3 mM against glutamine synthetase extracted from shoots of Japanese barnyard millet. Likewise, Omura *et al.* (1984*a*) failed to inhibit the enzyme from spinach leaves with 2.5 mM phosalacine; however when they incubated phosalacine with cell-free extracts from either spinach leaves or *B. subtilis,* PPT was liberated; this clearly demonstrated that the peptide must be hydrolysed by the plant (or bacterium) before a herbicidal effect can be produced.

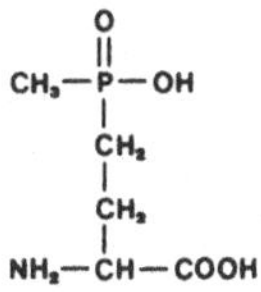

```
          O
          ||
    CH3—P—OH
          |
          CH2
          |
          CH2
          |
    NH2—CH—COOH
```

Phosphinothricin
(PPT)

Chemically synthesised PPT, called glufosinate, is a mixture of the D- and L-isomers (Köcher, 1983), but only the L-isomer is active (Manderscheid & Wild, 1986). Many variations on the phosphinothricin structure have been synthesised, but no improvement in herbicidal activity has been found (Maier & Lea, 1983). The 2-oxo analogue of phosphinothricin, 2-oxo-4-(methylphosphinyl)-butanoic acid (PPO) is herbicidal and exerts a similar effect to phosphinothricin itself. However, it does not inhibit GS *in vitro*, and is most probably converted enzymically by transamination to phosphinothricin after entering the plant cell (Lea *et al.*, 1984; Joy, K.W. and Lea, P.J. unpublished results).

Studies on the inhibition of isolated glutamine synthetases by PPT (Mandersheid & Wild, 1986; Colanduoni & Villafranca, 1986), have shown that it is also a two step process similar to that for the inhibition of sheep brain GS by MSO (Ronzio, Rowe & Meister, 1969): (i) under initial rate conditions PPT is a reversible inhibitor and competitive with glutamate; (ii) subsequent phosphorylation of PPT by ATP produces an analogue of the tetrahedral adduct of the substrates, and gives irreversible inhibition (Fig. 7).

Initial rate inhibition kinetics with PPT using pea seedling GS_2 gave the same diagnostic patterns in double reciprocal plots as for APBA (Fig. 3), showing that PPT is uncompetitive for ATP, competitive for glutamate and non-competitive for NH_3.

Inhibition of the first binding step has been widely examined (Bayer *et al.*, 1972; Fraser & Ridley, 1984; Leason *et al.*, 1982; Mandersheid & Wild, 1984). Table 2 shows a comparison of binding constants for glutamate ATP and PPT on isolated GS_1 and GS_2 from a range of crops and weeds. Many species lack the cytosolic GS_1 (McNally *et al.*, 1983), while others contain too little for kinetic analysis. PPT inhibited GS_2 with K_i values between 4 and 9 μM, but bound more tightly to GS_1 giving half those values. Although there is remarkably little variation in K_i values for the isolated enzymes between species, a 70-fold variation in the herbicidal susceptibility of some of these plants has been shown (Ridley & McNally, 1985) (Table 3).

More detailed studies of the complete process of inhibition have been carried out by Manderscheid & Wild (1986) using GS from wheat chloroplasts (GS_2) and root cytosol (GS_r), by Colanduoni & Villafranca (1986) using GS from *E.coli* and by

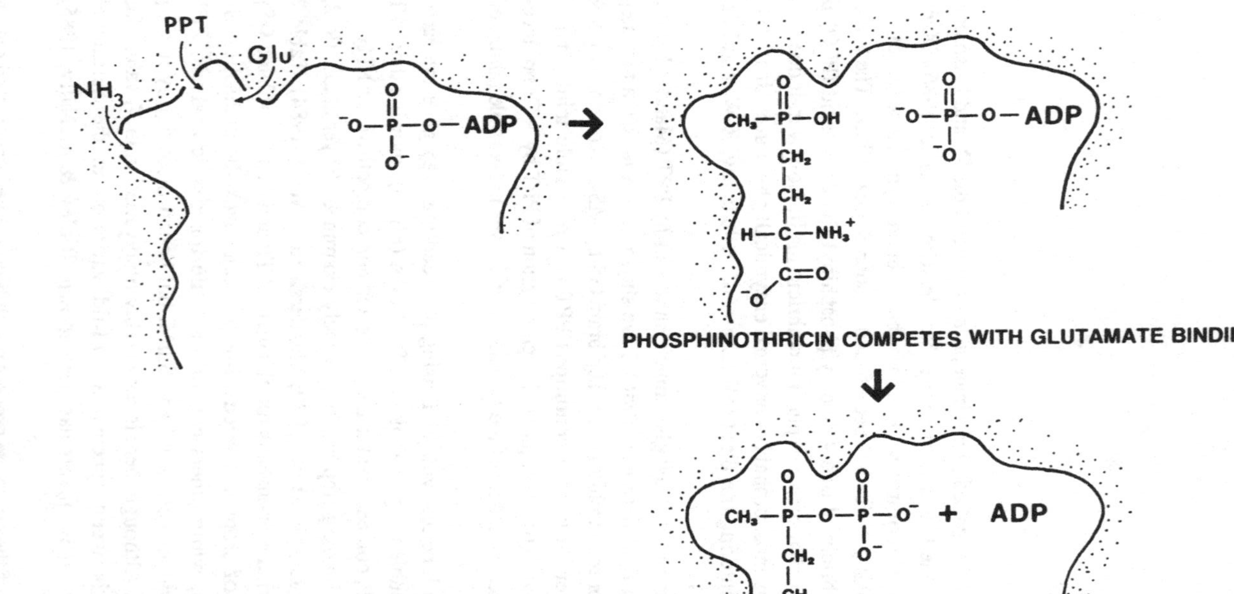

Fig. 7. An illustration of the binding reaction of phosphinothricin. ATP binds to GS initially, allowing PPT and Glu to compete for the same binding site. Phosphorylation of PPT leads to an 'irreversibly' bound mimic of the tetrahedral adduct of the substrates (see Fig. 5).

Table 2. *Comparison of* K_i *values under initial rate conditions for DL-phosphinothricin inhibition of leaf glutamine synthetases from a range of crops and weeds, and comparative* K_m *values for L-glutamate and ATP.*

Species		K_m glutamate (mM)		K_m ATP (mM)		K_i PPT (μM)		Refs
		GS_1	GS_2	GS_1	GS_2	GS_1	GS_2	
Barley		4.9	8.2	0.3	0.6	3.5	6.0	1
Wheat		ND	7.7	ND	0.9	ND	9.0	1
Maize		5.2	9.8	0.4	0.9	2.0	4.0	1
*Soya			4.1		0.3		8.0	1
*Sugar beet			8.4		1.3		6.5	1
*Oil seed rape			2.9		ND		6.4	2
*Spinach			6.7		1.8		6.1	3
*Asparagus			6.4		ND	(L-PPT)	2.6	4
Pea			11.4		0.3		2.9	5
*Wild Oat	(a)		5.5		0.6		5.0	1
	(b)		2.5		ND		4.5	2
*Sicklepod	(a)		7.3		0.9		8.5	1
	(b)		3.3		ND		7.6	2
Spiny Cockleburr		ND	4.0	ND	ND	ND	6.6	2
Couch grass		ND	8.2	ND	ND	ND	6.5	2
Goose grass		ND	4.2	ND	ND	ND	3.6	2
Green foxtail		1.8	3.2	ND	ND	1.7	5.2	2

* = plants lacking GS_1

ND = not determined (GS_1 is often present in too small an amount to assay)

References:

1. Acaster & Weitzman (1985)
2. Ridley & McNally (1985)
3. Ericson (1985)
4. Fraser & Ridley (1984)
5. S.M. Ridley & J.L. Howard, unpublished data (substrate concentrations when fixed, were: 1 mM ATP, 70 mM L-glutamate, and 0.2 mM NH_3)

Ridley (unpublished data) using pea GS_2. Pre-incubation of the enzymes with ATP and PPT (before addition of glutamate) resulted in a progressively increasing inhibition, but PPT preincubated without ATP did not inhibit. This indicates a progressive irreversible inhibition caused by the phosphorylation of PPT, as shown in Fig. 8 with two doses of PPT. The reaction differs from that found with APBA which, although it is also phosphorylated by ATP, does not exhibit the irreversible binding (Villafranca, Eads & Colanduoni, 1986). The essential difference between

Table 3. *Comparative herbicidal effects of PPT on a variety of plant species. Plants were sprayed with a range of herbicide rates between 0.016 and 4.0 kg ha⁻¹.*

Family	Species	Development stage when sprayed	LD_{50} (kg ha^{-1})
Composite	*Xanthium spinosum* (C3)	4 leaves	0.30
Cruciferae	*Brassica napus* (C3)	3+ leaves	0.35
Gramineae	*Elymus repens* (C3)	2+ leaves	3.8
	Avena fatua (C3)	3 leaves	2.5
	Setaria viridis (C4)		
	(a) mature	6 leaves	0.27
	(b) young	4 leaves	<0.125
Leguminosae	*Cassia obtusifolia* (C3)	3 leaves	8.5*
Rubiaceae	*Galium aparine* (C3)	2 whorls	1.9

*Value obtained by extrapolation
Data from Ridley & McNally (1985)

PPT and APBA is the presence of the methyl group on the phosphorus atom in PPT in place of an acid -OH in APBA, which completely alters the charge distribution around the phosphorus. Ronzio, Rowe & Meister (1969) used this type of pre-incubation experiment to compare MSO and methionine sulphone as part of their studies on the mechanism of inhibition.

Mandersheid & Wild (1986) could not reactivate the PPT-inhibited enzyme by dilution or gel filtration, but Colanduoni & Villafranca (1986) were able to reactivate the inhibitor–enzyme complex by acidification (pH 4.1) and addition of ammonium sulphate under conditions of high ionic strength.

In the two step reaction leading to irreversible inhibition of the two plant isoenzymes GS_r and GS_2, the rate limiting step in the inhibition of GS_r by PPT is the formation of the initial enzyme–inhibitor complex, whereas the rate-limiting step in the inhibition of GS_2 is the subsequent step (Mandersheid & Wild, 1986).

Inactivation by an analogue of the phosphorylated phosphinothricin, ACPPA. A close analogue of the phosphorylated PPT has been made at ICI Jealott's Hill, (3-amino-3-carboxypropyl)-(phosphonomethyl)phosphinic acid (ACPPA). This mimics the tightly bound product of the PPT + ATP reaction. ACPPA has been tested to see if this would substitute directly for the PPT inhibition product (Ridley S. & Lewis T, unpublished data).

Fig. 8. Time-dependent inactivation of pea GS2 when preincubated with 6 and 78 μM PPT in the presence and absence of ATP before adding glutamate.

The preincubation mixture contained (in 1.5 ml) 6 mM NH_2OH, 5 mM EDTA, 20 mM $MgSO_4$, 2 mM ATP (when present), 50 mM Tris-HCl, pH 7.6, and 6 or 78 μM PPT. At intervals 100 μl aliquots were removed to assay the activity that remained using 78 mM final concentration L-glutamate (giving a final vol. of 115 μl), and incubating for 10 min, 30 °C. Activity was recorded by measuring glutamyl hydroxamate production (Ferguson & Sims, 1971).

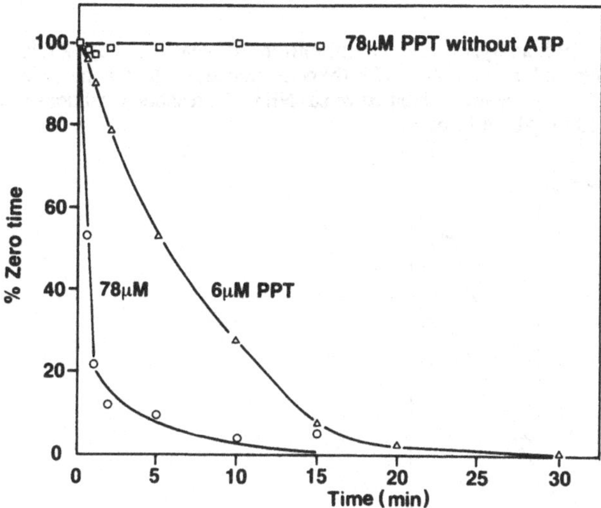

$$CH_3-\underset{\substack{|\\CH_2\\|\\CH_2\\|\\NH_2-CH-COOH}}{\overset{\substack{O\\||}}{P}}-CH_2-\underset{\substack{|\\OH}}{\overset{\substack{O\\||}}{P}}-OH$$

ACPPA

Inhibition of ACPPA under initial rate conditions against the three substrates using pea chloroplast GS_2 gave the diagnostic kinetic patterns shown in Fig. 9. This compound bound with a K_i of 2.2 mM which is five-fold tighter than glutamate on this same isoenzyme (K_m = 11 mM). It was non-competitive for glutamate and NH_3, and uncompetitive for ATP. Thus ACPPA displays reversible inhibition under initial rate conditions.

In spite of the likelihood that the phosphonomethyl moiety of ACPPA occupies the same site on the enzyme as the terminal phosphoryl moiety of ATP (see Fig. 4), the kinetics still indicate ATP binding to GS before ACPPA. The compound bound about ten-fold less tightly than PPT to pea GS_2 ($K_i = 2.9\ \mu M$), so the relative weakness of binding in comparison with PPT must indicate that a conformational change in the enzyme had already taken place and was unfavourable to the binding of a substrate analogue of the tetrahedral adduct.

Figure 9. Kinetic analyses for the inhibition of pea GS_2 by ACPPA. (a) Uncompetitive inhibition versus ATP; (b) non-competitive inhibition versus L-glutamate; (c) non-competitive inhibition versus NH_3. The reaction conditions were the same as used for APBA kinetics.

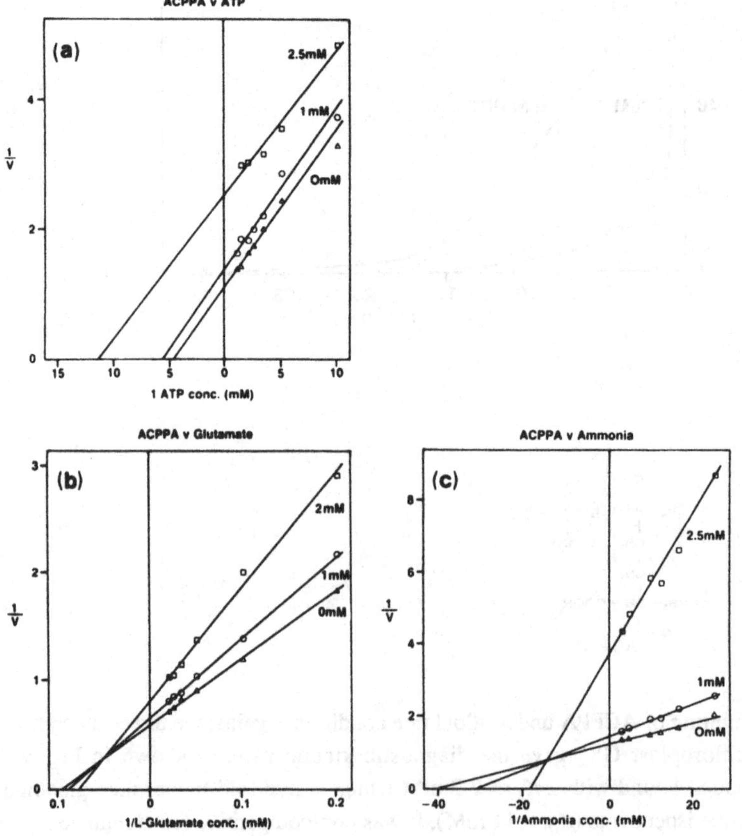

Time dependent irreversible inactivation by ACPPA The presence of a nucleotide might be expected to assist the binding of a substrate analogue such as ACPPA, because it is known that ADP remains bound to GS in the enzymic reaction (Krishnaswamy, Pamiljans & Meister, 1962), and when GS is inhibited by PPT and MSO after they have become phosphorylated (Colanduoni & Villafranca, 1986). When ACPPA (1 mM) was preincubated with ADP or ATP (prior to addition of glutamate), only the presence of ATP produced a progressive irreversible inhibition, whilst preincubation with ADP had no significant effect (Fig. 10).

This result is contrary to what has been reported by Meister's group for another tetrahedral adduct analogue, methionine sulphoximine phosphate, where both ADP and ATP enhanced the binding to sheep brain GS (Rowe *et al.*, 1969). The result also

Figure 10. Time-dependent inactivation of pea GS_2 when preincubated with 1 mM ACPPA in the presence of 2 mM ATP or 1 mM ADP before adding L-glutamate.

The preincubation mixture (1.5 ml) contained: 50 mM Tris-HCl, pH 7.6, 5 mM EDTA, 6 mM NH_4Cl, 20 mM $MgCl_2$, 2 mM ATP or 1 mM ADP, and 1 mM ACPPA, and incubated at 30 °C. 100 μl samples were removed at intervals and 55 mM L-Glu added, plus 1 mM ATP to the ADP pre-incubated sample (giving a final vol. of 110 μl), and incubated for 10 min, at 30 °C. Activity was recorded by the production of inorganic phosphate.

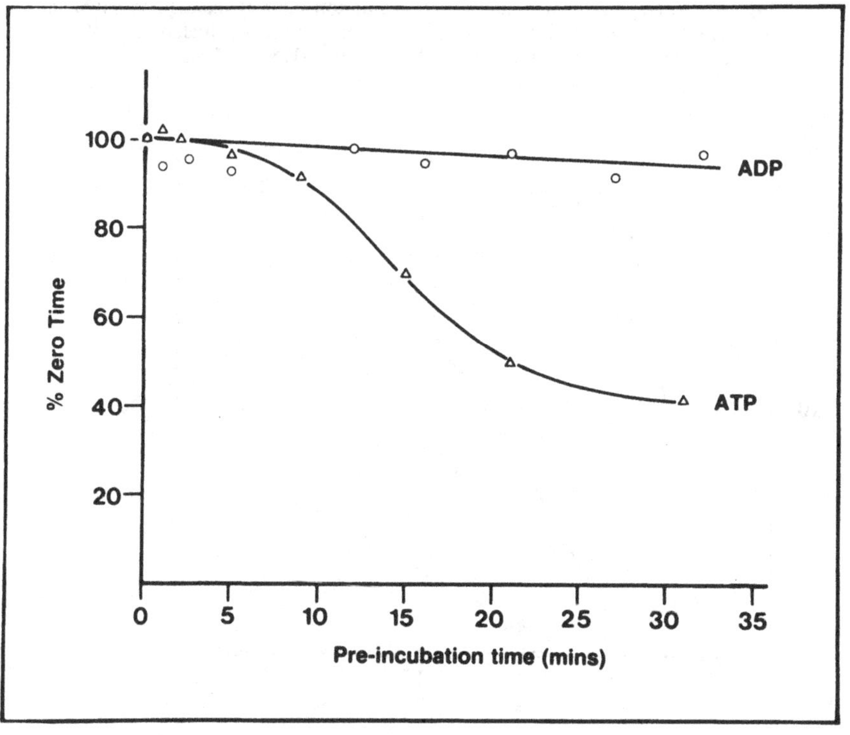

contrasts with that found on incubating the gamma-glutamyl-P analogue, 4-(phosphonacetyl)-L-α-aminobutyrate (PALAB) with ADP and pea seed GS (Wedler, Horn & Roby, 1980). In this case there was a progressive irreversible inhibition.

The result with ACPPA suggests it cannot directly substitute for the product of phosphorylation of PPT at the active site of the enzyme. Only ATP, but not ADP, assists the binding of ACPPA. This is either because ATP induces a direct conformational change in the enzyme, or because it further phosphorylates ACPPA, and thereby enhances binding. In spite of enhanced binding of ACPPA by ATP, the compound shows no herbicidal activity.

Action of PPT on plant metabolism Köcher (1983) carried out a detailed study on the effect of PPT (Hoe 39866) on the metabolism of different plant species. The rate of photosynthetic carbon assimilation of *Sorghum* was reduced to 10% of the control plants 8 h after spraying with PPT, the ammonia level increased to 576 μg N/g fresh weight. The ammonia level in both *Sorghum* and *Pisum* increased faster in plants when light was given immediately following spraying with PPT rather than after a 24-hour dark period.

That ammonia accumulation is dependent upon light, can be readily seen in Fig. 11.

Fig. 11. Time course of ammonia accumulation on leaf damage in soybean. Ammonia: ○ control, △ treated, 40 W/m^2; ● control, ▲ treated, 2.5 W/m^2. Damage: □ treated, 40 W/m^2; ■ treated, 2.5 W/m^2. (Köcher, 1983).

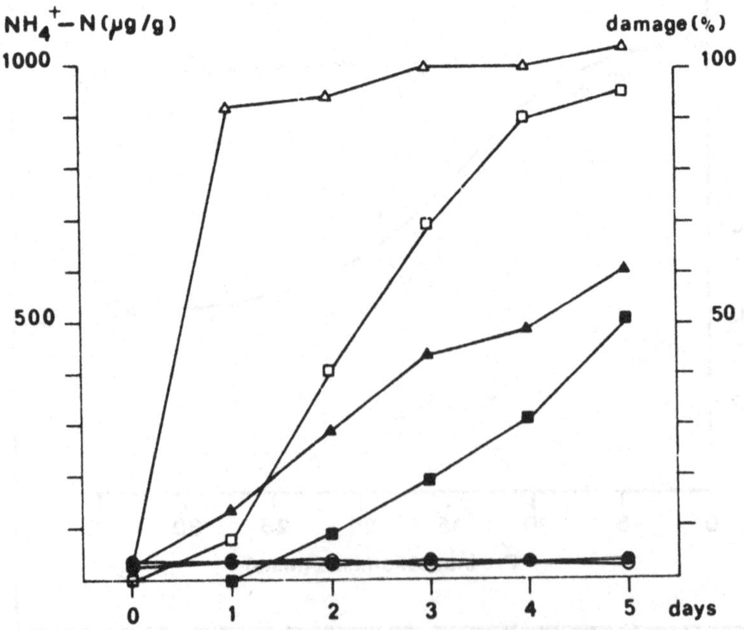

When soybean leaves were sprayed with PPT, at a light intensity of 40 W/m^2, the ammonia level increased slowly at a linear rate up to the end of the experiment. Leaf damage appeared at a slower rate in the low light treated plants. Köcher (1983) was also able to show that the increase in ammonia caused by PPT stimulated the leakage of K^+ ions from cotyledons of *Sinapsis*.

In a comparative study of the effects of MSO, PPT and PPO, Lea *et al.* (1984) showed that all three compounds caused an increase in ammonia levels and a decrease in free glutamine in a range of higher plants. PPT caused the evolution of ammonia in both nitrate reducing and nitrogen fixing cultures of cyanobacteria, indicating that the compound could both enter the cell and inhibit GS activity *in vivo* (Lea *et al.*, 1984).

3. Tabtoxinine-β-lactam

This inhibitor arises from the antibiotic tabtoxin, which is a toxic dipeptide produced by a number of phytopathogenic pathovars of *Pseudomonas syringae*. On hydrolysis, tabtoxin releases the active moiety, tabtoxinine-β-lactam (T-β-L) [2-amino-4-(3-hydroxy-2-oxo-azacyclobutan-3-yl-butanoic acid] and threonine (Taylor, Schnoes & Durbin, 1972). The antibiotic is the toxin responsible for wild fire disease in tobacco. Taylor *et al.* (1972) have also isolated a less abundant dipeptide of T-β-L and serine designated [2-serine]-tabtoxin.

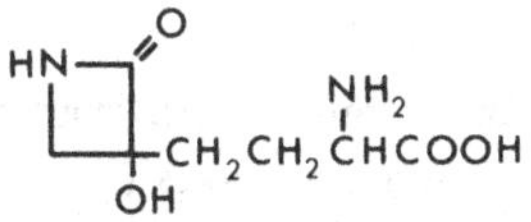

HN, O, OH, $CH_2CH_2CH(NH_2)COOH$

Tabtoxinine-ß-lactam

Tabtoxinine-β-lactam inhibits GS *in vitro*, whereas the dipeptide is not active. The open chain form, tabtoxinine, does not inhibit GS suggesting that the β-lactam ring is essential for inhibition, and some differences in the binding of T-β-L compared with MSO have been found (Langston-Unkefer *et al.*, 1987). Like MSO and PPT, the inhibition had an absolute requirement for ATP, and this led to a time-dependent irreversible inhibition, binding to the enzyme in an approximate ratio of one inhibitor per subunit. Preincubation of GS with T-β-L and ATP plus NH_4Cl produced more rapid inhibition, than preincubation with ATP alone (Langston-Unkeffer *et al.*, 1987; Thomas & Durbin, 1985). Leaves infected with the pathogen became chlorotic, which was considered to be the consequence of the accumulation of ammonia (Turner & Debbage, 1982), and in this respect is similar to other antibiotics that inhibit GS (see below). Oats resistant to pathogen attack, contain GS that is less sensitive to T-β-L inhibition *in vitro*(Knight *et al.*, 1988). Very recently T-β-L has been shown to increase the capacity of legume root nodules to fix nitrogen by the selective inhibition of one GS isoenzyme (Knight & Langston-Unkefer, 1988).

4. Oxetin

This antibiotic has been isolated from a *Streptomyces* species, and the structure elucidated as (2R,3S)-3-amino-2-oxetane carboxylic acid. It inhibited GS non-competitively with respect to glutamate giving a K_i value of 3.4 mM on GS_2 from spinach. In spite of this relatively high K_i, it is reported to have exhibited herbicidal activity (Omura *et al.*, 1984*c*).

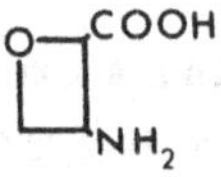

O COOH

NH_2

Oxetin

In the presence of only ATP and glutamate, GS catalyses a partial reaction in which glutamate cyclises to form 5-oxoproline, ADP and inorganic phosphate (Krishnaswamy, Pamiljans & Meister, 1962). Oxetin is a close analogue of 5-oxoproline, but the precise way in which it inhibits GS has not yet been researched.

5. 5-Hydroxylysine

This compound has been shown to be an inhibitor of GS from a cyanobacterium (Ladha, Rowell & Stewart, 1978), and from pea leaves (Leason *et al.*, 1982). Acaster & Weitzman (1985) found that 5-hydroxylysine inhibits both GS_1 and GS_2 isoforms from maize, with K_i values of 0.49 mM for GS_1 and 0.85 mM for GS_2. Like PPT, the inhibitor is almost twice as potent on GS_1 as on GS_2. These authors established from kinetic diagnoses that inhibition on both GS_1 and GS_2 from maize by 5-hydroxylysine with respect to glutamate was uncompetitive, but no further detailed kinetics have been carried out to reveal the mechanism of inhibition.

The mode of action of MSO and PPT as herbicides

It is obvious from the results discussed in the earlier sections that both MSO and PPT can inhibit GS *in vivo* and cause a dramatic build up in the level of ammonia within leaves in the light. It is clear that these inhibitors can cause the death of plants over a few days (Singh & Widholm, 1975; Dessauer & Hannah, 1978). It has also been reported that the compounds can inhibit the rate of photosynthetic CO_2 assimilation within a matter of minutes (Platt & Anthon, 1981; Martin *et al.*, 1983; Köcher, 1983; Archireddy *et al.*, 1983; Ikeda, Ogren & Hageman, 1984; Walker *et al.*, 1984*b*; Berger & Fock, 1985; Johansson & Larsson, 1986). A good example of this can be seen in Fig. 12 in which the effectiveness of PPT, PPO and MSO as inhibitors of CO_2 fixation in barley leaves can be readily seen.

It would seem at first sight that the build up of 'toxic' ammonia could cause both the death of plants and inhibit photosynthesis. However, Walker *et al.* (1984*b*) were the first to suggest that ammonia *per se* was not directly inhibiting photosynthesis. Wheat leaves supplied with 30 mM ammonium chloride accumulated more ammonia than leaves treated with the MSO, but showed less inhibition of photosynthesis. Similar results by Johansson & Larsson (1986) using *Lemna gibba* demonstrated that MSO caused the inhibition of photosynthesis at lower levels of internal ammonia, than were formed by the direct feeding of 10 mM ammonium salts which caused no inhibition. Such results would suggest that MSO has a direct effect upon the operation of the Calvin cycle or photosynthetic electron transport; however this could not be demonstrated in short term experiments with isolated chloroplasts (Anderson & Done, 1977; Muhitch & Fletcher, 1983; Köcher, 1983).

Until recently it has not been possible to separate the direct effect of a GS inhibitor from that of the ammonia formed from its action. However, there are now available a

Fig. 12. The effect of various inhibitors of glutamine synthetase on the rate of CO_2 assimilation fed to cut shoots of barley measured using an Infra Red Gas Analyser.

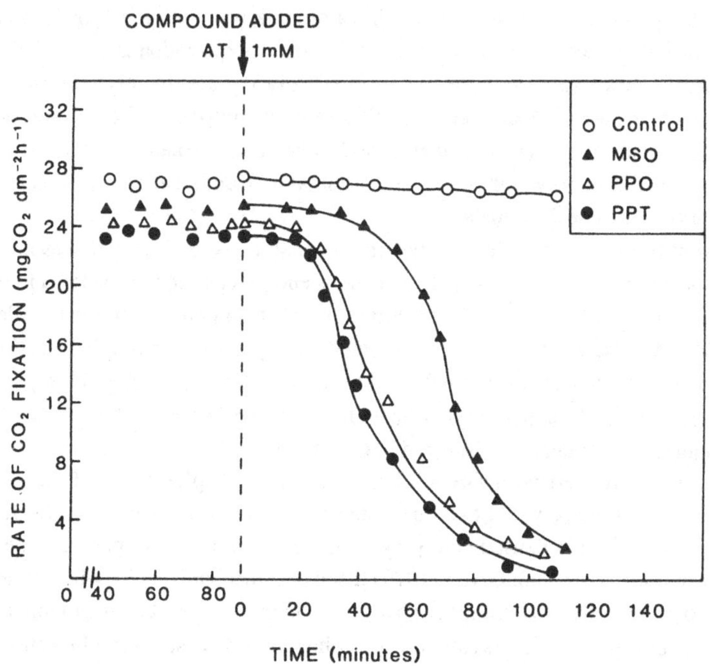

number of photorespiratory mutants of higher plants that are blocked in their ability to assimilate ammonia. Mutants lacking ferredoxin dependent glutamate synthase have been isolated from *Arabidopsis*, (Somerville & Ogren, 1980), barley (Kendall *et al.*, 1986; Blackwell *et al.*, 1988) and pea (Blackwell *et al.*, 1987*a*). Leaves of the mutant plants, when placed in air, exhibited a dramatic drop in photosynthetic capacity over a ten minute period (similar to the plants treated with PPT in Fig. 12), whilst accumulating low levels of ammonia. The preincubation of the leaves in glutamate or serine diminished the fall in photosynthetic rate.

Mutants of barley deficient in chloroplastic glutamine synthetase have been isolated both at Rothamsted (Wallsgrove *et al.*, 1987) and Lancaster (Blackwell *et al.*, 1987*b*). Such mutants when placed in air should be the equivalent of MSO or PPT treated plants. The photosynthetic rate of such mutant plants when transferred from 2% O_2 to air fell slowly over a 30-minute period, whilst the internal ammonia concentration of the leaves reached 60 μmol g^{-1} fresh weight over this period. The fall in photosynthetic rate may be partially reversed by prefeeding the leaves with glutamine. In a similar experiment using wheat leaves, Ikeda *et al.* (1984) showed that MSO did not inhibit photosynthesis under the non-photorespiratory conditions of 2% O_2. However, there was strong inhibition in air, which could be reversed by glutamine despite the presence of high concentrations of ammonia.

Taking into account the initial data of Walker *et al.* (1984*b*), Ikeda *et al.* (1984) and Johansson & Larsson (1986), along with results described above for the photorespiratory mutants, it would appear that ammonia was not the major cause of the immediate fall in photosynthetic rates. The more likely explanation is that sufficient amino donors were not available to convert glyoxylate to glycine in the photorespiratory nitrogen cycle, which would prevent the recycling of glycerate back into the chloroplast. It is interesting that in both MSO treated plants (Walker *et al.*, 1984*a*) and in the glutamate synthase deficient mutants (Kendall *et al.*, 1986) carbon is directed into the synthesis of malate.

The prevention of the Calvin cycle operating is analogous to placing leaves in the light in the absence of CO_2. Under such conditions energy cannot be dissipated in the formation of ATP and $NADPH_2$ and this normally results in photoinhibition (Powles, 1984), although some electrons may be directed to oxygen in the Mehler reaction. However extensive photoinhibition will lead to the formation of triplet chlorophyll, singlet oxygen $O_2{}^1$ and the hydroxyl radical $OH^•$, the role of these activated oxygen species in causing plant death is discussed in Chapter 3.

Like the plants treated with GS inhibitors, the mutant plants with lesions in ammonia assimilation die when they are exposed to air for a period of days. It is unlikely that this death is caused solely by an inability to carry out photosynthetic carbon assimilation. Although Givan (1979) discussed mechanisms by which plants avoid 'ammonia toxicity' it is not clear how the toxic effect operates. Ammonia can cause the uncoupling of photosynthetic phosphorylation in spinach chloroplasts

(Krogman *et al.*, 1959). It is probable that unprotonated NH_3 diffuses freely through the chloroplast and thylakoid membranes. The NH_3 is then able to take up a H^+ to form NH_4^+ (Kleiner, 1981), and in order to maintain an ionic balance there must also be an inward flux of anions. Following the influx of ions there is an osmotic uptake of water that causes the chloroplast and thylakoids to swell dramatically (Izawa & Good, 1966; Good, 1977). The change of shape of the chloroplasts and the formation of protrusions has been seen in tomato leaves following the onset of ammonia toxicity (Puritch & Barker, 1967). Under optimum conditions there should be a proton gradient across the thylakoid membranes of ΔpH 3.9. When chloroplasts were treated with 2 mM NH_4Cl this gradient dropped to 3.24 and the rate of CO_2 fixation and the ATP/ADP ratio fell by about 50% concurrently (Slovacek & Hind, 1981). Thus it is clear that low concentrations of ammonia can prevent photophosphorylation and thus limit ATP production without inhibiting electron flow. It has been recently shown that the addition of 5 mM ammonia to intact chloroplasts can stimulate photoinhibition by inhibiting both PSI and PSII (Krause & Laasch, 1987). Hydroxylamine (NH_2OH), which may be considered an analogue of ammonia has also been shown to exacerbate photoinhibition by maintaining the P_{680} in PSII in an oxidised state (Cleland, Critchley & Melis, 1987). At high concentrations (> 50 mM), ammonia can also bind to the water splitting, manganese containing site in photosystem II and thus also inhibit O_2 evolution and electron transport (Izawa, 1977).

In animal systems MSO also caused a depletion in the level of glutathione by inhibiting the first enzyme involved in its synthesis, (γ-glutamyl cysteine synthetase) an enzyme whose reaction mechanism is similar to that of GS (Meister, 1983). MSO and other derivatives have been tested as potential anticancer agents in humans (Meister, 1983; Dranoff *et al.*, 1985), because of their ability to reduce the level of glutathione in tumour cells. Glutathione plays an important role in the detoxification of activated oxygen species in higher plants (Smith *et al.*, 1985) and of course in the detoxification of applied herbicides (see chapter 9). Unfortunately there is little information available on the enzyme γ-glutamyl cysteine synthetase in higher plants (Rennenberg, 1982) and as far as we are aware no information is available on the action of MSO or PPT on the level of glutathione in plants.

The selection of resistance to GS inhibition

Confirming an earlier result by Braun (1955), Meins & Abrahams (1972) demonstrated that both methionine and glutamine reversed the inhibitory effect of MSO on the growth of *Chlorella,* probably by preventing the uptake of the inhibitor. A mutant line resistant to MSO, was shown to take up MSO at only 10% of the rate of wild type cells. Carlson (1973), utilising tobacco tissue culture cells, isolated lines resistant to MSO which were regenerated and shown to be resistant to infection by *Pseudomonas tabacci,* which forms 'halo blight' by the production of tabtoxin (see above). The resistant plants appeared to contain high levels of free methionine, but unfortunately no further data on the mechanism of resistance is available. Two groups of workers (Singh & Widholm, 1975; Dessauer & Hannah, 1978) have screened a range of seedlings for resistance to methionine sulphoximine. The screen was aimed at isolating mutants that overproduced methionine, but the amino acid did not reverse the growth inhibitory action of MSO and no resistant lines were selected. Utilising azide-mutated M_2 seed of barley that had previously been shown to produce a range of mutations in amino acid metabolism (see Kendall *et al.*, 1986), seedlings were screened for resistance to foliar applications of MSO (J.C. Turner & P.J. Lea unpublished results). The occasional seedling appeared to show resistance, but evidence of resistance in the progeny of such plants following self-fertilisation has not yet been obtained.

Using alfalfa suspension cell cultures Donn *et al.* (1984) were able to isolate lines that were 20–100 fold more resistant to PPT than wild type cells. Such cells were found to contain 3–7 fold higher levels of GS activity. Further studies showed that there had been a 4–11 fold amplification of the GS gene and 8 fold increases in the mRNA coding for GS were detected. A recent report has indicated that De Block *et al.* (1987) have isolated an enzyme from *Steptomyces* that detoxifies PPT by acetylation. The gene for the acetyltransferase has been transferred to tomato, tobacco and potato plants all of which can withstand ten times of the normal dose of the PPT herbicide 'Basta'.

Although there seems to be a major difference between the K_i for MSO between bacterial and plant GS, there is a remarkable consistancy between the K_i values obtained for MSO and PPT in the plant species so far studied (e.g. Table 2). It clearly would be of interest to isolate a mutant plant enzyme that was less sensitive to PPT inhibition. In a novel approach to this problem Dassarma, Tischer & Goodman (1986) have managed to insert a GS gene from alfalfa into a mutant of *E. coli* lacking GS. The transformed *E. coli* was able to grow normally using plant GS to assimilate ammonia. Using this transformed *E. coli* it should be possible to screen huge numbers of cells for resistance to PPT and thus isolate a mutated plant GS gene with an altered active site which binds PPT less effectively.

Note

Since the completion of this review and presentation of the lecture, two papers from the same group (Sauer, Wild & Ruehle, 1987; Wild, Sauer & Ruehle, 1987), have also shown that PPT is a strong inhibitor of photosynthesis under photorespiratory conditions only. The authors concluded that ammonia alone was not responsible for the fall in photosynthetic rates.

References

Acaster, M.A. & Weitzman, P.D.J. (1985). Kinetic analysis of glutamine synthetases from various plants. *FEBS Letters* **189**, 241–4.

Achireddy, N.R., Vann, D.R., Fletcher, J.S. & Beevers, L. (1983). The influence of methionine sulphoximine on photosynthesis and nitrogen metabolism in excised pepper (*Capsicum annum* L.) leaves. *Plant Science Letters* **32**, 73–8.

Almassy, R.J., Janson, C.A., Hamlin, R., Xuong, N-H. & Eisenberg. D. (1986). Novel subunit–subunit interaction in the structure of glutamine synthetase. *Nature* **323**, 304–9.

Anderson, J.W. & Done, J. (1977). A polarographic study of ammonia assimilation by isolated chloroplasts. *Plant Physiology* **60**, 504–8.

Andrews, M. (1986). The partitioning of nitrate assimilation between root and shoot of higher plants. *Plant Cell and Environment* **9**, 511–9.

Arima, Y. & Kumazawa, K. (1977). Evidence of ammonium assimilation via the glutamine synthetase – glutamate synthase system in rice seedling roots. *Plant Cell Physiology* **18**, 1121–9.

Bayer, E., Gugel, K.K., Haegele, K., Hagenmaier, H., Jessipow, S., Konig, W.A. & Zahner, H. (1972). Stoffwechselprodukte von Mikroorganismen. Phosphinothricin and Phosphinothricyl-Alanyl-Alanin. *Helv.Chim.Acta* **55**, 224–39.

Bentley, H.R., McDermott, E.E., Pace, J., Whitehead, J.K., Moran, T. (1949). Action of nitrogen trichloride (agene) on proteins: isolation of crystalline toxic factor. *Nature* **164**, 438–9.

Berger, M.G. & Fock, H-P. (1985). Comparative studies on the photorespiratory nitrogen metabolism in wheat and maize leaves. *Journal of Plant Physiology* **119**, 257–67.

Bergman, A., Gardestrom, P. & Ėricson, I. (1981). Release and refixation of ammonia during photorespiration. *Physiologia Plantarum* **53**, 528–32.

Blackwell, R.D. Murray, A.J.S. & Lea, P.J. (1987*a*). The isolation and characterisation of photorespiratory mutants of barley and pea. In *Progress in Photosynthesis Research* ed. J. Biggins, Vol. III. pp. 625–8. Dordecht, Martinus Nijhoff.

Blackwell, R.D., Murray, A.J.S. & Lea, P.J. (1987*b*). Inhibition of photosynthesis in barley with decreased levels of chloroplastic glutamine synthetase activity. *Journal of Experimental Botany* **38**, 1799–809.

Blackwell, R.D., Murray, A.J.S. & Lea, P.J. (1988). Photorespiratory amino donors, sucrose synthesis and the induction of CO_2 fixation in barley deficient in glutamine synthetase and/or glutamate synthase. *Journal of Experimental Botany* **39**, 845–58.

Braun, A.C. (1955). A study on the mode of action of the wildfire toxin. *Phytopathology* **45**, 659–64.

Breteler, H. & Siegerist, M. (1984). Effect of ammonium on nitrate utilisation by roots of dwarf bean. *Plant Physiology* **75**, 1009–103.

Carlson, P.S. (1973). Methionine sulphoximine-resistant mutants of tobacco. *Science* **180**, 1366–8.

Cleland, R.E., Critchley, C. & Melis, A. (1987). Alteration of electron flow around P_{680}: The effect on photoinhibition. In *Progress in Photosynthesis Research* ed J. Biggins, Vol.IV pp. 27–30. Dordrecht: Martinus Nijhoff.

Colanduoni, J.A. & Villafranca, J.J. (1986). Inhibition of *Escherichia coli* glutamine synthetase by phosphinothricin. *Bioorganic Chemistry* **14**, 163–9.

Cullimore, J.V. & Bennett, M. (1988). The molecular biology and biochemistry of plant glutamine synthetase from root nodules of *Phaseolus vulgaris* L. and other legumes. *Journal of Plant Physiology* **132**, 387–93.

Cullimore, J.V., Gebhardt, C., Saarelainen, R., Miflin, B.J., Idler, K.B. & Barker, R.F. (1984). Glutamine synthetase of *Phaseolus vulgaris*: organ specific expression of a multigene family. *Journal Molecular and Applied Genetics* **2**, 589–600.

Cullimore, J.V., Lara, M., Lea, P.J. & Miflin, B.J. (1983). Purification and properties of two forms of glutamine synthetase from the plant fraction of *Phaseolus* root nodules. *Plants* **157**, 245–53.

Cullimore, J.V. & Miflin, B.J. (1984). Immunological studies on glutamine synthetase using antisera raised to the two plant forms of the enzyme from *Phaseolus* root nodules. *Journal of Experimental Botany* **35**, 581–7.

Cullimore, J.V. & Sims, A.P. (1980). An association between photorespiration and protein catabolism: studies with *Chlamydomonas*. *Planta* **150**, 392–6.

Dassarma, S., Tischer, E. & Goodman, H.M. (1986). Plant glutamine synthetase complements a *glnA* mutation in *Escherichia coli*. *Science* **232**, 1241–4.

De Block, M., Botterman, J., Thoen, C., Gossele, V., Rao Movva, N., Thompson, C., Van Montagu, M. & Leemans, J. (1987). Engineering herbicide resistance in plants by expression of a detoxifying enzyme. *EMBO Journal* **6**, 2513–18.

Dessauer, D.W. & Hannah, L.C. (1978). Inhibition of cowpea seedling growth by methionine analogues. *Crop Science* **18**, 593–7.

Donn, G., Tischer, E., Smith, J.A. & Goodman, H.M. (1984). Herbicide-resistant alfalfa cells: an example of gene amplification in plants. *Journal Molecular Applied Genetics* **2**, 621–35.

Dranoff, G., Elion, G.B., Friedman, H.S., Campbell, G.L. & Bigner, D.D. (1985). Influence of glutamine on the growth of human glioma and medulloblastoma in culture. *Cancer Research* **45**, 4077–81.

Ericson, M.C. (1985). Purification and properties of glutamine synthetase from spinach leaves. *Plant Physiology* **79**, 923–7.

Fentem, P.A., Lea, P.J. & Stewart, G.R. (1983*a*). Action of inhibitors of ammonia assimilation on amino acid metabolism in *Hordeum vulgare* L. (cv Golden Promise). *Plant Physiology* **71**, 502–6.

Fentem, P.A., Lea, P.J. & Stewart, G.R. (1983*b*). Ammonia assimilation in the roots of nitrate and ammonia-grown *Hordeum vulgare* (cv Golden Promise). *Plant Physiology* **71**, 496–501.

Frantz, T.A. Peterson, D.M. & Durbin, R.D. (1982). Sources of ammonium in oat leaves treated with tabtoxin or methionine sulphoximine. *Plant Physiology* **69**, 345–8.

Fraser, A.R. & Ridley, S.M. (1984). Kinetics for glutamine synthetase inhibition by phosphinothricin and measurements of other enzyme activities *in situ* in isolated asparagus cells using a freeze–thaw technique. *Planta* **161**, 470–4.

Gebhardt, C., Oliver, J.E., Forde, B.G., Saarelainen, R. & Miflin, B.J. (1986). Primary structure and differential expression of glutamine synthetase in nodules roots and leaves of *Phaseolus vulgaris*. *EMBO Journal* **5**, 1429–1435.

Givan, C. (1979). Metabolic detoxification of ammonia in tissues of higher plants. *Phytochemistry* **18**, 375–82.

Good, N.E. (1977). Uncoupling of electron transport from phosphorylation in plants. In *Encyclopaedia of Plant Physiology* eds. A. Trebst & M. Avron, Vol.5, pp. 429-436. Berlin: Springer.

Hirel, B., Vidal, J. & Gadal, P. (1982). Evidence for a cytosolic dependent light induction of chloroplastic glutamine synthetase in etiolated rice leaves. *Planta* **155**, 17–23.

Hirel, B., Weatherley, C., Cretin, C., Bergounioux, C. & Gadal, P. (1984). Multiple subunit composition of chloroplast glutamine synthetase of *Nicotiana tabacum* L. *Plant Physiology* **74**, 448–50.

Ikeda, M., Ogren, W.L. & Hageman, R.H. (1984). Effect of methionine sulphoximine on photosynthetic carbon metabolism in wheat *(Triticum aestivum* cultivar Poland) leaves. *Plant Cell Physiology* **25**, 447–52.

Ito, O., Yoneyama, T. & Kumazawa, K. (1978). The effect of light on ammonium assimilation and glutamine metabolism in the cells isolated from spinach leaves. *Plant Cell Physiology* **19**, 1109–19.

Izawa, S. & Good, N.E. (1966). Effect of salts and electron transport inhibitors on the conformation of isolated chloroplasts. *Plant Physiology* **41**, 533–43.

Izawa, S. (1977). Inhibitors of electron transport. In *Encyclopaedia of Plant Physiology* eds. A. Trebst and M. Avron, Vol.5, pp. 266-282. Berlin: Springer.

Johansson, L. & Larsson, C.M. (1986). Relationship between inhibition of CO_2 fixation and glutamine synthetase inactivation in *Lemna gibba* L. treated with L-methione-D,L-sulphoximine (MSO). *Journal of Experimental Botany* **37**, 221–9.

Kendall, A.C., Wallsgrove, R.M., Hall, N.P., Turner, J.C. & Lea, P.J. (1986). Carbon and nitrogen metabolism in barley (*Hordeum vulgare*) mutants lacking ferredoxin glutamate synthase. *Planta* **168**, 316–23.

Keys, A.J., Bird, I.F., Cornelius, M.J., Lea, P.J., Wallsgrove, R.M. & Miflin, B.J. (1978). Photorespiratory nitrogen cycle. *Nature* **275**, 741–3.

Kleiner, D. (1981). The transport of NH_3 and NH_4^+ across biological membranes. *Biochimica et Biophysica Acta* **639**, 41–52.

Knight, T.J., Bush, D.R. & Langston-Unkefer, P.J. (1988). Oats tolerant of *Pseudomonas syringae pv. Tabaci* contain tabtoxinine-β-lactam-insensitive leaf glutamine synthetases. *Plant Physiology* **88**, 333–9.

Knight, T.J. & Langston-Unkefer, P.J. (1988). Enhancement of symbiotic dinitrogen-fixation by a toxin-releasing plant pathogen. *Science* **241**, 951–4.

Köcher, H. (1983). Influence of the light factor on physiological effects of the herbicide Hoe 39866. *Aspects of Applied Biology* **4**, 227–34.

Krause, G.H. & Laasch, H. (1987). Photoinhibition of photosynthesis. Studies on mechanisms of damage and protection in chloroplasts. In *Progress in Photosynthesis Research* ed. J. Biggins. Vol. IV. pp. 19–24. Dordrecht: Martinus Nijhoff.

Krishnaswamy, P.R., Pamiljans, V. & Meister, A. (1962). Studies on the mechanism of glutamine synthesis: evidence for the formation of enzyme-bound activated glutamic acid. *Journal of Biological Chemistry* **237**, 2932–40.

Krogman, D.W., Jagendorf, A.T. & Avron, M. (1959). Uncouplers of spinach chloroplast photosynthetic phosphorylation. *Plant Physiology* **34**, 272–7.

Kumar, P.A., Nair, T.V. & Abrol, Y.P. (1984). Effect of photorespiratory metabolites, inhibitors and methionine sulphoximine on the accumulation of ammonia in the leaves of mung bean and *Amaranthus*. *Plant Science Letters* **33**, 303–7.

Ladha, J.K., Rowell, P., & Steward, W.D.P. (1978). Effects of 5-hydroxylysine on acetylene reduction and NH_4^+ – assimilation in the cyanobacterium *Anabaena cylindrica*. *Biochem. Biophys. Res. Comm.* **83**, 688–96.

Langston-Unkefer, P.J., Robinson, A.C., Knight, T.J. & Durbin, R.D. (1987). Inactivation of pea seed glutamine synthetase by toxin, tabtoxinine-lactam. *Journal of Biological Chemistry*. **262**, 1608–13.

Lara, M., Porta, H., Padilla, J., Folch, J. & Sanchez, F. (1984). Heterogeneity of glutamine synthetase polypeptides in *Phaseolus vulgaris* L. *Plant Physiology* **76**, 1019–23.

Lea, P.J. & Joy, K.W. (1983) Amino acid interconversions in germinating seeds. In *Mobilisation of Reserves in Germination* eds. C. Nozzolillo, P.J. Lea & F.A. Loewus, pp. 77–109. Plenum: New York.

Lea, P.J., Joy, K.W., Ramos, J.L. & Guerrero, M.G. (1984). The action of 2-amino-4-(methylphosphinyl)-butanoic acid (phosphinothricin) and its 2-oxo-derivative on the metabolism of cyanobacteria and higher plants. *Phytochemistry* **23**, 1–6.

Lea, P. & Miflin, B.J. (1980) Transport and metabolism of asparagine and other compounds within plants. In *The Biochemistry of Plants*, Vol. 5, ed. B.J. Miflin, pp. 569–607. London: Academic Press.

Lea, P.J., Wallsgrove, R.M. & Miflin, B.J. (1984). The biosynthesis of amino acids in plants. In *The Chemistry and Biochemistry of the Amino Acids*, ed. G.C.Barrett, pp. 197–226. Chapman & Hall: London.

Leason, M., Cunliffe, D., Parkin, D., Lea, P.J. & Miflin, B.J. (1982). Inhibition of pea leaf glutamine synthetase by methionine sulphoximine, phosphinothricin and other glutamate analogues. *Phytochemistry* **21**, 855–7.

Lightfoot, D.A., Green, N.K. & Cullimore, J.V. (1988). The chloroplast located glutamine synthetase of *Phaseolus vulgaris* L.: Nucleotide sequence, expression in different organs and uptake into isolated chloroplasts. *Plant Molecular Biology* **11**, 191–202.

Loyola-Vargas, V.M. & de Jimenez, E.S. (1984). Differential role of glutamate dehydrogenase in nitrogen metabolism of maize tissues. *Plant Physiology*, **76**, 536–40.

Maier, L. & Lea, P.J. (1983). Synthesis and properties of phosphinothricin derivatives. *Phosphorus and Sulfur* **17**, 1–19.

Manderscheid, R. & Wild, A. (1986). Studies on the mechanism of inhibition by phosphinothricin of glutamine synthetase isolated from *Triticum aestivum* L. *J.Plant Physiol.* **123**, 135–42.

Mann, A.F., Fentem, P.A. & Stewart, G.R. (1979). Identification of two forms of glutamine synthetase in barley *(Hordeum vulgare) Biochemical Biophysical Research Communications* **88**, 515–21.

Manning. J.M., Moore, S. Rowe, W.B. & Meister, A. (1969). Identification of L-methionine S-sulfoximine as the diastereoisomer of L-methionine SR-sulfoximine that inhibits glutamine synthetase. *Biochemistry* **8**, 2681–5.

Martin, F., Winspear, M.J., Macfarlane, J.D. & Oaks, A. (1983). Effect of methionine sulphoximine on the accumulation of ammonia in C_3 and C_4 leaves. *Plant Physiology* **71**, 177–81.

Maurizi, M.R. & Ginsburg, A. (1982). Active site ligand stabilisation of quaternary structures of glutamine synthetase from *Escherichia coli*. *Journal of Biological Chemistry* **257**, 7246–51.

McNally, S.F., Hirel, B., Gadal, P., Mann, F. & Stewart, G.R. (1983). Glutamine synthetase of higher plants. Evidence for a specific isoform content related to their possible physiological role and their compartmentation within the leaf. *Plant Physiology* **72**, 22–5.

Meek, T.D. & Villafranca, J.J. (1980). Kinetic mechanisms of *Escherichia coli* glutamine synthetase. *Biochemistry* **19**, 5513–19.

Meins, F. & Abrahams, M.L. (1972). How methionine and glutamine prevent the inhibition of growth by methionine sulphoximine. *Biochimica & Biophysica Acta* **206**, 307–11.

Meister, A. (1973). Glutamine synthetase of mammals. In *The Enzymes*, ed. P.D.Boyer, Vol. X, pp. 699–754. New York: Academic Press.

Meister, A. (1980). Glutamine Synthetase. In *Glutamine: Metabolism, Enzymology, and Regulation*, ed. J. Mora & R. Palacios, pp.1–40. New York: Academic Press.

Meister, A. (1983). Selective modification of glutathione metabolism. *Science* **220**, 472–7.

Miflin, B.J. & Lea, P.J. (1976). The pathway of nitrogen assimilation in plants. *Phytochemistry* **15**, 873–85.

Muhitch, M.J. & Fletcher, J.S. (1983). Influence of methionine sulphoximine on photosynthesis in isolated chloroplasts. *Photosynthesis Research* **4**, 241–4.

Oaks, A. & Hirel, A. (1985). Nitrogen metabolism in roots. *Annual Review of Plant Physiology* **36**, 345–65.

Omura, S., Murata, M., Hanaki, H., Hinotozawa, K., Oiwa, R. & Tanaka, H. (1948*a*). Phosalacine, a new herbicidal antibiotic containing phosphinothricin. Fermentation, isolation, biological activity and mechanism of action. *Journal of Antibiotics* **37**, 829–35.

Omura, S., Hinotozawa, K., Imamura, N. & Murata, M. (1984*b*). The structure of phosalacine, a new herbicidal antibiotic containing phosphinothricin. *Journal of Antibiotics* **37**, 939–40.

Omura, S., Murata, M., Imamura, N., Iwai, Y., Tanaka, H., Furusaki, A. & Matsumoto, T. (1984*c*). Oxetin, a new antimetabolite from an actinomycete. Fermentation, isolation, structure and biological activity. *Journal of Antibiotics* **37**, 1324–32.

O'Neal, D. & Joy, K.W. (1973). Glutamine synthetase from pea leaves. *Archives of Biochemistry and Biophysics* **159**, 113–22.

Ortega, J.L., Campos, F., Sanchez, F.J. & Lara, M. (1986). Expression of two different glutamine synthetase polypeptides during root development in *Phaseolus vulgaris* L. *Plant Physiology* **80**, 1051–4.

Pace, J. & McDermott. (1952). Methionine sulphoximine and some enzyme systems involving glutamine. *Nature* **169**, 415–16.

Padilla, J.E., Campos, F., Conde, V., Lara, M. & Sanchez, F. (1987). Nodule specific glutamine synthetase is expressed before the onset of nitrogen fixation. *Plant Molecular Biology* **9**, 65–74.

Platt, S.G. & Anthon, G.E. (1981). Ammonia accumulation and inhibition of photosynthesis in methionine sulphoximine treated spinach. *Plant Physiology* **67**, 509–13.

Platt, S.G. & Rand, L. (1983). Methionine sulphoximine effects on C_4 plant discs: comparison with C_3 species. *Plant and Cell Physiology* **23**, 917–21.

Powles, S.B. (1984). Photoinhibition of photosynthesis induced by visible light. *Annual Review of Plant Physiology* **35**, 15–44.

Probyn, T.A. & Lewis, O.A.M. (1979). The route of nitrate-nitrogen assimilation in the root of *Datura stramonium*. *Journal of Experimental Botany* **30**, 299–305.

Pruess, D.L., Scannell, J.P., Ax, H.A., Kellett, M., Weiss, F., Demny, T.C. & Stempel, A. (1973). Antimetabolites produced by microorganisms. VII L-(N^5-phosphono)methionine-S-sulphoximinyl-L-alanyl-L-alanine. *Journal of Antibiotics* **26**, 261–5.

Puritch, G.S. & Barker, A.V. (1967). Structure and function of tomato leaf chloroplasts during ammonium toxicity. *Plant Physiology* **42**, 1229–38.

Pushkin, A.V., Antoniuk, L.P., Solovieva, N.A., Shubin, V.V., Evstigneeva, Z.G., Kretovich, W.L., Cherednikova, T.V., Tsuprun, V.L., Zograf, O.N., Kiselev, N.A. (1985). Glutamine synthetase of pea leaf and seed cytosol. Structure and properties. *Biochimica et Biophysica Acta* **828**, 336–50.

Ramos, J.L., Guerrero, M.G. & Losada, M. (1982). Photoproduction of ammonia from nitrate by *Anacistis nidulans* cells. *Biochimica et Biophysica Acta* **679**, 323–30.

Rennenberg. H. (1982). Glutathione metabolism and possible biological roles in higher plants. *Phytochemistry* **12**, 2771–81.

Rhodes, D., Deal, L., Haworth, P., Jamieson, G.C., Reuter, C.C. & Ericson, M.C. (1986). Amino acid metabolism of *Lemna minor* L. l. Responses to methionine sulphoximine. *Plant Physiology* **82**, 1057–65.

Ridley, S.M. & McNally, S.F. (1985). Effects of phosphinothricin on the isoenzymes of glutamine synthetase isolated from plant species which exhibit varying degrees of susceptibility to herbicide. *Plant Science* **39**, 31–6.

Robertson, J.G., Farnden, K.J.F. & Banks, J.M. (1975). Induction of glutamine synthetase during nodule development in lupin. *Australian Journal of Plant Physiology* **2**, 165–72.

Ronzio, R.A. & Meister, A. (1968). Phosphorylation of methionine sulfoximine by glutamine synthetase. *Proceedings of the National Academy of Sciences (USA)* **59**, 164–70.

Ronzio, R.A., Rowe, W.B. & Meister, A. (1969). Studies on the mechanism of inhibition of glutamine synthetase by methionine sulphoximine. *Biochemistry* **8**, 1066–75.

Rowe, W.B., Ronzio, R.A. & Meister, A. (1969). Inhibition of glutamine synthetase by methionine sulfoximine. Studies on methionine sulfoximine phosphate. *Biochemistry* **8**, 2674–80.

Sauer, H., Wild, A. & Ruehle, W. (1987). The effect of phosphinothricin (glufosinate) on photosynthesis. 2. The causes of the inhibition of photosynthesis *Zeitung Naturforschung* **42**, 270–8.

Scannell, J.P., Pruess, D.L., Demney, T.C., Ax, H-A., Weiss, F., Williams, T., and Stempel, A. (1975). L-(N^5-phosphono)methionine-S-sulphoximinyl-L-alanyl-L-alanine, and antimetabolite of L-glutamine produced by a *Streptomycete*. In *Chemistry and Biology of Peptides*, eds R. Walter and J. Meienhafer pp. 415–21. Ann. Arbor Science Publishers.

Singh, M. & Widholm, J.M. (1975). Inhibition of corn, soybean and wheat seedling growth by amino acid analogues. *Crop Science* **15**, 79–81.

Skokut, T.A., Wolk, C.P., Thomas, J., Meekes, J.C., Schatten, P.W. & Chuen, W.A. (1978). Initial organic products of assimilation of ^{13}N ammonium and ^{13}N nitrate by tobacco cells cultured on different sources of nitrogen. *Plant Physiology* **62**, 299–304.

Slovacek, R.E. & Hind, G. (1981). Correlation between photosynthesis and the transthylakoid proton gradient. *Biochimica et Biophysica Acta* **635**, 393–404.

Smith, I.K., Kendall, A.C., Keys, A.J., Turner, J.C. & Lea, P.J. (1985). The regulation of the biosynthesis of glutathione in leaves of barley (*Hordeum vulgare*, L.) *Plant Science* **41**, 11–17.

Snapp, S.S. & Vance, C.P. (1986). Asparagine synthesis in alfalfa root nodules. *Plant Physiology* **82**, 390–5.

Somerville, C.R. & Ogren, W.L. (1980). Inhibition of photosynthesis in mutants of *Arabidopsis* lacking glutamate synthase activity. *Nature* **286**, 257–9.

Stasiewicz, S. & Durham, V.L. (1979) Isolation and characterisation of two forms of glutamine synthetase from soybean hypocotyl. *Biochemical and Biophysical Research Communications* **87**, 627–34.

Stewart, C.R. (1979). The effect of ammonium, glutamine, methionine, sulphoximine and azaserine on asparagine synthesis in soybean leaves. *Plant Science Letters* **14**, 169–73.

Stewart, G.R. & Rhodes, D. (1976). Evidence for the assimilation of ammonia via the glutamine pathway in nitrate grown *Lemna* minor. *FEBS Letters* **64**, 296–9.

Ta, T.C. & Joy, K.W. (1986). Metabolism of some amino acids in relation to the photorespiratory nitrogen cycle of pea leaves. *Planta* **169**, 117–22.

Ta, T.C., Joy, K.W. & Ireland, R.J. (1984). Amino acid metabolism in pea leaves. Utilisation of nitrogen from amide and amino groups of [^{15}N]-asparagine. *Plant Physiology* **74**, 822–6.

Tachibana, K., Watanabe, T., Sekizawa, Y., Takematsu, T. (1986). Action mechanism of bialaphos. *Journal of Pesticide Science*, **11**, 27–31.

Taylor, P.A., Schnoes, H.K. & Durbin, R.D. (1972). Characterization of chlorosis-inducing toxins from plant pathogenic *Pseudomonas* sp. *Biochimica et Biophysica Acta* **286**, 107–17.

Thomas, M.D. & Durbin, R.D. (1985). Glutamine synthetase from *Pseudomonas syringae* pv. *tabaci;* Properties and inhibition by tabtoxinine-β-lactam. *J. General Microbiol.* **131**, 1061–7.

Tingey, S.V., Walker, E.L. & Coruzzi, G.M. (1987). Glutamine synthetase genes of pea encode distinct polypeptides which are differentially expressed in leaves, roots and nodules. *EMBO Journal* **6**, 1–10.

Tobin, A.K., Ridley, S.M. & Stewart, G.R. (1985). Changes in the activities of chloroplast and cytosolic isoenzymes of glutamine synthetase during normal leaf growth and plastid development in wheat. *Planta* **163**, 544–8.

Tolbert, N.E. (1980). Photorespiration. In *The Biochemistry of Plants* Vol. 2, ed. D.D. Davies, pp. 487–523. London: Academic Press.

Turner, J.G. & Debbage, J.M. (1982). Tabtoxin-induced symptoms are associated with the accumulation of ammonia formed during photorespiration. *Physiological Plant Pathology* **20**, 223–33.

Vézina, L-P., Hope, H.J. & Joy, K.W. (1987). Isoenzymes of glutamine synthetase in roots of pea and alfalfa. *Plant Physiology* **83**, 58–62.

Villafranca, J.J., Eads, C.D. & Colanduoni, J.A. (1986). Mechanistic investigations of the glutamine synthetase reaction. In *Mechanisms of Enzymatic Reactions Stereochemistry*, ed. P.A. Frey, pp. 47–58. New York: Elsevier Science Publishing Co., Inc.

Walker, K.A., Givan, C.V. & Keys, A.J. (1984a). Glutamic acid metabolism and the photorespiratory nitrogen cycle in wheat leaves. *Plant Physiology* **75**, 60–6.

Walker, K.A., Keys, A.J. & Givan, C.V. (1984*b*). Effect of L-methionine sulphoximine on the products of photosynthesis in wheat (*Triticum aestivum*) leaves. *Journal of Experimental Botany* **35**, 1800–10.

Wallsgrove, R.M., Keys, A.J., Bird, I.F., Cornelius, M.J., Lea, P.J. & Miflin, B.J. (1980). The location of glutamine synthetase in leaf cells and its role in the reassimilation of ammonia released in photorespiration. *Journal of Experimental Botany* **31**, 1005–17.

Wallsgrove, R.M., Keys, A.J., Lea, P.J. & Miflin, B.J. (1983). Photosynthesis, photorespiration and nitrogen metabolism. *Plant, Cell and Environment* **6**, 301–9.

Wallsgrove, R.M., Turner, J.C., Hall, N.P., Kendall, A.C. & Bright, S.W.J. (1987). Barley mutants lacking chloroplast glutamine synthetase – biochemical and genetic analysis. *Plant Physiology* **83**, 155–8.

Wedler, F.C. & Horn, B.R. (1976). Catalytic mechanisms of glutamine synthetase enzymes. *Journal of Biological Chemistry* **251**, 7530–8.

Wedler, F.C., Horn, B.R. & Roby, W.G. (1980). Interaction of a new γ-glutamyl-phosphate analog 4-(Phosphonacetyl(L-α–aminobutyrate with glutamine synthetase (EC 6.3.1.2) enzymes from *Escherichia coli*, plant and mammalian sources. *Archives of Biochemistry and Biophysics*, **202**, 482–90.

Wild, A., Sauer, H. & Ruehle, W. (1987). The effect of phosphinothricin (glufosinate) on photosynthesis. 1. Inhibition of photosynthesis and accumulation of ammonia in *Sinapsis alba*. *Zeitung Naturforschung* **42**, 263–9.

Woo, K.C. & Canvin, D.T. (1980). Effect of ammonia, nitrite, glutamate and inhibitors of nitrogen metabolism on photosynthetic carbon fixation in isolated spinach leaf cells. *Canadian Journal of Botany*, **48**, 511–16.

Woo, K.C. & Osmond, C.B. (1982). Stimulation of ammonia and 2-oxoglutarate-dependent O_2 evolution in isolated chloroplasts by dicarboxylates and the role of the chloroplast in photorespiratory nitrogen recycling. *Plant Physiology*, **69**, 591–6.

Woolley, D.W., Pringle, R.B. & Braun, A.C. (1952). Isolation of the phytopathogenic toxin of *Pseudomonas tabacci*, an antagonist of methionine. *Journal of Biological Chemistry*, **197**, 409–17.

Yamaya, T., Oaks, A., Rhodes, D. & Matsumoto, H. (1986). Synthesis of [^{15}N]-glutamate from [^{15}N]-H_4^+ and [^{15}N]-glycine by mitochondria isolated from pea and corn shoots. *Plant Physiology*, **81**, 754–7.

W. J. OWEN

Metabolism of herbicides – detoxification as a basis of selectivity

Introduction

Herbicides recommended for the selective control of unwanted weeds have been developed to exploit a difference in phytotoxicity between species adequate to kill competing weeds without significantly reducing crop yields. In some cases the margin of selectivity may be quite modest and can be rendered inadequate when the timing of application coincides with unfavourable climatic conditions, as was reported to be the case for some of the phenylureas in winter wheat in Autumn 1983. There are a number of factors which can contribute to herbicide selectivity including soil placement, rates of absorption and subsequent translocation, localisation (both within the plant and at the sub-cellular level) and transformation to products of modified phytotoxicity. In addition studies made with atrazine-resistant biotypes of such weed species as *Chenopodium album* and *Amaranthus hybridus* L. (Steinback, Pfister & Arntzen, 1982) have highlighted the importance of differences in sensitivity of the target site, in this case the 32 kD protein component of photosystem II. It is likely that other examples of reduced target site sensitivity will be encountered when biotypes resistant to herbicides with different modes of action emerge. Differences in target site sensitivity as a basis for selectivity has been considered previously by Gressel (1985) and is also referred to elsewhere in this volume (Chapters 2, 11 and 12).

The selective properties of herbicides often result from a complex interaction of a number of the factors listed above though there are many examples where one dominant factor has been implicated. This is certainly true in the case of herbicide uptake and movement (Hess, 1985) and differential metabolism to less-phytotoxic products, the subject of the present Chapter. The mechanisms by which plants transform foreign compounds have been adequately reviewed previously (Shimabukuro, Lamoureux & Frear, 1982; Cole, 1983; Shimabukuro, 1985), including progress using isolated enzyme preparations (Lamoureux & Frear, 1979). More recent reviews (Hathway, 1986; Cole, Edwards & Owen, 1987) have specifically concentrated on metabolism as the basis for the selectivity of particular herbicides. Whereas Cole *et al.* (1987) placed emphasis on more recently introduced compounds the intention in the present review is to draw on examples, where appropriate, from both new and older compounds in order to cover a wide spectrum of the various biotransformations known to contribute to metabolic selectivity. These include oxidations resulting in aromatic ring and alkyl hydroxylation, *N*-dealkylation, *O*-dealkylation and sulphoxidation, hydrolytic reactions, deamination and conjugation with carbohydrate residues (glycosidation) or the tripeptide glutathione. Examples of compounds for which differential metabolism has been partly implicated in species selectivity have been cited previously (Jensen, 1982).

Apart from contributing to our overall understanding of herbicide selectivity the metabolic fate of herbicides is also worthy of study from a number of other standpoints. A detailed knowledge of the metabolic capabilities of weed species in particular may be used to advantage in the design of inactive herbicide progenitors which are bioactivated in the target plants. In addition, metabolism studies lead to increased knowledge of the enzymology of detoxification processes and the identification of detoxification genes suitable for gene transfer studies. These latter aspects are discussed in Chapters 11 and 12 respectively.

Mechanisms of detoxification implicated in species selectivity

Aryl hydroxylation

Because of the aromatic nature of many, if not most, herbicides it is not surprising that ring hydroxylation represents a very important detoxification reaction in higher plants (see e.g. Cole, 1983). However, species differ in their ability to aryl hydroxylate particular parent structures, presumably because of variation in the specificities or levels of expression of the mixed-function oxidases (mfos) responsible for these reactions. Such is the case for bentazon (1.1, Fig.1) for which a 200-fold selectivity margin between rice and *Cyperus serotinus* is attributable to rapid

Fig. 1. Pathway of bentazon metabolism in plants

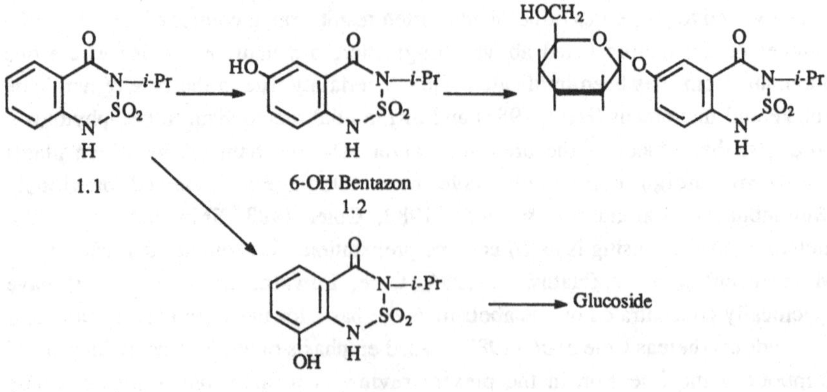

HOCH$_2$
O
N—*i*-Pr
SO$_2$
N
H
1.1
HO
6-OH Bentazon
1.2
8-OH Bentazon
OH
Glucoside

conversion in rice to 6-hydroxy bentazon (1.2), followed by glucose conjugation (Mine, Miyakado & Matsunaka, 1975). Metabolism in *C. serotinus* on the other hand is ineffective as is the case in other susceptible weed species such as *Cirsium arvense, Eleocharis kuroguwai, Sagittaria pygmaea* and *Solanum nigrum* (Mahoney & Penner, 1975*b*; Mine *et al.*, 1975; Penner, 1975). Rapid production of water soluble glycosidic metabolites also occurs in tolerant maize, navy-bean (*Phaseolus vulgaris*), *Echinochloa crus-galli* (Mahoney & Penner, 1975*a*; Mine *et al.*, 1975) and wheat (Retzlaff & Hamm, 1976) whereas in soybean both the 6- and 8-hydroxy (1.3) derivatives are formed (Otto *et al.*, 1979). Differences in tolerance to bentazon between soybean cultivars have also been attributed to relative rates of ring hydroxylation (Wax, Bernard & Hayes, 1974; Hayes & Wax, 1975) though metabolites were not characterised in detail.

Ring hydroxylation also plays an important role in the selectivity of diclofop-methyl (2.1, Fig. 2) between resistant cereals and susceptible grass weeds. Subsequent to rapid hydrolysis of the ester the resulting phytotoxic diclofop acid (2.2) is inactivated by hydroxylation of the dichlorophenyl ring to give 2.3 followed by glycosidation (2.4) (Gorbach, Kuenzler & Asshauer, 1977; Shimabukuro, Walsh & Hoerauf 1979; Jacobson, Shimabukuro & McMichael, 1985). Hydroxylation occurred at all three available positions on the ring though the 5'-hydroxy derivative predominated. In contrast aryl hydroxylation is of minor importance in susceptible oat and wild oat *(Avena fatua)* in which diclofop is primarily converted to the neutral glycosyl ester (2.5) (Shimabukuro *et al.*, 1979). Polar conjugates of diclofop have also been recorded as the major metabolites of diclofop-methyl in other susceptible species, notably *Echinochloa crus-galli* (Boldt & Putnam, 1981) and maize (Gorecka, Shimabukuro & Walsh, 1981). Though glucose esters of acidic herbicides might be expected to accumulate in cell vacuoles, as for glycosidic conjugates, the susceptibility of oat and *Avena fatua* to diclofop-methyl implies that in this case compartmentalisation of metabolites is inadequate to confer tolerance. Susceptibility presumably results from subsequent liberation of diclofop-acid from the glucose ester, which receives support from the observation (Jacobson *et al.*, 1985) that the acetylated derivative of the diclofop glucose ester was as inhibitory as diclofop-methyl to growth of oat roots. Metabolism of diclofop-methyl in cell suspension and callus cultures respectively of oat and *A. fatua* (Dusky, Davis & Shimabukuro, 1982) also occurred predominantly by glucose ester formation whereas ring hydroxylation was the major route in suspension cultures of diploid wheat (*Triticum monococcum*) (Dusky, Davis & Shimabukuro, 1980).

Despite the remarkably high biological activity of the more recently introduced sulphonylurea herbicides the margin of selectivity between cereals and sensitive weeds can be very large, approaching some 4000-fold in the case of chlorsulfuron (3.1, Fig. 3) (Sweetser, Schow & Hutchison, 1982). Neither herbicide uptake nor translocation by tolerant (wheat, barley, *A.fatua* and *Poa annua)* and susceptible (cotton, soybean, mustard and sugar beet) species correlated with selectivity which was attributed to a striking difference in ability of young plants to oxidise chlorsulfuron to the 5'-hydroxy derivative (3.2) which accumulated in tolerant species as the *O*-glucoside (Sweetser *et al.*, 1982). The observation that 5'-hydroxy chlorsulfuron was also a potent inhibitor of acetolactate synthase (Dr. H. Brown, E.I. du pont de Nemours & Co., personal communication), the target enzyme of the sulphonylureas, suggests that in this case subsequent glycosidation rather than hydroxylation constitutes the important detoxification step. Differential metabolism of chlorsulfuron has also been implicated as a mechanism of selectivity between wheat and barley (tolerant), *Viola bicolor* (moderately susceptible) and sugar beet and *Matricaria chamomilla*

Fig. 2. Major routes of diclofop-methyl metabolism in resistant wheat and susceptible oat and wild oat.

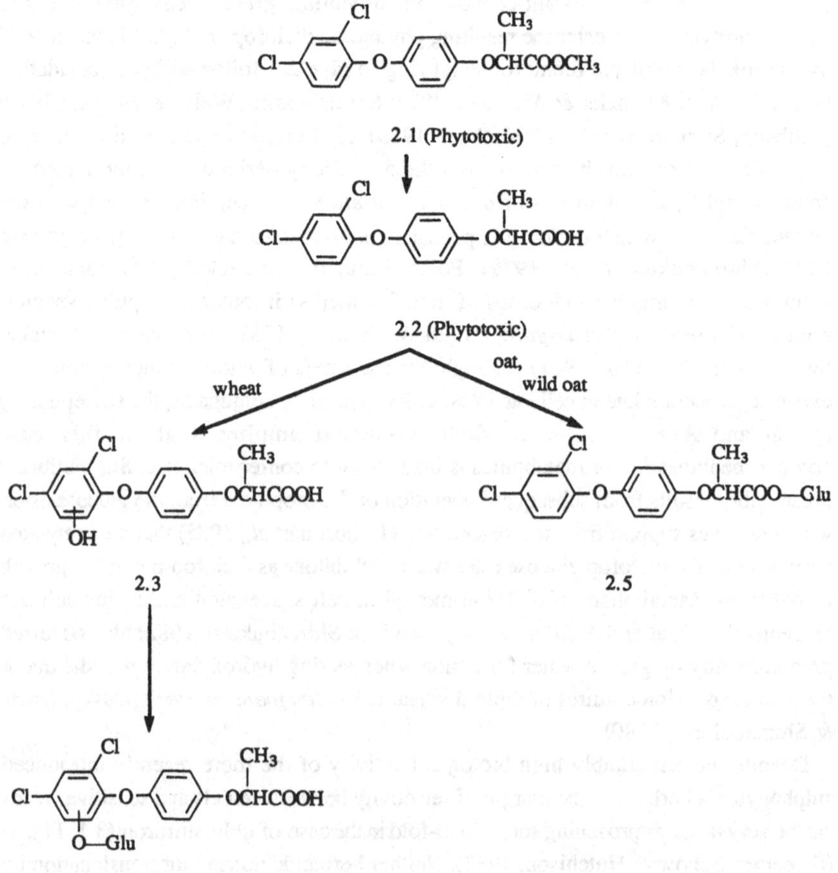

2.1 (Phytotoxic)

2.2 (Phytotoxic)

wheat

oat, wild oat

2.3

2.5

2.4

(susceptible) (Muller, Kang & Haruska, 1984) though products were not characterised.

The phenoxyalkanoic acid herbicide 2,4-D undergoes aromatic hydroxylation in many plant species. Hydroxylation occurs preferentially at the 4-position, by a mechanism generally referred to as the 'NIH' shift (Guroff *et al.*, 1967), to give a mixture of 4-OH-2,5-D and 4-OH-2,3-D in which the former product usually, though not always, predominates. However, elimination of chlorine may also occur concomitant with hydroxylation as evidenced by the identification of 4-OH-2-chlorophenoxyacetic acid (Feung, Hamilton & Mumma, 1975) and 2-OH-4-chlorophenoxyacetic acid (Bristol, Ghanuni & Oleson, 1977). Unsubstituted positions on the ring are also hydroxylated to varying degrees to give 3-OH-2,4-D (Bristol *et al.*, 1977). Though 3-OH-2,4-D retains auxin activity the major products of 2,4-D hydroxylation, 4-OH-2,5-D and 4-OH-2,4-D, are both inactive and consequently their formation constitutes an effective detoxification mechanism (Feung, Hamilton & Witham, 1971; Hamilton *et al.*, 1971). From these and other data it appears that aryl hydroxylation via the 'NIH' shift is restricted to the para position. Consideration is given to this mechanism here on the basis of it having been recorded only for 2,4-D. It has not been reported in plants for any other pesticide with the same or similar chlorine substitution pattern, as exemplified by diclofop-methyl (Fig. 2). The 'NIH' shift is not even reported for 2-methyl-4-chlorophenoxyacetic acid (MCPA) which is hydroxylated predominantly at the ring-methyl substituent (Cole & Loughman, 1983, 1984, 1985). Species which are 2,4-D-tolerant and utilise the 'NIH' shift mechanism

Fig. 3. Metabolism of chlorsulfuron in plants

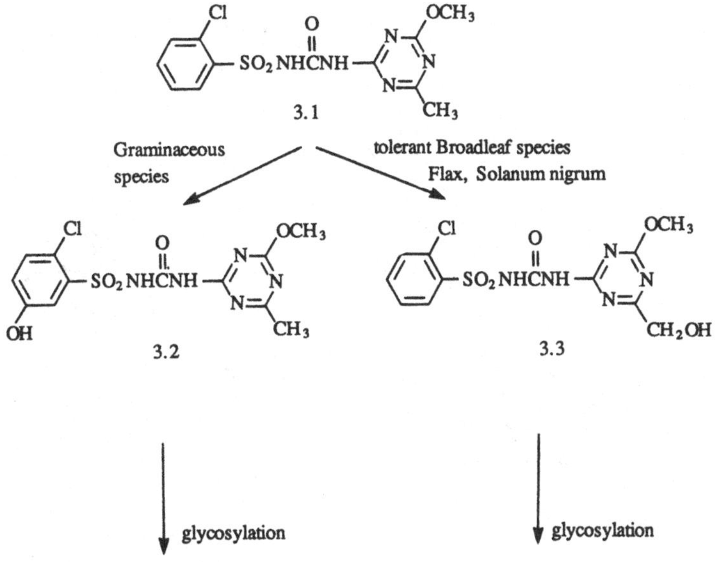

Cl
O
OCH_3
SO_2NHCNH
N
N
N
3.1
CH_3
Graminaceous species
tolerant Broadleaf species
Flax, Solanum nigrum
Cl
O
OCH_3
SO_2NHCNH
OH
CH_3
3.2
Cl
O
OCH_3
SO_2NHCNH
CH_2OH
3.3
glycosylation
glycosylation

include barley, oat and wheat (Bristol *et al.*, 1977; Feung *et al.*, 1978). The pattern of hydroxylated metabolites in a cell suspension culture of hexaploid wheat reflected that obtained in intact plants (Scheel & Sandermann, 1981). Hydroxylation of 2,4-D is also an important mechanism in cucumber (Chkanikov *et al.*, 1977) and leguminous species such as soybean (Feung *et al.*, 1978). However, the major metabolites in soybean were amino acid amide conjugates, notably 2,4-D-glutamic acid, although the proportion of phenolic derivatives increased with time (Feung *et al.*, 1971; Feung, Hamilton & Mumma, 1973). Consequently it may be that the amide conjugates, rather than 2,4-D itself, are the preferred substrates for hydroxylation. In contrast to other cereals rice callus was a poor hydroxylating tissue and accumulated a glycosyl ester of 2,4-D (Feung, Hamilton & Mumma, 1976) which was also the major initial product in cucumber (Chkanikov *et al.*, 1977).

Alkyl hydroxylation

Oxidation of alkyl substituents is also considered to be mediated by microsomal mfo enzymes. For both *N*- and *O*-substitutions initial alkyl hydroxylation results in the formation of alcohol intermediates that are generally unstable and, unless rapidly glycosylated, decompose to liberate formaldehyde in the case of a methyl group. These processes are referred to as *N*-dealkylation and *O*-dealkylation respectively. Further oxidation of *C*-alkyl derived alcoholic functions to

Fig. 4. Proposed metabolic activation in wild oat versus degradation in wheat, corn, and wild oat for the *meta* and *para* isomers of Assert.

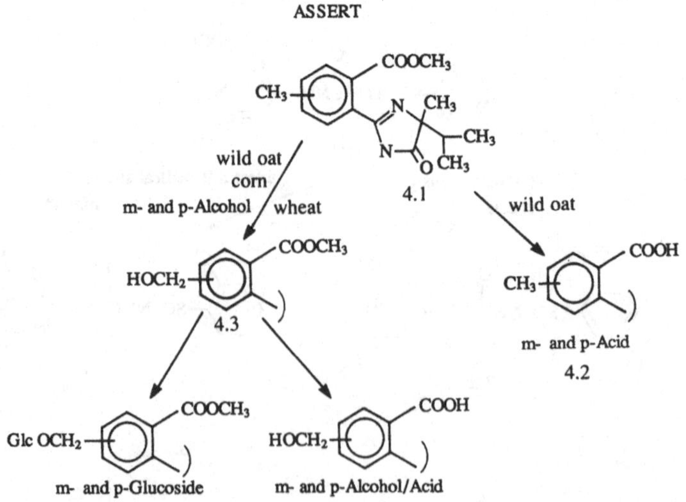

ASSERT

$COOCH_3$ CH_3 N CH_3 CH_3 N O CH_3

4.1

wild oat corn m- and p-Alcohol wheat

wild oat

$HOCH_2$ $COOCH_3$

4.3

CH_3 COOH

m- and p-Acid

4.2

Glc OCH_2 $COOCH_3$

m- and p-Glucoside

$HOCH_2$ COOH

m- and p-Alcohol/Acid

corresponding carboxylic acids is generally slow and is probably of little relevance to selectivity. Rates of ester conjugation may also be sufficiently rapid to compete with decarboxylation.

Hydroxylation of *C*-alkyl groups is of common occurrence in plants but notable species differences exist which contribute to selectivity. In contrast to graminaceous crops the tolerance of certain broadleaf species to chlorsulfuron results from hydroxylation of the 4-methyl substituent on the triazine ring (to produce 3.3) and subsequent glycosylation (Fig. 3) rather than 5' ring hydroxylation (Hutchison, Shapiro & Sweetser, 1984). The herbicidal nature of 3.3 (Sweetser, 1985) implies that for this metabolic route also the glycosylation step is essential for detoxification. A particularly interesting feature of the sulphonylureas is that comparatively subtle modifications to the parent structure can result in targetting for selective use in different crops. Thus DPX-F 384 (Londax) is used for weed control in rice whereas chlorsulfuron is non-selective in this crop (Ray, 1985). Tolerance of rice to Londax apparently correlated with an ability to *O*-demethylate a pyrimidine methoxy substituent to the corresponding hydroxy derivative (Dr H. Brown, E.I. du Pont de Nemours & Co., personal communication). Differential metabolism also provides an explanation for the selectivity of the imidazolinone aryl-carboxylate herbicides which also inhibit the acetolactate synthase enzyme. Thus Assert (4.1, Fig. 4) which is comprised of a mixture of *m*- and *p*-methyl substituted isomers, was largely de-esterified to the phytotoxic acids (4.2) in susceptible *A. fatua* whereas ring-methyl hydroxylation to the *m*- and *p*-benzyl alcohols (4.3) followed by glycosylation occurred in tolerant maize and wheat (Brown, Chiu & Miller, 1987).

Substituted phenylureas such as monuron and diuron (Fig. 5) are susceptible to metabolic attack by successive *N*-dealkylation. Whereas *N*-didealkylated products are inactive monodemethylated derivatives possess residual phytotoxicity so that removal of both *N*-methyl substitutions is usually required for complete detoxification. Plant species differ widely in their abilities to carry out one or both of these *N*-demethylation steps which has been correlated with degree of susceptibility (Geissbuhler, 1969). With the notable exception of cotton (Frear, 1968*a;* Frear, Swanson & Tanaka, 1969; Ryan *et al.*, 1981) the ability of crop species to detoxify monuron and diuron is limited so that these herbicides have little selective use in agriculture. Selectivity is markedly enhanced however by the replacement of the *para* chlorine with a methyl group as in chlortoluron (Fig. 5) which is widely used for weed control in winter wheat. This structural modification offered an additional mode of oxidative attack and extensive ring-methyl hydroxylation was considered to be the basis of the chlortoluron-tolerance of wheat and barley (Ryan *et al.*, 1981; Ryan & Owen, 1982). Hydroxylation of the ring methyl also occurred in cotton though in this species *N*-didealkylation was prominent. Susceptible species such as *Lolium perenne, A. fatua* and blackgrass (*Alopecurus myosuroides*) possessed only a limited ability to metabolise chlortoluron which was essentially restricted to *N*-monodealkylation (Ryan *et al.*, 1981). The position of the ring methyl substitution can also have a profound effect on selectivity. Thus the chlortoluron isomer, CGA 43057 (Fig. 5) in which the methyl group is *meta* located, is non-selective this being attributed to the poor ability of plants to oxidise the methyl group in this position (Ryan *et al.*, 1981). Rapid *N*-didealkylation resulted in cotton expressing tolerance to this methyl-phenylurea also.

The margin of selectivity of chlortoluron is inadequate for certain cultivars of wheat and barley. Studies carried out in our own laboratory (Ryan & Owen, 1983) have indicated that detoxification by ring methyl oxidation represented a minor metabolic route in leaf discs of susceptible cultivars in which metabolism was generally slow and restricted to *N*-monodealkylation. This metabolite profile resembled that of susceptible graminaceous weed species. The distinction was less apparent in a similar study using intact plants (Cabanne, Gaillardon & Scalla, 1985) though different

Fig. 5. Structures of some phenylurea herbicides.

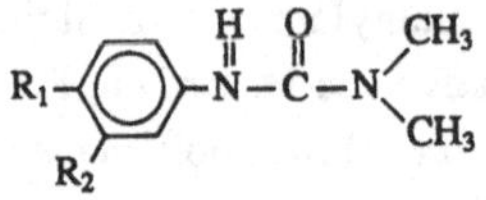

Monuron	:	$R_1 = Cl, R_2 = H$
Diuron	:	$R_1 = Cl, R_2 = Cl$
Chlortoluron	:	$R_1 = CH_3, R_2 = Cl$
CGA 43057	:	$R_1 = Cl, R_2 = CH_3$

Table 1. *Liberation of $^{14}CO_2$ from (methoxy-^{14}C) metoxuron-treated leaf discs of resistant and susceptible wheat plants*

Time after treatment (h)	$^{14}CO_2$ trapped (d.p.m./g fresh weight)	
	Resistant wheat (Cappelle)	Suseptible wheat (Maris Nimrod)
2	12,800	5,000
6	33,100	6,000
9	41,500	8,000
12	51,500	7,200

Leaf discs treated with (methoxy-^{14}C) metoxuron were incubated for the prescribed times in Warburg flasks, the released $^{14}CO_2$ being trapped by means of a piece of convoluted filter paper saturated with 6 M KOH situated in the centre well. The amount of radioactivity trapped was subsequently determined by scintillation counting.

cultivars were used. However, wheat varieties show parallel differences in response to the related phenylurea metoxuron and in this case a difference in relative contributions to metabolism of *N*-demethylation versus *O*-demethylation of the ring methoxy substituent was the major factor that correlated with varietal reaction (Emami-Saravi, 1979). The relative rates of *O*-demethylation, measured as loss of $^{14}CO_2$ from leaf discs treated with [methoxy-^{14}C]metoxuron, in tolerant and susceptible cultivars differed by as much as 7-fold over a 12 h period (Table 1). The phenolic derivative formed as a product of *O*-demethylation of metoxuron was demonstrated to be inactive as a photosystem II inhibitor (Emami-Saravi, 1979). Despite the rapid loss of radioactivity from [methoxy-^{14}C]metoxuron observed in leaf disc experiments attempts in our laboratory to demonstrate metoxuron *O*-demethylase activity in isolated sub-cellular preparations have met with little success. However, we were able to demonstrate an *N*-demethylase active in the conversion of metoxuron to the mono-methyl derivative, which was enriched in the microsomal fraction of wheat. The enzyme from tolerant and susceptible cultivars showed little difference in specific activity which was in the order of 6.8 nmoles/30 min/mg protein. This value compared favourably with that of 2.16 nmoles/30 min/mg protein quoted for a monuron *N*-demethylase in microsomes prepared from etiolated cotton hypocotyls (Frear, Swanson & Tanaka, 1969; Frear, Swanson & Tanaka, 1972). The latter authors provided evidence implicating involvement of cytochrome P-450 in the oxidative *N*-demethylation of monuron. Such evidence in the case of xenobiotic metabolism in plant tissues is rare, other documented examples being a *p*-chloro-N-methylaniline *N*-demethylase from castor bean endosperm (Young & Beevers, 1976) and avocado (McPherson *et al.*, 1975*a,b*); Dohn & Krieger, 1984) an aldrin epoxidase from pear roots (Earl & Kennedy, 1975), a *p*-nitro anisole *O*-demethylase from avocado mesocarp (McPherson *et al.*, 1975*a,b*) and a 2,4-D aryl-hydroxylase from cucumber leaves (Makeev, Makoveichuk & Chkanikov, 1977). More recently Sweetser (1985) has speculated on the involvement of an inducible cytochrome P-450 dependent mixed function oxidase involved in the hydroxylation of chlorsulfuron in wheat based on the incorporation of one atom of ^{18}O from a molecular oxygen into a hydroxylated metabolite and the sensitivity of chlorsulfuron metabolism to inhibitors of protein biosynthesis. Difficulties associated with the isolation of plant microsomal preparations active in the utilisation of other xenobiotic substrates are currently restricting progress on the identification of mixed function oxidases appropriate as candidates for gene transfer studies.

Compounds which function as inhibitors of cytochrome P-450 linked mixed function oxidases offer potential for use as synergists of herbicides that are detoxified in weed species via initial microsomal hydroxylation. Thus the enzyme activated inhibitor 1-aminobenzotriazole (ABT) blocked the oxidation of chlortoluron and isoproturon in wheat (Cabanne *et al.*, 1985*b*) and enhanced the toxicity of both herbicides (Gaillardon *et al.*, 1985). Interestingly ABT was found to preferentially inhibit the ring-methyl oxidation versus N-demethylation of chlortoluron but no such selectivity was observed in a study of the influence of this inhibitor on the fate of chlortoluron in suspension cultured cells of maize and cotton (Cole & Owen, 1987). Similarly the methyl hydroxylation of MCPA in potato tuber slices was also susceptible to inhibition by ABT (Cole & Loughman, 1985). More potent inhibition of oxidative metabolism of chlortoluron was achieved in this laboratory by the use of the plant growth regulator tetcyclasis (see chemical glossary). At a concentration of 10 μM metabolism of chlortoluron was inhibited by 60–70% in both cotton and maize cells following a 24 h pre-treatment (Fig. 6) with both ring-methyl oxidation and *N*-demethylation being similarly susceptible (Cole & Owen, 1987). These latter observations raise the possibility that *C*- and *N*-alkyl hydroxylation of chlortoluron may be mediated by the same species of mixed function oxidase. An alternative explanation however could be provided by a general lack of specificity in the action of tetcyclasis which receives support from its recorded ability to interfere with mixed function oxidases involved in kaurene oxidation (Graebe, 1982) and in the biosynthesis of plant sterols (Grossman, Rademacher & Jung, 1983) and xanthophylls (Sandmann & Bramley, 1985).

In addition to the examples referred to above *O*-dealkylation leading to detoxification of 2,4-D by side-chain removal as indicated by release of glyoxylic acid from side-chain labelled 2,4-D, has been implicated as being important to selectivity between certain plant species. *O*-demethylation of the dichloroanisole derivative generated by initial decarboxylation of the side-chain offers an alternative mechanism. Though these pathways occur to only a minor extent in many plants, in the case of red currant

Fig. 6. Effect of tecyclasis on the metabolism of chlortoluron in suspension-cultured cells.

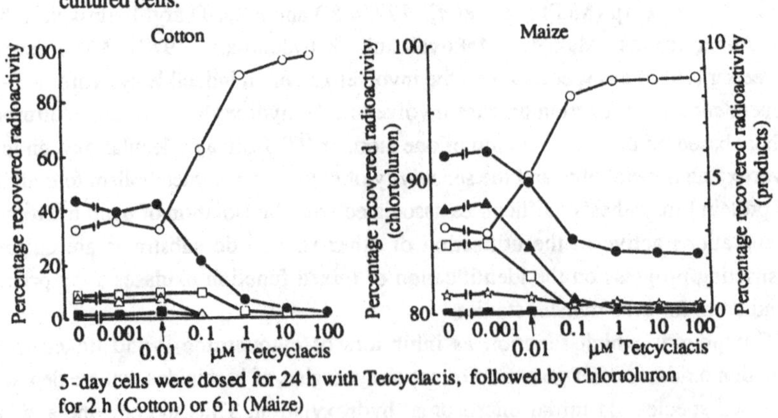

5-day cells were dosed for 24 h with Tetcyclacis, followed by Chlortoluron for 2 h (Cotton) or 6 h (Maize)

○ Chlortoluron ● mono-demethylated △ didemethylated ▲ CH_3-hydroxylated
□ CH_3-hydroxylated/monodemethylated ■ CH_3-hydroxylated/didemethylated
☆ polar conjugates

(Ribes sativum Syme) it constitutes a major degradative mechanism and is responsible for conferring tolerance (Luckwill and Lloyd-Jones, 1960). In contrast, susceptible blackcurrent *(Ribes nigrum* L.) had a poor ability to degrade the side-chain.

N-dealkylation also plays a role in the metabolism and selectivity of *s*-triazine herbicides such as atrazine (Jenson, Bandeen & Machado, 1977; Gressel, Shimabukuro and Duysen, 1983), simazine (Akinyemiju, Dickmann & Leavitt, 1983), simetryne (Matsumoto & Ishizuka, 1980) and terbutryne (Edwards & Owen, unpublished observation). In species in which this pathway was prominent the major products identified were mono-dealkylated triazines which retain phytotoxicity. However, in a study of atrazine metabolism in *Chenopodium album, C. strictum* and *Amaranthus ponelli* Khan, Warwick & Marriage (1985) identified the bound residue from resistant biotypes as *N*-desethylatrazine and suggested that sequestration of phytotoxic metabolites contributed to tolerance to the herbicide.

Sulphoxidation

Sulphide groups or other oxidisable sulphur moieties are of common occurrence in pesticide molecules and have been shown to be sulphoxidised by numerous plant species (Metcalf *et al.*, 1966; Muller & Payot, 1966; Bowman, Beroza & Harding, 1969; Andrawes, Bagley & Herrett, 1971). The precise mechanism of sulphur-oxidation in plants is not clear. In mammalian systems, sulphide substrates are oxidised primarily by microsomal cytochrome P-450 dependent monoxygenases (Nakatsugawra & Morelli, 1976) though an FAD-dependent *S*-oxygenase has also been identified in microsomal preparations of pig liver (Poulsen, Hyslop & Ziegler, 1979). Some sulphur oxidations have been demonstrated to be mediated by horseradish peroxidase. Thus this enzyme catalysed the sulphoxidation of a heterocyclic sulphide, chlorpromazine, though the carbamate insecticide, mesurol, an aromatic-alkyl sulphide was not a substrate (Lamoureux & Frear, 1979). Products of sulphoxidation are more water soluble and tend to be more chemically reactive, pesticidally active (see e.g. EPTC, Chapter 10) and short lived in the environment than the precursor sulphides. Consequently, plant sulphoxidation is a very significant factor in the biochemistry and physiology of pesticides.

In addition to its role in the bioactivation of various pesticides including herbicides where species differences can account for selectivity (Casida, Gray & Tilles, 1974), sulphoxidation of alkylsulphide groups is often an important initial step in detoxification. Thus the sulphoxides generated by the oxidation of the methylsulphide ring substituents of terbutryne (Edwards & Owen, unpublished observations) and metribuzin (Fig. 9) (Frear, Swanson & Mansager, 1985) are substrates for glutathione *S*-transferases which catalyse their conversion to non-phytotoxic peptide conjugates. Indeed, the 2-methylthio-*s*-triazines apparently inhibit glutathione *S*-transferases in maize (Frear & Swanson, 1970) and can only be detoxified by this route following sulphoxidation. In the case of the thiocarbamate herbicide EPTC Horvath & Pulay

(1980) obtained evidence implicating the sulphone derivative as the true carbamoylating agent rather than the sulphoxide. Further metabolism of a glutathione conjugate of the fungicide pentachloronitrobenzene (PCNB) in suspension cultured cells of peanut (Lamoureux & Rusness 1980; Rusness & Lamoureux, 1980; Lamoureux *et al.*, 1981) resulted in the formation of phenylmethylsulphides. Such metabolities arose as a result of the concerted action of a C-S lyase and an S-adenosyl-L[methyl]-methionine-dependent methyl transferase on a cysteine conjugate derived from the product of the glutatione *S*-transferase reaction. The same authors noted that phenylmethylsulphides became oxidised *in vivo* to the sulphoxides. Blair *et al.* (1984) recently reported the ability of suspension cultures of cotton, carrot and tobacco to mediate the enzymic sulphoxidation of a model aromatic alkyl-sulphide, *p*-chlorophenylmethylsulphide (PCPMS). Rates of PCPMS sulphoxidation to the sulphoxide were greatest in cotton and poorest in tobacco which served to illustrate the existence of species differences with respect to this mode of oxidative attack also.

Hydrolysis

Hydrolysis of carboxylic acid ester groups of various xenobiotics is of widespread occurrence in plants. In the context of herbicides, cleavage of carboxylic esters more commonly results in bioactivation (see e.g. Cole *et al.*, 1987) rather than detoxification and examples of herbicide selectivity attributed to species differences in capacity for ester hydrolysis are discussed elsewhere in this volume (Chapter 10). In general, the substrate specificity of plant carboxylesterases appears quite broad as indicated by the ability of wheat to cleave the methyl ester of diclofop (Fig. 2), the 2-ethylhexyl ester of a phthalate plasticiser (Krell & Sandermann, 1986) and the ethoxybutyl ester of the pyridinyloxyacetic acid herbicide triclopyr (Lewer and Owen, 1987). More detailed information will be required in order to assess whether significant species differences in specificity exist which may be exploitable in the design of selective herbicides.

The amide linkage is somewhat similar to that in esters and is attacked in plants by amidases. However, there has been speculation that some carboxylamidases may be carboxylesterases whose specificity extends to amide groups and partially purified carboxylesterases have been clearly demonstrated to utilise amide substrates (Hassall, 1982). The possibility that these observations might arise from the presence in such extracts of two or more similar enzymes has also been considered and Junge & Krisch (1975) obtained data which implied partial separation of esterase and amidase activities of pig liver. Amino acid amide conjugates of 2,4-D and the related 2,4,5-T have been identified for some time as prominent metabolities of these phenoxyalkanoic acid herbicides in some plant species, notably legumes such as soybean (Feung *et al.*, 1971; Feung *et al.*, 1973; Chkanikov *et al.*, 1972, 1977; Arjmand, Hamilton & Mumma, 1978; Scheel & Sandermann, 1981). Of the 20 or so amino acid conjugates that are possible 2,4-D metabolites 2,4-D-aspartate and 2,4-D-glutamate generally predominate. Recent work in this laboratory has led to the tentative identification of aspartate and glutamate amide conjugates of triclopyr as the major products formed from this pyridinyloxyacetic acid derivative in soybean suspension cultures and intact plants of *Stellaria media* (Lewer & Owen, 1987). In contrast aspartate and glutamate conjugates of the naturally occurring auxin IAA were minor metabolites in *Parthenocissus tricuspidata* crown gall callus tissue in which the glycine, alanine and valine conjugates predominated (Feung, Hamilton & Mumma 1976).

Studies on the fate of amino acid conjugates of 2,4-D in soybean callus have suggested that these metabolites are readily hydrolysed in plant tissue (Feung *et al.*, 1973; Owen, Hamilton & Mumma, unpublished observation) which probably accounts for their observed biological activity (Feung, Mumma & Hamilton, 1974; Feung, Hamilton & Mumma, 1977). The implication from these data is that conversion of 2,4-D to amide conjugates does not constitute an effective detoxification mechanism. In the case of IAA their formation is probably related to translocation and/or the regulation of localised concentrations of active auxin. Whether plant species differ significantly in ability to hydrolyse amide conjugates has not been extensively investigated though in a screen of the herbicidal properties of 2,4-D amino acid conjugates (Feung *et al.*, 1977) green beans, pea, sunflower and soybean responded similarly. Studies carried out in this laboratory indicated a faster rate of hydrolysis of triclopyr-aspartate (applied to leaves as the dibenzyl ester) in young plants of *Chenopodium album* compared to *S. media* (Lewer & Owen, unpublished observation). This difference may contribute to the greater susceptibility of *C. album* to triclopyr.

In contrast, available evidence suggests that arylamide bonds are generally quite resistant to hydrolytic attack in plant tissues. A notable exception is the substituted propionanilide derivative, propanil, which was rapidly hydrolysed in resistant rice plants but not in susceptible weed species such as *Echinochloa crus-galli* (Frear & Still, 1968). Hydrolysis of propanil has been attributed to a specific arylacylamidase enzyme (Frear & Still, 1968; Hoagland, 1978) the activity of which was sixty times greater in rice compared to *E. crus-galli*. Similar enzyme activities have been extracted from some twenty other species of higher plant including tulip (Hoagland & Graf, 1972), *Taraxacum officinale* (Hoagland, 1975) and lettuce (Hoagland, Graf & Handel, 1974) which was the best source. A more recent report (Oyamada *et al.*, 1986) of hydrolysis of the related napththoxypropionanilide derivative, naproanilide, in rice plants suggested that the specificity of the rice arylacylamidase extends beyond propanil. A product of this cleavage reaction was the phytotoxic napthoxypropionic acid (NOP). Tolerance of rice to naproanilide was attributed to rapid detoxification of NOP by hydroxylation and glycosidation. Amide hydrolysis was also rapid in the susceptible *Sagittaria pygmaea* Miq. but unlike rice further metabolism of NOP in this species produced mainly methyl and glucose esters rather than ring hydroxy derivatives. Since the methyl ester was reported to be phytotoxic (Oyamada *et al.*,

1986) and the glucose ester represented a potential source of NOP, this difference in metabolic profile between the two plant species was considered to be a possible mechanism for the herbicidal selectivity of naproanilide.

Nitrile hydrolysis also occurs readily in plants and proceeds via the corresponding amide to the carboxylic acid which may be subsequently decarboxylated. This metabolic route has been established for bromoxynil in wheat (Buckland, Collins & Pullin, 1973) and for cyanazine in wheat, potato and maize (Beynon, Stoydin & Wright, 1972*a,b*). Both these herbicides were metabolised by a number of competing pathways rendering interpretation of the importance of the hydrolysis of the cyano group in the detoxification difficult. There is little evidence to suggest that nitrile hydrolysis was a major route in tolerant maize but was insignificant in the moderately tolerant weed *Amaranthus retroflexus L.* (Waring, Edwards, Owen unpublished observation) in which a cyanazine glutathione conjugate was the dominant product. It is not yet clear whether this latter species lacks ability to hydrolyse nitrile groups or whether the opportunity to utilise this route is reduced through competition from a very efficient glutathione conjugation mechanism.

A further competing detoxification mechanism in the case of the chloro-*s*-triazines is the rapid nucleophilic displacement of the 2-chloro substituent in the triazine ring to give non-phytotoxic 2-hydroxy derivatives. These reactions are not enzymic but are catalysed by the naturally-occurring cyclic hydroxamate, 2,4-dihydroxy-3-keto-7-methoxy-1,4-benzoxazine (benzoxazinone), primarily in the roots of tolerant plants such as maize (Roth and Knüsli, 1961; Hamilton & Moreland, 1962; Hamilton, 1964). Species such as cotton, pea, sorghum and Johnson grass (*Sorghum halepense*) which do not contain benzoxazinone produced insignificant amounts of hydroxy-triazine derivatives (Hamilton, 1964). Thus in maize and other grass species the various metabolic options available for detoxification of atrazine and other chlorotriazines include *N*-dealkylation, benzoxazinone-catalysed hydrolysis of the C-2 chlorine or its nucleophilic displacement by glutathione (see below) (Jensen, Stephenson & Hunt, 1977). Rapid conversion to hydroxy-atrazine alone confers tolerance in *Coix lachryma-jobi* in which glutathione conjugation plays a very minor role in detoxification. In contrast the ability of susceptible panicoid grasses such as *Echinochloa crus-galli* and *Setaria glauca* to carry out any of the above biotransformations was very limited (Jensen *et al.*, 1977).

Glutathione conjugation

The role of glutathione *S*-transferases in herbicide detoxification and selectivity has been considered in a number of previous reviews, including Lamoureux & Frear (1979) and Cole *et al.* (1987). The structures of herbicides and their derivatives which serve as substrates for these enzymes in various plant species are shown in Fig. 7. With exceptions, glutathione conjugation products represent the major metabolites in maize and other related panicoid species demonstrating field tolerance to chlorotriazines such as atrazine. The operation of both the glutathione conjugation and C-2 chlorine hydrolysis pathways and, to a lesser extent, *N*-dealkylation in maize renders this species particularly tolerant of atrazine. Tolerance in sorghum on the other hand is conferred predominantly by glutathione-*S*-transferase activity (Lamoureux *et al.* 1973). Not all maize cultivars are atrazine tolerant and the susceptibility of line GT112 has been attributed to a low glutathione-*S*-transferase activity (Shimabukuro *et al.* 1971). Activity of the enzyme was similarly deficient in the susceptible panicoid grasses *E. crus-galli* and *S. glauca* (Jensen *et al.* 1977).

Glutathione conjugation is also important in the metabolism of other herbicides including the nitrodiphenylethers fluorodifen (Frear & Swanson, 1973; Shimabukuro *et al.* 1973; Diesperger & Sandermann, 1979) acifluorfen (Frear, Swanson & Mansager, 1983) and fomesafen (Evans, Cavell & Hignett, 1987) the chloracetanilides propachlor (Lamoureux, Stafford & Tanaka, 1971; Burkholder, 1978), alachlor (Mozer, Tiemeier & Jaworski, 1983) and metolachlor (Blattman *et al.*, 1986, Edwards & Owen 1986; Fuerst & Gronwald, 1986;) and the sulphonylurea derivative DPX-F6025 ('classic') (Dr H. Brown E.I. du Pont de Nemours and Co., personal communication). The latter compound has been developed for broadleaf

Fig. 7. Herbicide substrates for plant glutathione-*S*-transferases.

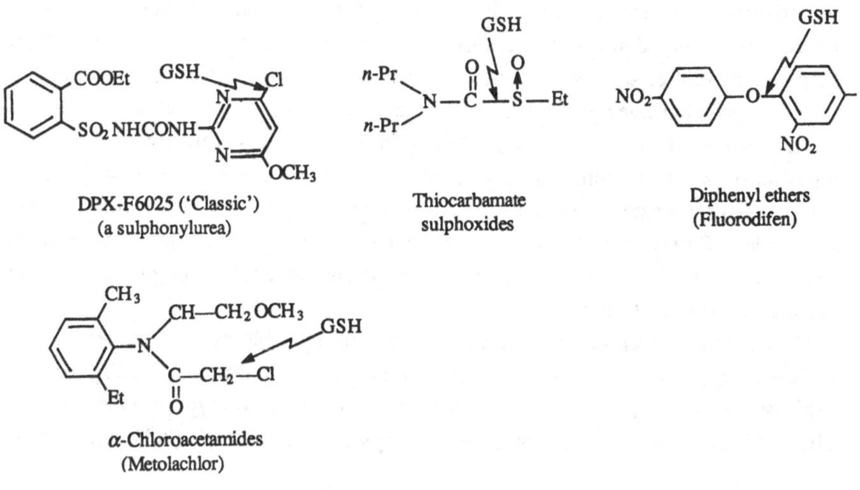

weed control in tolerant soybean which rapidly detoxifies 'classic' by nucleophilic displacement of the pyrimidine chlorine by homoglutathione. The action of glutathione *S*-transferases on nitrodiphenylethers results in fission of the aromatic ether bond. Thus fluorodifen was cleaved to *p*-nitrophenol and *S*-(2-nitro-4-trifluoromethylphenyl)glutathione (Lamoureux & Rusness, 1983). Similarly Frear, Swanson & Mansager (1983) identified 2-chloro-4-trifluoromethylphenol and *S*-(3-carboxy-4-nitrophenyl)-homoglutathione as initial products of acifluorfen metabolism in soybean. Activity of glutathione *S*-transferase towards fluorodifen was higher in resistant species such as cotton, maize, peanut, pea, soybean and okra than in susceptible cucumber, tomato and marrow (Frear & Swanson, 1973). Thus fluorodifen selectivity appeared to be based on distribution and concentration of the transferase. On purification and further characterisation the glutathione *S*-transferase of pea was found to be very specific for fluorodifen and did not utilise related nitrodiphenylethers such as nitrofen (Frear & Swanson, 1973). However, nitrofen has also been shown to be degraded by ether cleavage in plants (Hawton & Stobbe, 1971) which suggests the existence of species differences in specificity of glutathione *S*-transferases towards nitrodiphenylethers. Interestingly the pea transferase did not utilise atrazine as a substrate and this illustrates the difficulties in predicting the detoxification potential of a plant species with other substrates which undergo similar modes of metabolism. Studies on the metabolism of herbicides which serve as substrates for glutathione *S*-transferases have thus established the existence of isoenzymic forms which are specific for particular herbicide types. Thus isoenzyme multiplicity offered a possible explanation for the difference in response to atrazine and EPTC between maize (tolerant to both herbicides) and *Panicum miliaceum* (atrazine-tolerant only) despite the fact that both herbicides undergo similar routes of detoxification by glutathione conjugation (Ezra & Stephenson, 1985). In this comparison, however, other explanations such as the requirement for sulphoxidation to activate EPTC and the reduced intracellular concentration of glutathione in *P. miliaceum* compared to maize were also tenable. More recent studies using maize cell suspension cultures suggested that the isoenzymes responsible for the detoxification of the chloracetanilide metolachlor were distinct from the enzyme which conjugated atrazine (Edwards & Owen, 1986). These findings were subsequently confirmed by analysis of whole protein extracts of maize leaves by chromatofocussing which separated three peaks of GST activity differing in their activity towards the two herbicides (Edwards & Owen, 1987). This complexity apparently results from multiple genes which may also give rise to various molecular weight forms of the enzyme (Moore *et al.*, 1986).

Once formed tripeptide conjugates are generally unstable in plant tissues and are metabolised further. Thus in maize *S*-(metolachlor)-glutathione (8.2, Fig. 8) was rapidly converted, by the concerted action of carboxypeptidase and γ-glutamyltranspeptidase enzymes, to the cysteine conjugate (8.3) which was subsequently oxidised to the thiolactic acid (8.4) (Blattman *et al.*, 1986; Owen & Donzel, unpublished observation). Similar pathways have been recorded for the fate of glutathione and homoglutathione conjugates of other herbicides though malonylation of the cysteine conjugates as occurs in fluorodifen metabolism (Lamoureux & Rusness, 1983), may be of more common occurrence than conversion to thiolactic derivatives.

Though glutathione *S*-transferases are involved in the transformation of chloracetanilides their role in the selectivity of these herbicides is less certain. Thus the observation that propachlor served as a substrate for the transferases isolated from 10 agriculturally important plant species (Burkholder, 1978) suggested that certain types of glutathione *S*-transferase activity may be widely distributed in higher plants. However rates of conjugation may explain the differing sensitivities of cell suspension cultures of maize and rice to metolachlor. Formation of the glutathione conjugate was found to be some 8-fold faster in maize compared to rice (Table 2). Interestingly,

Fig. 8. Further metabolism of the glutathione conjugate of metolachlor.

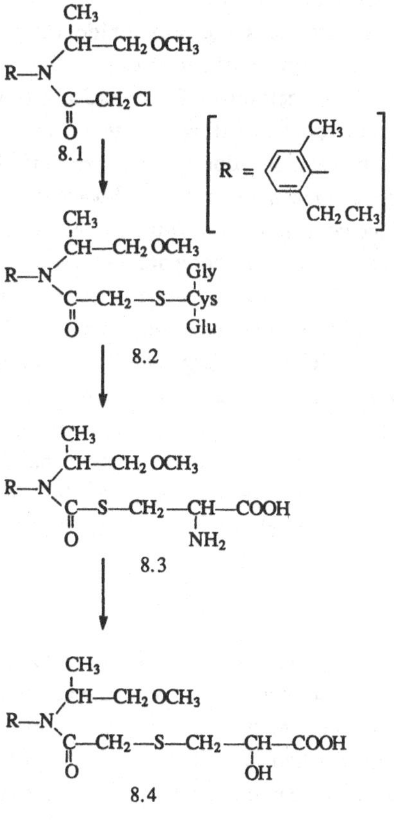

CH_3
$CH—CH_2OCH_3$
R—N
$C—CH_2Cl$
O

8.1

[R = (ring with CH_3 and CH_2CH_3)]

CH_3
$CH—CH_2OCH_3$
R—N Gly
$C—CH_2—S—Cys$
O Glu

8.2

CH_3
$CH—CH_2OCH_3$
R—N
$C—S—CH_2—CH—COOH$
O NH_2

8.3

CH_3
$CH—CH_2OCH_3$
R—N
$C—CH_2—S—CH_2—CH—COOH$
O OH

8.4

Table 2. *Metabolism of metolachlor in suspension cultures of maize and rice*

		Major metabolite fractions (% total recovered ^{14}C)			
Species	Time	Metolachlor (Met)	Met-GSH[a]	Met-Cys[b]	Met-thiolactic acid
Maize	1 h	33.6	42.1	4.8	–
	8 h	6.0	47.7	18.4	2.5
Rice	1 h	70.2	5.1	8.2	–
8 h	45.0	21.8	14.6	–	

[a] glutathione conjugate; [b] cysteine conjugate

similar amounts of the cysteine conjugate were found in both cultures which suggested that initial conjugation was the deficient step in rice rather than carboxypeptidase or γ-glutamyltranspeptidase activities.

In addition to the above parent herbicides the sulphoxide metabolities of dithiocarbamates EPTC, 2-methylmercapto-*s*-triazines (e.g. terbutryn) and 3-methylmercapto-1,2,4-triazine-5-ones (e.g. metribuzin, Fig. 9) are also substrates for plant glutathione *S*-transferases and species differences in capacity for their conjugation can contribute to selectivity. Detoxification of metribuzin (Fig. 9) in plants occurs by three competing pathways but a recent report (Frear, Swanson & Mansager, 1985) suggested that homoglutathione conjugation following sulphoxidation may be the most important route in soybean, though no comparative studies correlating metabolism with plant tolerance were recorded. Varying proportions of the sulphoxide, depending on species and soybean cultivar (see Cole *et al.*, 1987), may also be hydrolysed to the ketone (9.7) which can undergo *N*-acylation to the malonyl derivative (9.8) or deamination to 9.9. Metribuzin metabolism in the susceptible weed species *Sesbania exaltata* did not result in the appearance of significant radioactivity in the polar metabolite fraction, in contrast to soybean, suggesting that glutathione conjugation was of minor importance in this species (Hargroder & Rogers, 1974). Differential uptake and translocation were also considered to contribute to the difference in response.

Reductive deamination

Reductive reactions do not generally play a major role in pesticide detoxification by plant tissues. The most commonly encountered reaction of this type is the reduction of aromatic nitro groups to the corresponding anilines and the observed reduction of the NO_2 group of botran (2,6-dichlor-4-nitroaniline) by axenic callus cultures of soybean (quoted by Lamoureux & Frear, 1979) demonstrated the ability of plant tissue *per se* to mediate such biotransformations. Whereas such reactions apparently do not occur extensively in higher plants, the reductive deamination of metribuzin to 9.2 (Fig. 9) represents an important detoxification mechanism which has been considered to be the basis of the metribuzin-tolerance of some soybean cultivars (Fedtke & Schmidt, 1983). Thus deamination in leaves of tolerant 'Bragg' and 'Tracy M' was considerably more rapid than that observed in susceptible 'Tracy' and 'Coker 156'. The deaminase enzyme activity was apparently located in the peroxisome fraction and required a reduced flavin cofactor under anaerobic (N_2) conditions (Fedtke & Schmidt, 1983). It was possible to correlate rates of deaminase reaction with extent of photosynthetic CO_2 fixation in the various cultivars studied. A similar enzyme system has been implicated in the detoxification of the related triazinone metamitron in tolerant sugar beet (Fedtke & Schmidt, 1979). Rates of deamination of metamitron were correlated with increasing tolerance among sugar beet and five weed species. The reaction product was less inhibitory in the Hill reaction of isolated chloroplasts than was metamitron itself (Fedtke & Schmidt, 1979) indicating the importance of deamination in selectivity.

Glycosidation

Production of *O*-*β*-*D*-glucosides in plants almost invariably follows transformations of herbicides to hydroxylated derivatives though the extent of conjugation may vary between both species and plant tissues. In addition the ability of plants to synthesise the glucose esters of acidic herbicides has been referred to above (p. 173). Available evidence suggests that because of their inherent instability, formation of such esters, even when extensive, does not constitute an effective detoxification mechanism. In general, glycosidation may contribute to detoxification by virtue of the enhanced water solubility of the products which facilitates their disposal in the vacuole. However in the case of chlorsulfuron metabolism in tolerant plant species glycosidation resulted in the conversion of phytotoxic hydroxylated metabolites to inactive glycosides (p. 174) and in suspension cultures of soybean and wheat an *O*-glycosyl transferase was directly responsible for detoxification of pentachlorophenol (Schmitt *et al.*, 1985). Though less common, formation of *N*-glycosides also occurs in plants and direct *N*-glucosidation of chloramben represents a significant detoxification pathway in tolerant species (Stoller & Wax, 1968; Frear *et al.*, 1978). The *N*-glucosyl transferase which catalyses this reaction has been isolated, and partially characterised, from soybean (Frear, 1968*b*) and tomato (Frear *et al.*, 1983). In contrast, the rate of *N*-glucosidation was much reduced in susceptible species in which *O*-glucose ester formation was more prominent (Frear *et al.*, 1978). *N*-glycoside formation (9.3, Fig.9) also offers an additional metabolic route for detoxification of metribuzin. Smith & Wilkinson (1974) presented evidence implicating the importance of *N*-glucosidation in the tolerance of soybean cultivar 'Bragg' to metribuzin, whereas in susceptible 'Coker 102' and 'Semmes' the dione

(9.7) was the major metabolite identified. Intraspecies variation of metribuzin metabolism in soybean has also been recorded between tetraploid and diploid lines (Abusteit *et al.*, 1985). Metabolism was three to four-fold more rapid in tetraploid than in dipolid lines and the nature of the metabolites suggested that *N*-glucosidation was the major detoxification route employed. A parallel distinction in rates of metribuzin conjugation was obtained in suspension cultures of diploid and tetraploid soybean

Fig. 9. Metabolism of metribuzin in plants.

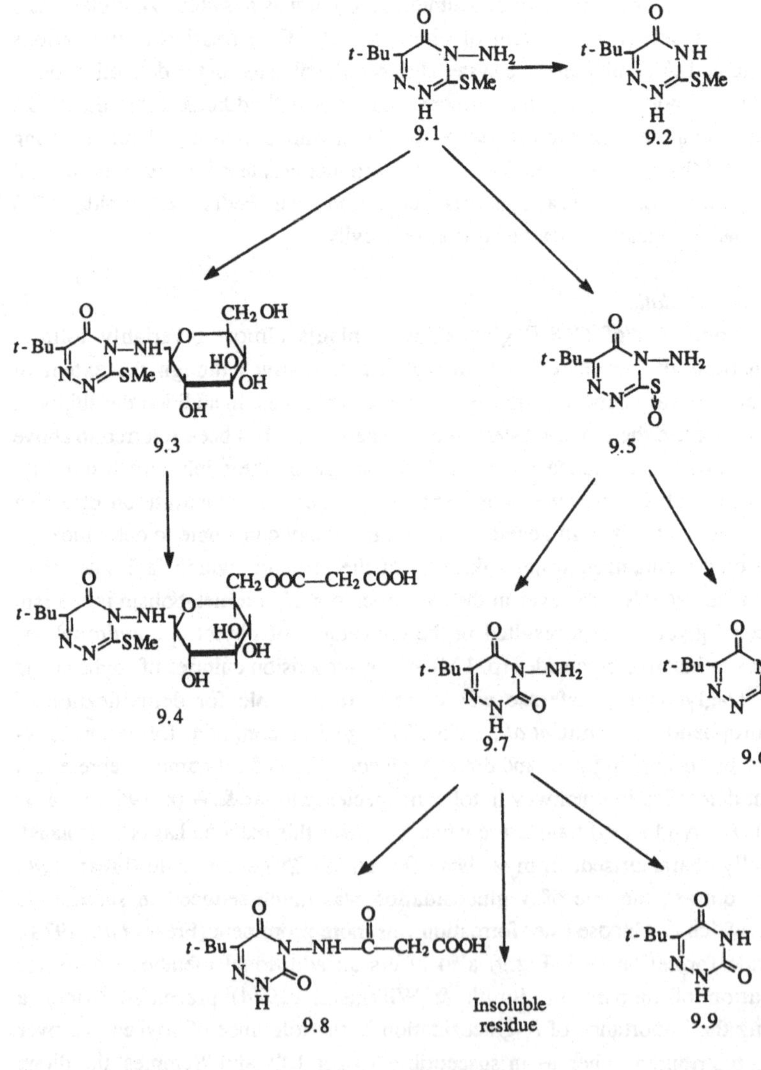

lines which illustrates the potential usefulness of plant cell cultures in herbicide metabolism and selectivity studies. In summary, there would appear to be little doubt that differential metabolism of metribuzin contributes to selectivity between soybeancultivars but tolerant cultivars may differ with respect to the importance of the three detoxification routes (Fig. 9).

As was the case with cysteine conjugates, glycosides need not represent terminal herbicide metabolites and are frequently subjected to further modification. Thus the *β-D*-glucosyl conjugate of pentachlorophenol was acylated to the *O*-(malonyl)-*β-D*-glucoside (Schmitt *et al.*, 1985) in suspension cultures of soybean and wheat as was the *N*-glucoside of metribuzin (9.4) in tomato (Frear *et al., 1983*).

Concluding Remarks

Differential metabolism is clearly important in determining selectivity of many herbicides both between species and crop varieties. However, it is important to remember that metabolism may be relevant to selectivity of a particular herbicide in certain crop/weed situations only. In other cases additional factors such as uptake and translocation may be as or more important. Plant cell cultures are proving a useful additional tool for metabolism studies and selective metabolism phenomena established in whole plants generally extrapolate to suspension cultures though results from this laboratory have indicated some exceptions. Finally, a thorough knowledge of the metabolic fate of pesticides in plants might lead to the development of new herbicides which exploit species differences in detoxification. The design of inactive herbicide progenitors which are activated by biotransformation might also result from a study of accumulated metabolism data.

References

Absteit, O., Corbin, F.J., Schmitt, D.P., Burton, J.W., Worsham, A.D. & Thompson, L. (1985). Absorption, translocation and metabolism of metribuzin in diploid and tetraploid soybean, *(Glycine max)* plants and cell cultures. *Weed Science* **33**, 618–28.

Akinyemiju, O.A., Dickmann, D.I. & Leavitt, R.A. (1983). Distribution and metabolism of simazine in simazine-tolerant and -intolerant poplar (*Populus sp.*) clones. *Weed Science*, **31**, 775–8.

Andrawes, N.R., Bagley, W.P. & Herrett, R.A. (1971). Metabolism of 2-methyl-2-(methylthio)propionaldehyde-*O*-(methylcarbamoyl)oxime (Temik, aldicarb pesticide) in potato plants. *Journal of Agricultural and Food Chemistry*, **19**, 731–7.

Arjmand, M., Hamilton, R.H. & Mumma, R.O. (1978). Metabolism of 2,4,5-trichlorophenoxyacetic acid. Evidence for amino acid conjugates in soybean callus tissue. *Journal of Agricultural and Food Chemistry*, **26**, 1125–28.

Beynon, K.I., Stoydin, G. & Wright, A.N. (1972a). The breakdown of the triazine herbicide cyanazine in soils and maize. *Pesticide Science*, **3**, 293–305.

Beynon, K.I., Stoydin, G. & Wright, A.N. (1972b). The breakdown of the triazine herbicide cyanazine in wheat and potatoes grown under indoor conditions in treated soils. *Pesticide Science*, **3**, 379–87.

Blair, L.C., Slife, F.W., Felsot, A. & Plewa, M.J. (1984). Rates of sulfide oxidation in cotton, carrot and tobacco cultured plant cells measured with a model aromatic alkyl-sulfide. *Pesticide Biochemistry and Physiology*, **21**, 291–300.

Blattmann , P., Gross, D., Kriemler, H.P. & Ramsteiner, K. (1986). Identification of thiolactic acid-type conjugates as major degradation products in the glutathione-dependent metabolism of the α-chloroacetamide herbicides metolachlor, dimetachlor and pretilachlor. Presented at the 6th IUPAC International Congress of Pesticide Chemistry, Ottawa, August 1986.

Boldt, P.F. & Putnam, A.R. (1981). Selectivity mechanisms for foliar applications of diclofop-methyl. II. Metabolism. *Weed Science*, **29**, 237–41.

Bowman, M.C., Beroza, M. & Harding, J.A. (1969). Determination of phorate and five of its metabolites in corn. *Journal of Agricultural and Food Chemistry* , **17**, 138–42.

Bristol, D.W., Ghanuni, A.M. & Oleson, A.E. (1977). Metabolism of 2,4-Dichlorophenoxyacetic acid by wheat cell suspension cultures. *Journal of Agricultural and Food Chemistry*, **25**, 1309–14.

Brown, M.A., Chiu, T.Y. & Miller, P. (1987). Hydrolytic activation versus oxidative degradation of assert herbicide, an imidazolinone aryl-carboxylate, in susceptible wild oat versus tolerant corn and wheat. *Pesticide Biochemistry and Physiology*, **27**, 24–9.

Buckland, J.L., Collins, R.F. & Pullin, E.M. (1973). Metabolism of bromoxynil octanoate in growing wheat. *Pesticide Science*, **4**, 149–62.

Burkholder, R.R.S. (1978). Plant glutathione S-transferase. M.S. Thesis, North Dakota State University, Fargo, North Dakota.

Cabanne, F., Gaillardon, P. & Scalla, R. (1985*a*). Phytotoxicity and metabolism of chlortoluron in two wheat varieties. *Pesticide Biochemistry and Physiology* , **23**, 212–20.

Cabanne, F., Gaillardon, P., Scalla, R. & Durst, F. (1985*b*). Aminobenzotriazole as a synergist of urea herbicides. *Proceedings of the 1985 British Crop Protection Conference – Weeds*, **3**, 1163–70.

Casida, J.E., Gray, R.A. & Tilles, H. (1974). Thiocarbamate sulfoxides: potent, selective and biodegradable herbicides. *Science*, **184**, 573–4.

Chkanikov, D.I., Makeev, A.M., Pavlova, N.N. & Dubovoi, V.P. (1972). *N*-(2,4-dichlorophenoxyacetyl)-L-glutamic acid: A new metabolite of 2,4-D. *Soviet Plant Physiology* **19**, 364–9.

Chkanikov, D.I., Pavlova, N.N., Makeev, A.M., Nazarova, T.A. & Makoveichuk, A.Yu. (1977). Paths of detoxification and immobilization of 2,4-D in cucumber plants. *Soviet Plant Physiology*, **24**, 457–63.

Cole, D.J. (1983). Oxidation of xenobiotics in plants. In *Progress in Pesticide Biochemistry and Toxicology*, vol. 3, ed. D.H. Hutson & T.R. Roberts, pp. 199–254. Chichester: Wiley.

Cole, D.J., Edwards, R. & Owen, W.J. (1987). The role of metabolism in herbicide selectivity. In *Progress in Pesticide Biochemistry and Toxicology*, vol. 6, ed. D. Hutson & T.R. Roberts, pp. 57–104. Chichester: Wiley.

Cole, D.J. & Loughman, B.C. (1983). The metabolic fate of (4-chloro-2-methylphenoxy) acetic acid in higher plants I. General results. *Journal of Experimental Botany*, **34**, 1299–310.

Cole, D.J. & Loughman, B.C. (1984). Transformation of phenoxyacetic acid herbicides by *Phaseolus vulgaris L.* callus cells. *Plant Cell Reports*, **3**, 5–7.

Cole, D.J. & Loughman, B.C. (1985). Factors affecting the hydroxylation and glycosylation of (4-chloro-2-methylphenoxy) acetic acid in *Solanum tuberosum* tuber tissue. *Physiologie Vegetale*, **23**, 879–86.

Cole, D.J. & Owen, W.J. (1987). Influence of monoxygenase inhibitors on the metabolism of the herbicides chlortoluron and metolachlor in cell suspension cultures. *Plant Science*, **50**, 13–20.

Diesperger, H. & Sandermann, H. (1979). Soluble and microsomal glutathione *S*-transferase activities in pea seedlings (*Pisum sativum* L.). *Planta*, **146**, 643–8.

Dohn, D.R. & Krieger, R.L. (1984). *N*-demethylation of *p*-chloro-*N*-methylaniline catalysed by subcellular fractions from the avocado pear (*Persea americana*). *Archives of Biochemistry and Biophysics*, **231**, 416–23.

Dusky, J.A., Davis, D.G. & Shimabukuro, R.H. (1980). Metabolism of diclofopmethyl in cell suspensions of diploid wheat (*Triticum monococcum* L.) *Physiologia Plantarum*, **49**, 151–6.

Dusky, J.A., Davis, D.G. & Shimabukuro, R.H. (1982). Metabolism of diclofopmethyl in cell cultures of *Avena sativa* and *Avena fatua*. *Physiologia Plantarum*, **54**, 490–4.

Earl, J.W. & Kennedy, I.R. (1975). Aldrin epoxidase from pear roots. *Phytochemistry*, **14**, 1507–12.

Edwards, R. & Owen, W.J. (1986). Comparison of glutathione *S*-transferases of *Zea mays* responsible for herbicide detoxification in plants and suspension-cultured cells. *Planta*, **169**, 208–15.

Edwards, R. & Owen, W.J. (1987). Isoenzymes of glutathione *S*-transferase in *Zea mays*. *Biochemical Society Transactions*, **15**, 1184.

Emami-Saravi, R. (1979). Studies on the varietal susceptibility in winter wheat to substituted phenylurea herbicides. PhD. Thesis, University of London.

Evans, J.D.H.L., Cavell, B.D. & Hignett, R.R. (1987). Fomesafen-Metabolism as a basis for its selectivity in soya. *Proceedings of the 1987 British Crop Protection Conference - Weeds*, **1**, 345–52.

Ezra, G. & Stephenson, G.R. (1985). Comparative metabolism of atrazine and EPTC in proso millet *(Panicum miliaceum* L.) and corn. *Pesticide Biochemistry and Physiology*, **24**, 207–12.

Fedtke, C. & Schmidt, R.R. (1979). Characterization of the metamitron deaminating enzyme activity from sugar beet *(Beta vulgaris* L.) leaves. *Zeitschrift fur Naturforschung*, **34C**, 948–50.

Fedtke, C., & Schmidt, R.R. (1983). Behaviour of metribuzin in tolerant and susceptible soybean varieties. In: *Pesticide Chemistry: Human Welfare and the Environment* (ed.-in-Chief J. Miyamoto & P.C. Kearney) vol.3 *Mode of Action Metabolism and Toxicology*, eds. S. Matsunaka, D.H. Hutson and S.D. Murphy, pp. 177–82 Oxford: Pergamon.

Feung, C-S., Hamilton, R.H. & Mumma, R.O. (1973). Metabolism of 2,4-dichlorophenoxyacetic acid V. Identification of metabolites in soybean callus tissue cultures. *Journal of Agricultural and Food Chemistry*, **21**, 637–40.

Feung, C-S., Hamilton, R.H. & Mumma, R.O. (1975). Metabolism of 2,4-dichlorophenoxyacetic acid VII. Comparison of metabolites from five species of plant callus tissue cultures. *Journal of Agricultural and Food Chemistry*, **23**, 373–6.

Feung, C-S., Hamilton, R.H. & Mumma, R.O. (1976). Metabolism of 2,4-dichlorophenoxyacetic acid X. Identification of metabolites in root callus tissues. *Journal of Agricultural and Food Chemistry*, **24**, 1013–5.

Feung, C-S., Hamilton, R.H. & Mumma, R.O. (1977). Metabolism of 2,4-dichlorophenoxyacetic acid XI. Herbicidal properties of amino acid conjugates. *Journal of Agricultural and Food Chemistry*, **25**, 898–900.

Feung, C-S. Hamilton, R.H. Witham, F.H. (1971). Metabolism of 2,4-dichlorophenoxyacetic acid by soybean cotyledon callus culture tissues. *Journal of Agricultural and Food Chemistry*, **19**, 475–9.

Feung, C.-S., Loerch, S.L., Hamilton, R.H. & Mumma, R.O. (1978). Comparative metabolic fate of 2,4-dichlorophenoxyacetic acid in plants and plant tissue culture. *Journal of Agricultural and Food Chemistry*, **26**, 1064–7.

Feung, C-S., Mumma, R.O. & Hamilton, R.H. (1974). Metabolism of 2,4-dichlorophenoxyacetic acid VI. Biological properties of amino acid conjugates. *Journal of Agricultural and Food Chemistry*, **22**, 307–9.

Frear, D.S. (1968*a*). Microsomal *N*-demethylation by a cotton leaf oxidase system, of 3-(4-chlorophenyl)-1,1-dimethylurea (monuron). *Science*, **162**, 674–5.

Frear, D.S. (1968*b*) Herbicide metabolism in plants, I. Purification and properties of UDP-glucose : arylamine *N*-glucosyl transferase from soybean. *Phytochemistry*, **7**, 381–90.

Frear, D.S., Mansager, E.R., Swanson, H.R. & Tanaka, F.S. (1983). Metribuzin metabolism in tomato: isolation and identification of *N*-glucoside conjugates. *Pesticide Biochemistry and Physiology*, **19**, 270–81.

Frear, D.S. & Still, G.G. (1968). The metabolism of 3,4-dichloropropionanilide in plants. Partial purification and properties of arylacylamidase from rice. *Phytochemistry*, **7**, 913–20.

Frear, D.S. & Swanson, H.R. (1970). Biosynthesis of *S*-(4-ethylamino-6-isopropylamino-2-*s*-triazino)glutathione: Partial purification and properties of a glutathione *S*-transferase from corn. *Phytochemistry*, **9**, 2123–32.

Frear, D.S. & Swanson, H.R. (1973). Metabolism of substituted diphenyl ether herbicides in plants I. Enzymic cleavage of fluorodifen in peas (*Pisum sativum*). *Pesticide Biochemistry and Physiology*, **3**, 473–82.

Frear, D.S., Swanson, H.R. & Mansager, E.R. (1983). Acifluorfen metabolism in soybean: diphenyl ether bond cleavage and the formation of homo-glutathione, cysteine and glucose conjugates. *Pesticide Biochemistry and Physiology*, **20**, 299–310.

Frear, D.S., Swanson, H.R. & Mansager, E.R. (1985). Alternate pathways of metribuzin metabolism in soybean; formation of *N*-glucoside and homo-glutathione conjugates. *Pesticide Biochemistry and Physiology*, **23**, 56–65.

Frear, D.S., Swanson, H.R. Mansager, E.R. & Wien, R.G. (1978). Chloramben metabolism in plants: isolation and identification of the glucose ester. *Journal of Agricultural and Food Chemistry*, **26**, 1347–51.

Frear, D.S., Swanson, H.R. & Tanaka, F.S. (1969). *N*-demethylation of substituted 3-(phenyl)-l-methylureas: isolation and characterization of a microsomal mixed function oxidase from cotton. *Phytochemistry*, **8**, 2157–69.

Frear, D.S., Swanson, H.R. & Tanaka, F.S. (1972). Herbicide metabolism in plants. In *Recent Advances in Phytochemistry*, vol. 5, ed. V.C. Runeckles & T.C. Tso, pp. 225–46. London, New York: Academic Press.

Fuerst, E.P. & Gronwald, J.W. (1986). Induction of rapid metabolism of metolachlor in sorghum *(Sorghum bicolor)* shoots by CGA-92194 and other antidotes. *Weed Science*, **34**, 354–61.

Gaillardon, P., Cabanne, F., Scalla, R. & Durst, F. (1985). Effect of mixed function oxidase inhibitors on the toxicity of chlortoluron and isoproturon to wheat. *Weed Research*, **25**, 397–402.

Geissbuhler, H. (1969). The substituted ureas. In *Degradation of Herbicides*, ed. P.C. Kearney & D.D. Kaufman, pp. 79–111. New York: Marcel Dekker, Inc.

Gorbach, S.G., Kuenzler, K. & Asshauer, J. (1977). On the metabolism of Hoe 23408 OH in wheat. *Journal of Agricultural and Food Chemistry*, **25**, 507–11.

Gorecka, K., Shimabukuro, R.H. & Walsh, W.C. (1981). Aryl hydroxylation: a selective mechanism for the herbicides diclofop-methyl and clofop-isobutyl in gramineous weeds. *Physiologia Plantarum* , **53**, 55–63.

Graebe, J.C. (1982). Giberellin biosynthesis in cell-free systems from higher plants. In *Plant Growth Substances*, ed. P.F. Wareing, pp. 71–80. London: Academic Press.

Gressel, J. (1985). Herbicide tolerance and resistance: alteration of site of activity. In *Weed Physiology Vol. II. Herbicide Physiology*, ed. S.O. Duke, pp. 159–89. Boca Raton: CRC Press.

Gressel, J., Shimabukuro, R.H. & Duysen, M.E. (1983). *N*-dealkylation of atrazine and simazine in *Senecio vulgaris* biotypes: A major degradation pathway. *Pesticide Biochemistry and Physiology*, **19**, 361–70.

Grossman, K., Rademacher, W., & Jung, J. (1983). Effects of NDA, a new plant growth retardant, on cell culture growth of *Zea mays*. *Journal of Plant Growth Regulation*, **2**, 19–29.

Guroff, G., Daly, J.W., Jerina, D.M., Renson, J., Witkop, B. & Udenfriend, S. (1967). Hydroxylation-induced migration: The NIH shift. *Science*,**157**, 1524–30.

Hamilton, R.H. (1964). Tolerance of several grass species to 2-chloro-*s*-triazine herbicides in relation to degradation and content of benzoxazinone derivatives. *Journal of Agricultural and Food Chemistry*, **12**, 14–7.

Hamilton, R.H. Hurter, J., Hall, J.K. & Ercegovich, C.D. (1971). Metabolism of phenoxyacetic acids: metabolism of 2,4-D and 2,4,5-T by bean plants. *Journal of Agricultural and Food Chemistry*, **19**, 480–3.

Hamilton, R.H. & Moreland, D.E. (1962). Simazine: degradation by corn seedlings. *Science*, **135**, 373–4.

Hargroder, T.G. & Rogers, R.L. (1974). Behaviour and fate of metribuzin in soybean and hemp sesbania. *Weed Science*, **22**, 238–45.

Hassall, K.A. (1982). *The Chemistry of Pesticides: Their Metabolism, Mode of Action and Uses in Crop Protection.* London: The MacMillan Press.

Hathway, D.E. (1986). Herbicide selectivity. *Biological Reviews*, **61**, 435–86.

Hawton, D. & Strobbe, E.H. (1971). Fate of nitrofen in rape, redroot pigweed and green foxtail. *Weed Science*,**19**, 555–8.

Hayes, R.M. & Wax, L.M. (1975). Differential intraspecific responses of soybean cultivars to bentazon. *Weed Science*, **23**, 516–21.

Hess, F.D. (1985). Herbicide absorption and translocation and their relationship to plant tolerances and susceptibility. In *Weed Physiology Vol. 2. Herbicide Physiology*, ed. S.O. Duke. pp. 191–214. Boca Raton: CRC Press.

Hoagland, R.E. (1975). The hydrolysis of 3', 4'-dichloropropionanilide by an aryl acylamidase from *Taraxacum officinale*. *Phytochemistry*, **14**, 383–6.

Hoagland, R.E. (1978). Isolation and some properties of aryl acylamidase from red rice (*Oryza sativa* L.) that metabolizes 3,4-dichloropropionanilide (Propanil). *Plant Cell Physiology*, ***19***, 1019–27.

Hoagland, R.E. & Graf, G. (1972). An aryl acylamidase from tulip, which hydrolyses 3,4-dichloropropionanilide. *Phytochemistry*, **11**, 521–7.

Hoagland, R.E., Graf, G. & Handel, E.D. (1974). Hydrolysis of 3,4-dichloropropionanilide by plant aryl acylamidases. *Weed Research*, **14**, 371–4.

Horvath, L. & Pulay, A. (1980). Metabolism of EPTC in germinating corn: Sulfone as the true carbamoylating agent. *Pesticide Biochemistry and Physiology*, **14**, 265–70.

Hutchison, J.M., Shapiro, R. & Sweetser, P.B. (1984). Metabolism of chlorsulfuron by tolerant broadleaves. *Pesticide Biochemistry and Physiology*, **22**, 243–7.

Jacobson, A., Shimabukuro, R.H. & McMichael, C. (1985). Response of wheat and oat seedlings to root-applied diclofop-methyl and 2,4-D. *Pesticide Biochemistry and Physiology*, **24**, 61–7.

Jensen, K.I.N. (1982). The roles of uptake, translocation and metabolism in the differential intraspecific responses to herbicides. In *Herbicide Resistance in Plants*, ed. H.M. LeBaron & J. Gressel, pp. 133–62. New York: J. Wiley & Sons.

Jensen, K.I.N., Bandeen, J.D. & Machado, V.S. (1977). Studies on the differential tolerance of two lamb's-quarters selections to triazine herbicides. *Canadian Journal of Plant Science*, **57**, 1169–77.

Jensen, K.I.N., Stephenson, G.R. & Hunt, L.A. (1977). Detoxification of atrazine in three Graminae Subfamilies. *Weed Science*, **25**, 212–20.

Khan, S.U., Warwick, S.I. & Marriage, P.B. (1985). Atrazine metabolism in resistant and susceptible biotypes of *Chenopodium album* L., *Chenopodium strictum* Roth., and *Amaranthus powellii* S. Wats. *Weed Research*, **25**, 33–7.

Krell, H-W. & Sandermann, H. (1986). Metabolism of the persistent plasticizer chemical bis(2-ethylhexyl)phthalate in cell suspension cultures of wheat *(Triticum aestivum.* L.): Discrepancy from the intact plant. *Journal of Agricultural and Food Chemistry*, **34**, 194–8.

Lamoureux, G.L. & Frear, D.S. (1979). Pesticide metabolism in higher plants: *in vitro* enzyme studies. In *Xenobiotic metabolism – in vitro methods*, American Chemical Society Symposium Series, Vol. 97, ed. G. D. Paulson, D.S. Frear & E.P. Marks, pp. 72–128. Washington D.C.: American Chemical Society.

Lamoureux, G.L., Gouot, J-H., Davis, D.G. & Rusness, D.G. (1981). Pentachloronitrobenzene metabolism in peanut 3. Metabolism in peanut cell suspension cultures. *Journal of Agricultural and Food Chemistry*, **29**, 996–1002.

Lamoureux, G.L. & Rusness, D.G. (1980). *In vitro* metabolism of pentachloronitrobenzene to pentachloromethylthiobenzene by onion: characterisation of glutathione *S*-transferase, cysteine C-S lyase and *S*-adenosylmethionine methyl transferase activities. *Pesticide Biochemistry and Physiology*, **14**, 50–61.

Lamoureux, G.L. & Rusness, D.G. (1983). Malonylcysteine conjugates as end-products of glutathione conjugate metabolism in plants. *Proceedings of the 5th IUPAC International Congress on Pesticide Chemistry 1982*, ed. J. Miyamoto & P.C. Kearney, Vol. 3, pp. 295–300. Oxford: Pergamon.

Lamoureux, G.L., Stafford, L.E., Shimabukuro, R.H. & Zaylskie, R.G. (1973). Atrazine metabolism in sorghum: catabolism of the glutathione conjugate of atrazine. *Journal of Agricultural and Food Chemistry*, **21**, 1020–30.

Lamoureux, G.L., Stafford, L.E. & Tanaka, F.S. (1971). Metabolism of 2-Chloro-*N*-Isopropylacetanilide (Propachlor) in the leaves of corn, sorghum, sugarcane and barley. *Journal of Agricultural and Food Chemistry*, **19**, 346–50.

Lewer, P. & Owen, W.J. (1987). Triclopyr: An investigation of the basis for its species selectivity. *Proceedings of the 1987 British Crop Protection Conference – Weeds*, **1**, 353–60.

Luckwill, L.C. & Lloyd-Jones, C.P. (1960). Metabolism of plant growth regulators I. 2,4-dichlorophenoxyacetic acid in leaves of red and blackcurrant. *Annals of Applied Biology*, **48**, 613–25.

Mahoney, M.D. & Penner, D. (1975*a*). Bentazon translocation and metabolism in soybean and navy bean. *Weed Science*, **23**, 265–71.

Mahoney, M.D. & Penner, D. (1975*b*). The basis for bentazon selectivity in navy bean, cocklebur and black nightshade. *Weed Science*, **23**, 272–6.

Makeev, A.M., Makoveichuk, A. Yu & Chkanikov, D.I. (1977). Microsomal hydroxylation of 2,4-D in plants. *Doklady Botanical Science*, **223**, 36–8.

Matsumoto, H. & Ishizuka, K. (1980). Herbicidal selectivity of foliar-applied simetryne: Its absorption, translocation and metabolism in gramineous plants. *Weed Research (Jap.)*, **25**, 185–93.

McPherson, F.J., Markham, A., Bridges, J.W., Hartman, G.C. & Parke, D.V. (1975*a*). A comparison of the properties *in vitro* of biphenyl 2- and 4-hydroxylase in the mesocarp from Avocado Pear (*Persea americana*) and Syrian hamster hepatic tissue. *Biochemical Society Transactions*, **3**, 281–3.

McPherson, F.J., Markham, A., Bridges, J.W., Hartman, G.C. & Parke, D.V. (1975*b*). Effects of pre-incubation *in vitro* with 3,4-benzopyrene and phenobarbitone on the drug metabolising systems of the microsomal and soluble fractions of the Avocado Pear (*Persea americana*). *Biochemical Society Transactions*, **3**, 283–5.

Metcalf, R.L., Fukuto, T.R., Collins, C., Borck, K., Burk, H.T., Reynolds, H.T. & Osman, M.F. (1966). Metabolism of 2-methyl-2-(methylthio)propionaldehyde-*O*-(methylcarbamoyl)-oxime in plant and insect. *Journal of Agricultural and Food Chemistry*, **14**, 579–84.

Mine, A., Miyakado, M. & Matsunaka, S. (1975). The mechanism of bentazon selectivity. *Pesticide Biochemistry & Physiology*, **5**, 566–74.

Moore, R.E., Davies, M.S., O'Connell, K.M., Harding, E.I., Wiegand, R.C. & Tiemeier, D.C. (1986). Cloning and expression of a cDNA encoding a maize glutathione *S*-transferase in *E.coli*. *Nucleic Acids Research*, **14**, 7227–35.

Mozer, T.J., Tiemeier, D.C. & Jaworski, E.G. (1983). Purification and characterisation of corn glutathione *S*-transferase. *Biochemistry*, **22**, 1068–72.

Muller, F., Kang, B.H. & Maruska, F.T. (1984). Fate of chlorsulfuron in cultivated plants and weeds and reasons for selectivity. *Mededelingen-Faculteit Landbouwwetenschappen Rijksuniversiteit Gent*, *49/36*, 1091–8.

Muller, P.W. & Payot, P.H. (1966). Fate of ^{14}C-labelled triazine in plants. In *Isotopes in Weed Research*. Vienna: International Atomic Energy Agency.

Nagatsugawa, T. & Morelli, M.A. (1976). Microsomal oxidation and insecticide metabolism. In *Insect Biochemistry and Physiology*, ed. C.F. Wilkinson, pp. 61–114. New York: Plenum.

Otto, S., Beutel, P., Drescher, N. & Huber, R. (1979). Investigation into the degradation of bentazon in plant and soil. In *IUPAC: Advances in Pesticide Science*, part 3, ed. H. Geissbuhler, pp. 551–6. Oxford: Pergamon Press.

Oyamada, M., Tanaka, T., Takasawa, Y. & Takematsu, T. (1986). Metabolic fate of the herbicide naproanilide in rice plants (*Oryza sativa* L.) and *Sagittaria pygmaea* Miq. *Journal of Pesticide Science*, **11**, 197–203.

Penner, D. (1975) Bentazone selectivity between soybean and Canada thistle. *Weed Research*, **15**, 259–62.

Poulsen, L.L., Hyslop, R.M. & Ziegler, D.M. (1979). *S*-oxygenation of *N*-substituted thioureas catalysed by the pig liver microsomal FAD containing monoxygenase. *Archives of Biochemistry and Biophysics*, **198**, 78–88.

Ray, T.B. (1985). The site of action of sulfonylurea herbicides. *Proceedings of the 1985 British Crop Protection Conference – Weeds*, **1**, 131–8.

Retzlaff, G. & Hamm, R. (1976). The relationship between CO_2 assimilation and the metabolism of bentazone in wheat plants. *Weed Research*, **16**, 263–6.

Roth, W. & Knüsli, E. (1961). Beitrag Zur Kenntnis der Resistenzphänomene einzelner Pflanzen gegenüber den phyto-toxischen Wirkstoff Simazin. *Experientia*, **17**, 312–3.

Rusness, D.G. & Lamoureux, G.L. (1980). Pentachloronitrobenzene metabolism in peanut 2. Characterization of chloroform-soluble metabolites *in vivo*. *Journal of Agricultural and Food Chemistry*, **28**, 1070–7.

Ryan, P.J., Gross, D., Owen, W.J. & Laanio, T.L. (1981). The metabolism of chlortoluron, diuron and CGA43057 in tolerant and susceptible plants. *Pesticide Biochemistry and Physiology*, **16**, 213–21.

Ryan, P.J. & Owen, W.J. (1982). The mechanism of selectivity of chlortoluron between cereals and grass weeds. *Proceedings of the 1982 British Crop Protection Conference – Weeds*, **1**, 317–24.

Ryan, P.J. & Owen, W.J. (1983). Metabolism of chlortuluron in tolerant and sensitive cereal varieties. *Aspects of Applied Biology*, **3**, 63–72.

Sandmann, G. & Bramley, P.M. (1985). The *in vitro* biosynthesis of b-cryptoxanthin and related xanthophylls with *Aphanocapsa* membranes. *Biochimica et Biophysica Acta*, **843**, 73–7.

Scheel, D. & Sandermann, H. (1981). Metabolism of 2,4-dichlorophenoxyacetic acid in cell suspension cultures of soybean (*Triticum aestivum* L.) I. General Results. *Planta*, **152**, 248–52.

Schmitt, R., Kaul, J., V.D. Trenck, T., Schaller, E. & Sandermann, H. (1985). b-D-Glucosyl and *O*-Malonyl-b-D-glucosyl conjugates of pentachlorophenol in soybean and wheat: Identification and enzymatic synthesis. *Pesticide Biochemistry and Physiology*, **24**, 77–85.

Shimabukuro, R.H. (1985). Detoxication of herbicides. In *Weed Physiology Vol 2. Herbicide Physiology*, ed. S.O. Duke, pp. 215–40. Boca Raton: CRC Press.

Shimabukuro, R.H., Frear, D.S., Swanson, H.R. & Walsh, W.C. (1971). Glutathione conjugation: An enzymic basis for atrazine resistance in corn. *Plant Physiology*, **47**, 10–14.

Shimabukuro, R.H., Lamoureux, G.L. & Frear, D.S. (1982). Pesticide metabolism in plants – reactions and mechanisms. In *Biodegradation of Pesticides*, ed. F. Matsumura & C.R. Krishna Murti, pp. 21–66. New York: Plenum.

Shimbukuro, R.H., Lamoureux, G.L., Swanson, H.R., Walsh, W.C., Stafford, L.E. & Frear, D.S. (1973). Metabolism of substituted diphenyl ether herbicides in plants III. Identification of a new fluorodifen metabolite, *S* (2-nitro-4-trifluoromethylphenyl)-glutathione in peanut. *Pesticide Biochemistry & Physiology*, **3**, 483–94.

Shimabukuro, R.H. Walsh, W.C. & Hoerauf, R.A. (1979). Metabolism and selectivity of diclofop-methyl in wild oat and wheat. *Journal of Agricultural and Food Chemistry*, **27**, 615–23.

Smith, A.E. & Wilkinson, R.E. (1974). Differential absorption, translocation and metabolism of metribuzin [4-amino-6-*tert*-butyl-3-(methylthio)-*as*-triazine-5(4H)-one] by soybean cultivars. *Physiologia Plantarum*, **32**, 253–7.

Steinback, K.E., Pfister, K. & Arntzen, C.J. (1982). Identification of the receptor site for triazine herbicides in chloroplast thylakoid membranes. In *Biochemical Responses Induced by Herbicides*, American Chemical Society Symposium Series, Vol 181, ed. D.E. Moreland, J.B. St. John & F.D. Hess, pp. 37–55. Washington D.C.: American Chemical Society.

Stoller, E.W. & Wax, L.M. (1968). Amiben metabolism and selectivity. *Weed Science*, **16**, 283–8.

Sweetser, P.B. (1985). Safening of sulphonylurea herbicides to cereal crops: Mode of herbicide antidote action. *Proceedings of the 1985 British Crop Protection Conference – Weeds*, **3**, 1147–54.

Sweetser, P.B., Schow, G.S. & Hutchison, J.M. (1982). Metabolism of chlorsulfuron by plants: biological basis for selectivity of a new herbicide for cereals. *Pesticide Biochemistry and Physiology*, **17**, 18–23.

Wax, L.M., Bernard, R.L. & Hayes, R.M. (1974). Response of soybean cultivars to bentazon, bromoxynil, chloroxuron and 2,4-DB. *Weed Science*, **22**, 35–41.

Young, O. & Beevers, H. (1976). Mixed function oxidases from germinating castor bean endosperm. *Phytochemistry*, **15**, 379–85.

M. W. KERR

Bioactivated Herbicides

Introduction

Most herbicides on entering the target plant undergo some metabolic transformation. This generally results in a loss of biological activity but there are a few instances where the parent molecule becomes activated in the plant.

Although most proherbicides or those which require bioactivation were almost certainly discovered by chance, it may be interesting to consider whether this feature has any significance or is just a scientific curiosity. Possible advantages may be summarised:

1. the proherbicide may have superior physical properties from the point of view of penetration or stability.
2. the bioactivation mechanism may alter the selectivity of the compound in a useful manner.
3. the reactivity of compounds may be toned down by the presence of protecting groups, allowing better distribution, especially if the activating mechanism is located near the site of herbicidal action. An extreme case of this might be a suicide inhibitor of an enzyme, although no herbicides have been reported to have this mechanism.
4. delayed release of the active herbicide may be a desirable feature allowing better distribution and a more prolonged effect.

These are some of the possible advantages of proherbicides, but the knowledge that bioactivation is occurring is essential if the mode of action and structure:activity relationships of an herbicide are to be clearly understood. For example, one explanation for the situation with the imidazolinones where activity against the target enzyme is up to a thousand fold less than that against intact cells could involve some form of bioactivation.

Bioactivated herbicides

a) Ester hydrolysis

It is widely assumed that esters of carboxylic acids will be readily hydrolysed in a plant, to the (presumably) active acid. The esterification should improve stability and penetration. In the case of a series of wild oat herbicides synthesised at Shell Research, Sittingbourne, such as SUFFIX (benzoylprop-ethyl), and MATAVEN (flamprop methyl), their remarkable selectivity is the result of differences in the rate of hydrolysis and subsequent inactivation by sugar conjugation (Fig. 1) (Jeffcoat & Harries, 1973) (Jeffcoat, Harries & Thomas, 1977). A similar

Fig. 1.

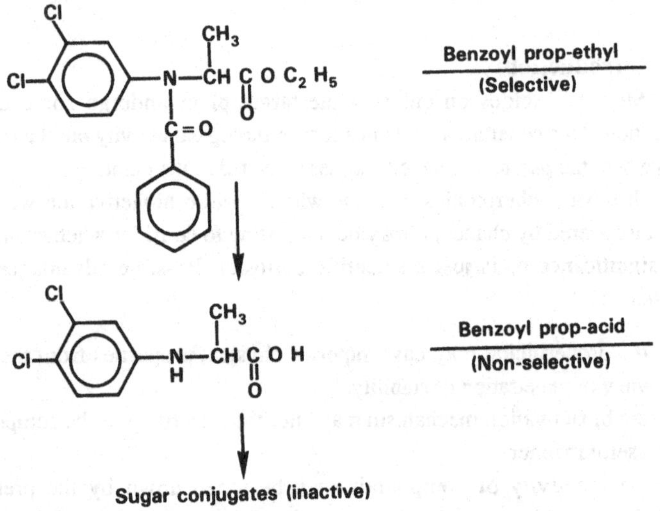

Benzoyl prop-ethyl (Selective)

Benzoyl prop-acid (Non-selective)

Sugar conjugates (inactive)

Fig. 2

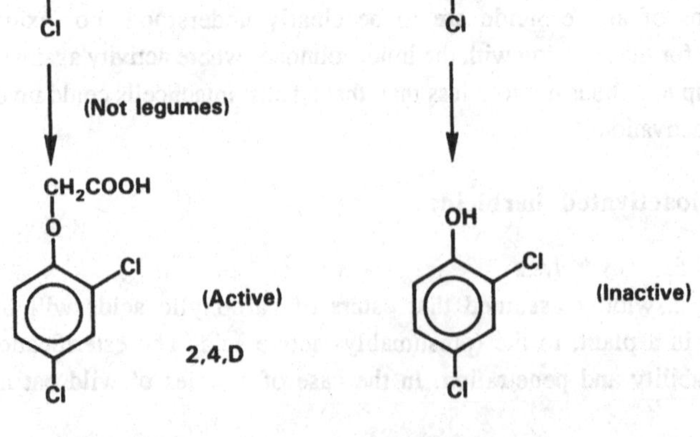

(Inactive) 2,4,DB

(Inactive)

(Not legumes)

(Active) 2,4,D

(Inactive)

situation obtains in the case of the new imidazolinone carboxylate ASSERT (Brown, Chiu & Miller, 1987) which also kills wild oats in cereal crops. In both cases the free acid is herbicidal but non-selective. The grass killer, diclofop-methyl, requires activation by ester hydrolysis but subsequent metabolism may be by two mechanisms. Aryl hydroxylation followed by irreversible conjugation occurs in resistant species but in susceptible plants a reversible ester conjugate is formed (Shimabukuro, Walsh & Hoerauf, 1979). An example of a biosynthesised proherbicide? The antagonism of dichlofop-methyl by auxin herbicides has been explained by its delayed release and slower transport. This allows 2,4-D for example, to reach the target tissue more rapidly and induce detoxifying conjugation enzymes (Taylor & Loader, 1986). In this case the effect is deleterious, but one could imagine the situation where, for instance, a safener was being used and it would be advantageous to delay the arrival of the toxicant at the target tissue to allow the safener time to induce the resistance mechanism.

b) β-Oxidation

One of the earliest examples of bioactivation was that of the longer chain analogues of the auxin herbicides 2,4-D and MCPA. It was shown (Wain, 1955) that 2,4-dichlorophenoxybutyric acid (2,4-DB) is converted to the active, 2,4-D in certain species (Fig. 2) by the β-oxidation mechanism which normally degrades fatty acids. Longer chain derivatives with even numbers of carbon atoms would also be converted to 2,4-D but those with odd numbers are converted to the inactive phenol. This mechanism introduces additional selectivity since 2,4-DB is not activated in legumes and so may be used safely in leguminous crops. Similarly, MCPB is activated in annual nettle (*Urtica urens*) and creeping thistle (*Cirsium arvense*) but not in clover (*Trifolium repens*) or celery (*Apium graveolens*).

The β-oxidation mechanism is also effective on longer chain analogues of indole acetic acid but does not appear to have been utilised in any subsequent herbicide design.

c) S-Oxidation

The thiocarbamate herbicide EPTC is subject to S-oxidation to the sulphoxide (Fig. 3) and the sulphone (Casida, Gray & Tilles, 1974) by a P450 oxidase. When applied to the roots of germinating oat seeds, the sulphoxide was more phytotoxic than EPTC which was more active than the sulphone. The sulphoxide was reported to be chemically unstable, however, and so was unsuitable as an herbicide. Additionally, detoxification of EPTC is by glutathione conjugation of this sulphoxide. Since the rate of conjugation is known to vary between species, the activity of the compound may be determined to a large extent by the balance between activation and detoxification. Some doubt was cast on these results (Dutka & Kömives, 1982) by later work involving direct injection of EPTC and its oxidation products into maize stems. This gave precisely the reverse order of activity sulphone > EPTC >

Fig. 3

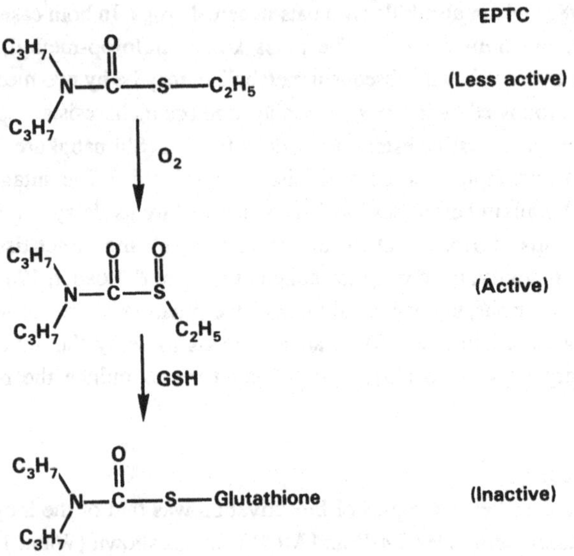

EPTC

$(C_3H_7)_2N{-}C(=O){-}S{-}C_2H_5$ (Less active)

↓ O_2

$(C_3H_7)_2N{-}C(=O){-}S(=O){-}C_2H_5$ (Active)

↓ GSH

$(C_3H_7)_2N{-}C(=O){-}S{-}$Glutathione (Inactive)

Fig. 4.

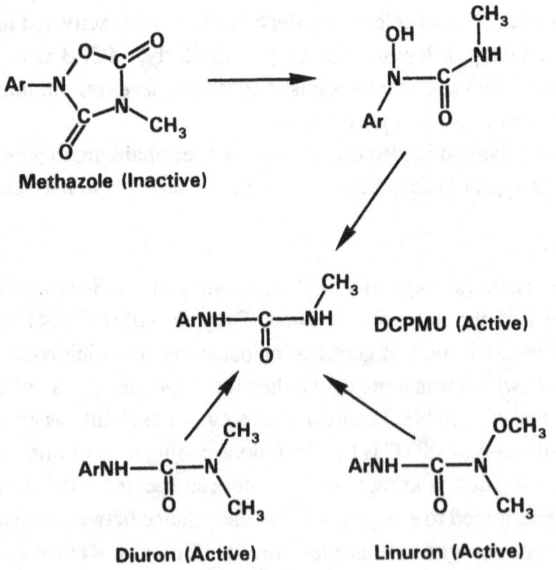

Methazole (Inactive) → Ar–N(OH)–C(=O)–NH–CH_3 → ArNH–C(=O)–NH–CH_3 DCPMU (Active)

ArNH–C(=O)–N(CH_3)$_2$ Diuron (Active) → DCPMU

ArNH–C(=O)–N(OCH_3)(CH_3) Linuron (Active) → DCPMU

sulphoxide. This highlights the difficulty of attributing values for intrinsic activity to compounds whose precise target site has not been identified. EPTC is known to interfere with lipid metabolism, and the oxidation products may act by carbamoylating SH groups, but the picture is still far from clear.

Many other herbicides containing SCH_3 groups undergo S-oxidation, such as other thiocarbamates (Casida *et al.*, 1974) and SCH_3-triazines (Bedford, Crawford & Hutson, 1975) but their activity compared with the parent compounds has not been studied.

d) Ring cleavage

The cyclic urea herbicide, methazole, is not intrinsically active but is converted to the photosynthesis (PS2) inhibitor DCPMU (Fig. 4) (Suzuki & Casida, 1981). Interestingly, two other urea herbicides, diuron and linuron, which are intrinsically active, are also metabolised to DCPMU which is equally active. Further N-demethylation of DCPMU gives the inactive dichlorophenyl urea. Another cyclic urea, buthidazole, with an imidazolidinone ring instead of the oxadiazolidinedione ring of methazole, also undergoes ring cleavage to an active N-methyl urea (Yu, Atallah & Whitacre, 1980). Both methazole and buthidazole may be metabolised by an alternative route involving N-demethylation prior to ring cleavage which does not produce an active intermediate (Yu *et al.*, 1980). The ratio of the two pathways varies between plant species, corn having only the inactive pathway whereas lucerne has both but this does not seem to confer the expected selectivity as both species are tolerant.

e) N-Demethylation

The pyridazinone herbicide metflurazon and its N-demethylated analogue, norflurazon, inhibit carotenoid biosynthesis causing bleaching. They also inhibit PS2 although this may not contribute to their herbicidal effect. Recent experiments with the unicellular green alga, *Chlorella fusca* (Tantawy, Braumann & Grimme, 1984) suggest that metflurazon is of low activity but is bioactivated by N-demethylation to norflurazon (Fig. 5) by a P450 catalysed mechanism. The alga adapts to metflurazon treatment by suppressing its N-demethylation mechanism and thus becomes resistant but is still just as susceptible to norflurazon. Interestingly, the N-demethylation decreases the PS2 inhibition activity by a factor of five. Further N-demethylation yields the inactive amine SAN9774. This mechanism would be in agreement with data from higher plants where susceptibility to metflurazon correlates with a balance between activation and inactivation which permits the accumulation of norflurazon.

As mentioned in (d), N-demethylation of dimethylureas does not significantly alter their PS2 inhibition activity, but N-demethylation of the N-dimethyl triazines, trietazine and ipazine, to simazine and atrazine respectively, increases their PS2 activity by two orders of magnitude (Good, 1961).

Fig. 5.

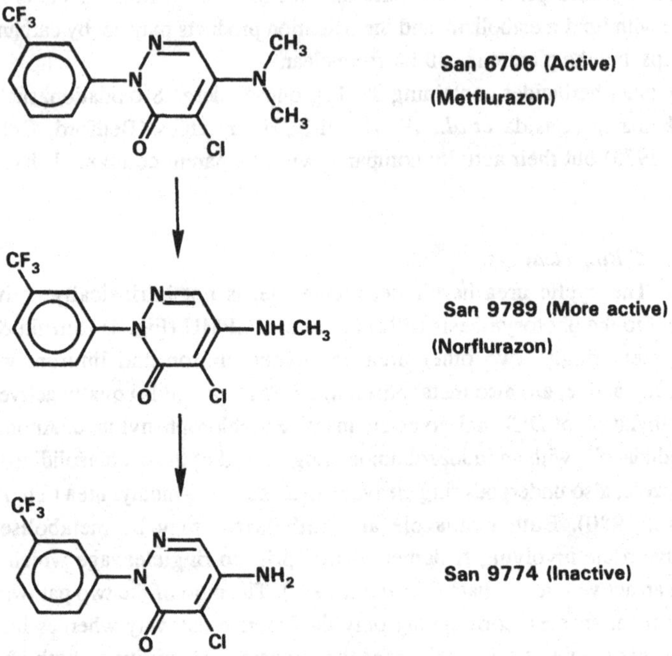

San 6706 (Active)
(Metflurazon)

San 9789 (More active)
(Norflurazon)

San 9774 (Inactive)

Fig. 6.

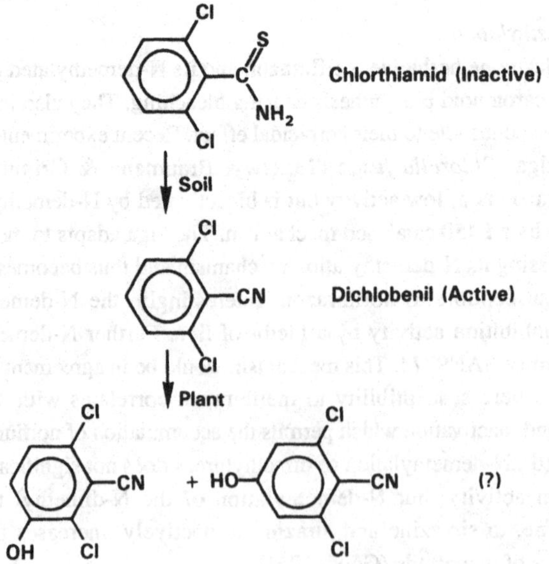

Chlorthiamid (Inactive)

Soil

Dichlobenil (Active)

Plant

(?)

f) Ring hydroxylation

This is usually regarded as being part of a detoxification mechanism. The herbicide chlorthiamid is converted to dichlobenil in the soil (Fig. 6) and so this is not strictly bioactivation, although chlorthiamid has the advantage of being less volatile than dichlobenil. In the plant dichlobenil is hydroxylated in the 3 or 4 position to yield compounds which are potent uncouplers of oxidative and photosynthetic phosphorylation and inhibitors of PS2 (Moreland, Hussey & Farmer, 1974) Dichlobenil has little effect on these reactions but is believed to interfere with cellulose biosynthesis. It is not clear what contribution these metabolites make to the herbicidal activity of dichlobenil.

g) Glutathione conjugation

Tridiphane is an herbicide which controls grass and annual broadleaved weeds in cereal crops by inhibiting meristematic activity. It also synergises the activity of atrazine by inhibiting glutathione conjugation, which is the principal mechanism for atrazine detoxification. Tridiphane itself undergoes glutathione conjugation (Fig. 7) (Lamoureux & Rusness, 1986) to yield a more active inhibitor of glutathione conjugation. This synergism is more effective in giant foxtail (*Setaria faberi*), the problem weed, than in maize (*Zea mays*). This is another reaction which is normally regarded as part of a detoxification mechanism but in this case is causing increased activity. These are some of the better-known examples of bioactivation although there are others including probably many yet to be elucidated. Where the mode of action is unclear it is quite difficult to determine whether the parent molecule or one of its metabolites is the active species.

Bialaphos is a naturally occurring tripeptide which exerts its herbicidal effects by inhibiting glutamine synthetase. Since one of the residues is the amino acid herbicide, gluphosinate, it seems likely that the tripeptide is being bioactivated by hydrolysis to this more active inhibitor of the same target enzyme.

A novel bioactivation reaction

Some years ago the compound WL46287 (Fig. 8) passed through the herbicide screens at Sittingbourne and showed an unusual selectivity pattern. Linseed (*Linum ustittatissimum*) was effectively killed in both the pre and post emergence tests (GID_{50} = 0.1kg/ha) but there was little effect on the other 32 species tested. Whilst this selectivity pattern has no obvious commercial potential, it was argued that there were two likely explanations for the effect:- either all species except linseed detoxify the material or only linseed converts WL46287 to an active herbicide.

If the second case obtains then a compound with more general herbicidal properties might be discovered. With this in mind, an investigation of the metabolism of WL46287 was undertaken.

Fig. 7.

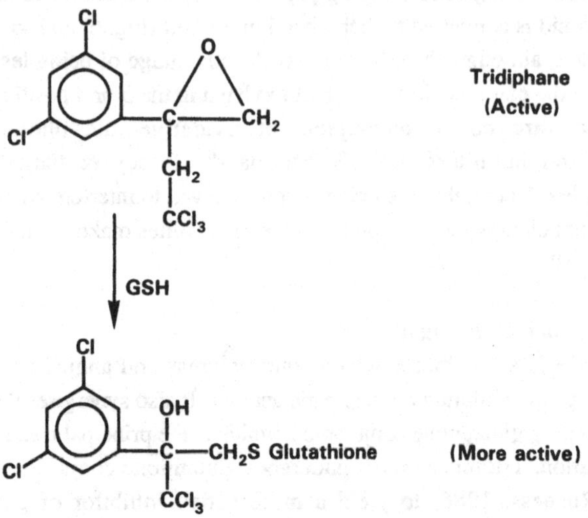

Cl, O, C, CH_2, Cl, CH_2, CCl_3 — Tridiphane (Active)

↓ GSH

Cl, OH, C—CH_2S Glutathione, Cl, CCl_3 — (More active)

Fig. 8.

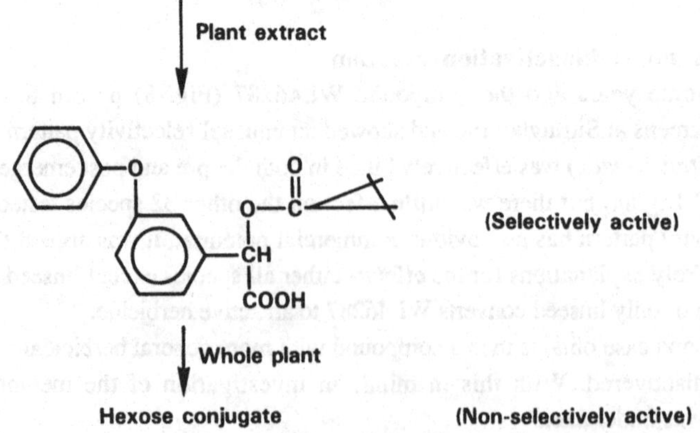

O, O, O—C, CH, $COSCH_3$ — WL 46287 (Selectively active)

↓ Plant extract

O, O, O—C, CH, COOH — (Selectively active)

↓ Whole plant

Hexose conjugate — (Non-selectively active)

Using radiolabelled material (^{14}C in the cyclopropyl or benzyl moiety), it was soon possible to eliminate differential uptake rates as the basis for selectivity. Using centrifuged juice from a variety of plant species, it was possible to demonstrate the conversion of WL46287 to a more polar metabolite. This was identified as the mandelic acid WL84855 (Fig. 8). The rate of conversion to this metabolite was about three times faster in linseed juice than in that from other species, suggesting that this might be the general toxicant. Unfortunately, when WL84855 was tested by applying to the roots of linseed, mustard (*Sinapis alba*) and barnyardgrass (*Echinochloa crus-galli*) seedlings, it showed the same selectivity towards linseed and the same dose: response curve as the parent material.

If plant sections or whole seedlings were used instead of plant juice, a further metabolite of WL46287 was identified which was even more polar than WL84855. This was tentatively identified as a hexose conjugate on the basis of its chemical and enzymic hydrolysis to the free acid, WL84855.

This metabolite was extracted from treated linseed tissue in sufficient quantity to allow the treatment of linseed, mustard and barnyardgrass seedlings via the roots (Fig. 9). Under these conditions, the growth of all three species was inhibited, indicating that this hexose conjugate was the general toxicant.

Since the mode of action was unknown, it was not clear whether the conjugate itself was active or was serving as a delivery mechanism. Unfortunately the conjugate proved to be extremely unstable and difficult to work with and so the project had to be abandoned.

An attempt to design a bioactivated herbicide

The well-known herbicides, paraquat and diquat, are reduced by photosystem I of the chloroplast electron transport chain and then react with oxygen generating toxic reduced oxygen species. Many other compounds have this property and can catalyse this cyclic Mehler reaction *in vitro* but are not effective herbicides. This may be because they are too reactive or unstable to reach the site of action, or because material is lost during catalysis by side reactions.

The compound 3,4,dihydroxyphenylalanine (3,4 DOPA) will catalyse a Mehler reaction *in vitro* and, being an amino acid, might be expected to have good transport

Fig. 9.

Concentration ($M \times 10^{-6}$)	Fresh weight (% of untreated) Linseed	Mustard	Barnyard grass
0.5	39	74	75
1.0	46	64	19
7.5	22	23	4

properties. However, it undergoes spontaneous oxidation and polymerisation in air to yield coloured melanin-type compounds (Fig. 10). The close analogue, 2,4 DOPA is quite stable but does not catalyse a Mehler reaction.

It was recently reported (Morrison & Cohen, 1983) that 2,4 DOPA is converted to 2,4,5 TOPA by tyrosinase (phenoloxidase). In plants, this enzyme is located exclusively in the chloroplast which is also the site of action for a photosystem I herbicide. 2,4,5 TOPA is known to be a very efficient generator of toxic oxygen species when exposed to air (Cohen & Heikkila, 1974). It seemed possible, therefore, to utilise 2,4 DOPA as an inactive precursor of the putative herbicide, 2,4,5 TOPA.

Using an *in vitro* chloroplast assay, it was demonstrated that 3,4 DOPA would catalyse a Mehler reaction but 2,4 DOPA would not. 2,4,5 TOPA gave, initially, a much more rapid reaction, some four times the maximum achieved with 3,4 DOPA but the rate decayed quite sharply. This may well have been due to the competing cyclisation reaction (Fig. 10). This cyclisation can be blocked by N-acetylation albeit with some loss of catalytic efficiency.

Incubation of 2,4 DOPA with chloroplasts results in slight conversion to 2,4,5

Fig. 10.

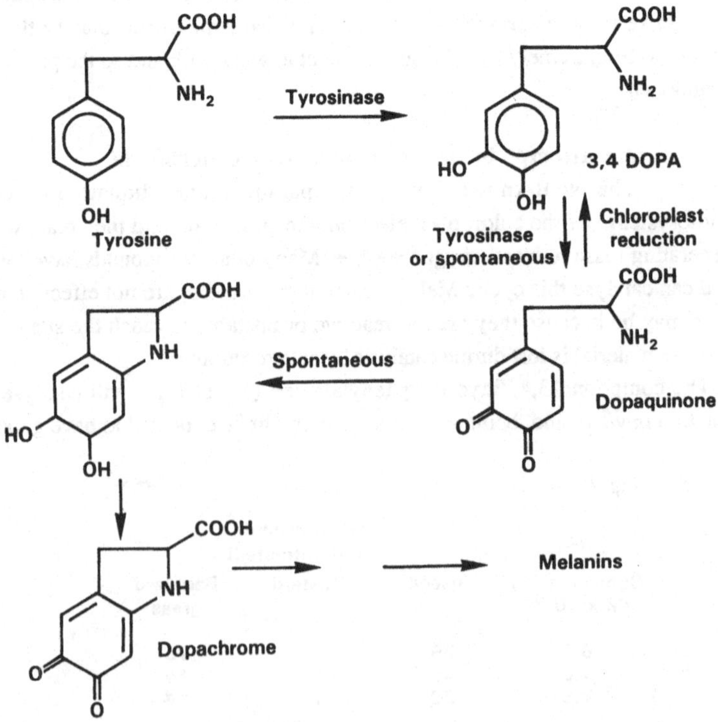

Table 1. Oxygen uptake catalysed by 2.4 DOPA in pea chloroplasts (arbitrary units)

pretreatment	dark	light
2 hours	4.0	11.0
2 hours + tyrosinase	4.0	16.5
4 hours	4.5	14.0
4 hours + tyrosinase	4.5	23.0

TOPA as judged by its ability to catalyse oxygen uptake (see Table 1). If tyrosinase was added, the apparent conversion rate was increased, although it was still very small. It may be that the washed thylakoids used in the reaction had lost their endogenous tyrosinase, or the enzyme, which is normally latent, was not effectively activated.

2,4 DOPA showed no activity in the herbicide screens but whether this was due to inadequate conversion or rapid loss of the 2,4,5 TOPA by cyclisation was not clear. N-acetylation would prevent cyclisation but since both OH groups were also acetylated in the derivatisation and they could not be selectively hydrolysed, it was not possible to carry out any *in vitro* assays with this compound. Triacetyl 2,4 DOPA was also inactive in the herbicide screens and so the project was terminated.

Although this attempt was unsuccessful, it did serve to emphasise the subtlety of the bioactivation of paraquat. This compound is not active until reduced by functional chloroplasts at its site of action. It is therefore more stable than redox catalysts such as 3,4 DOPA which, being already in a reduced state, can react spontaneously with air and only require photosystem I to complete the catalytic cycle.

Acknowledgements

The investigation of the bioactivation of WL46287 was carried out by Drs R.Baxter, M.T.Clark, and A.N.Wright.

2,4-DOPA and its triacetyl derivative were synthesised by Mr D. Brough.

References

Bedford, C.T., Crawford, M.J., & Hutson, D.T.(1975). Sulphoxidation of *CYANATRYN*, a 2-mercapto-*sym*-triazine herbicide. *Chemosphere*, **5**, 311–316.

Brown, M.A. Chiu, T.U, & Miller, P. (1987). Hydrolytic activation versus oxidative degradation of ASSERT herbicide, an imidazolinone aryl-carboxylate, in susceptible wild oat versus tolerant corn and wheat. *Pesticide Biochemistry and Physiology* **27**, 24–29.

Casida, J.E., Gray, R.A. & Tilles, H. (1974). Thiocarbamate sulphoxides: potent selective and biodegradable herbicides. *Science* **184**, 573–574.

Cohen, G. & Heikkila R.E. (1974). The generation of hydrogen peroxide, superoxide radical, and hydroxyl radical by 6-hydroxydopamine, dialuric acid, and related cytotoxic agents. *Journal of Biological Chemistry* **249**, 2447–2452.

Dutka, F. & Kömives, T. (1982). On the mode of action of EPTC and its antidotes in *Human Welfare and the Environment*, eds. in chief Miyamoto J. & Kearney, P.C., **3**, 213–218 Pergamon Press, Oxford.

Good, N.E. (1961). Inhibitors of the Hill reaction. *Plant Physiology* **36** 788–803.

Jeffcoat, B. & Harries, W.N. (1973). Selectivity and mode of action of ethyl (±)-2-(N-benzoyl-3, 4-dichloroanilino) propionate in the control of *Avena fatua* in cereals. *Pesticide Science*, **4**, 891–899.

Jeffcoat, B., Harries, W.N. & Thomas, D.B. (1977). Factors affecting the choice of flamprop methyl (±)-2-(N-(3-chloro-4-fluorophenyl) benzamido) propionate, for the control of *Avena* species in wheat. *Pesticide Science*, **8**, 1–12.

Lamoureux, G.L. & Rusness, D.G. (1986). Tridiphane (2-(3,5-dichlorophenyl)-2-(2,2,2-trichloroethyl) oxirane) an atrazine synergist; enzymatic conversion to a potent glutathione-s-transferase inhibitor. *Pesticide Biochemistry and Physiology*, **26**, 323–342.

Moreland, D.E., Hussey, G.G. & Farmer, F.S. (1974). Comparative effects of Dichlobenil and its phenolic alteration products on photo-and oxidative phosphorylation. *Pesticide Biochemistry and Physiology* , **4**, 356–364.

Morrison, M.E. & Cohen, G. (1983). Novel substrates for tyrosinase serve as precursors of 6-hydroxydopamine and 6-hydroxydopa. *Biochemistry*, **22**, 19–21.

Shimabukuro, R.H., Walsh, W.C. & Hoerauf, R.A. (1979). Metabolism and selectivity of diclofop-methyl in wild oat and wheat. *Journal of Agricultural Food Chemistry*, **27**, 615–623.

Suzuki, T. & Casida, J.E. (1981). Metabolites of diuron, linuron, and methazole formed by liver microsomal enzymes and spinach plants. *Journal of Agricultural Food Chemistry*, **29**, 1027–1033

Tantawy, M.M., Braumann, T. & Grimme, L.H. (1984). Uptake and metabolism of the phenylpyridazinone herbicide metflurazon during the bleaching and regeneration process of the green alga; *Chlorella fusca*. *Pesticide Biochemistry and Physiology*, **22**, 224–231.

Taylor, H.F. & Loader, M.P.C. (1986). Antagonism *in vivo* between auxin herbicides and auxin-inhibitor herbicides. *Proceedings of the 5th Congress of FESPP* 3.05.

Wain, R.L. (1955). Herbicidal selectivity through specific action of plants on compounds applied. *Journal of Agricultural Food Chemistry*, **3**, 128–130.

Yu, C.C., Atallah, Y.H. & Whitacre, D.M. (1980). Metabolism of the herbicide buthidazole in corn seedlings and alfalfa plants. *Journal of Agricultural Food Chemistry*, **28**, 1090–1095.

P.D.PUTWAIN & H.A. COLLIN

Mechanisms involved in the evolution of herbicide resistance in weeds

Introduction

Occurrence and spread of resistance to triazine herbicides

The first incidences of resistance in weed species to triazine herbicides occurred in 1968 in the State of Washington USA. During the 1970s and 1980s, notably in North America and mainly western Europe, and to a lesser extent in Israel in the 1980s there has been an irregular but relatively steady addition to the occurrences of new species becoming resistant mainly to triazine herbicides (but also to some others). Figure 1 gives an indication of the total number of species in different countries which are occurrences of resistance not previously recorded. Worldwide 49 species of 33 genera have become resistant to triazine herbicides (Le Baron pers. comm. 1987) and a further 11 species have evolved resistance to other herbicides.

A record of the occurrence of a herbicide-resistant biotype will often conceal the fact that numerous populations have independently evolved in many different locations over a period of just a few years. For example, in Hungary, *Amaranthus retroflexus* resistant to s-triazines occurred in scores of locations and 75% of the maize growing area has become infested by the resistant biotype (Hartmann, 1979). In the USA, evolution of resistance to s-triazines in *A. hybridus* first occurred in Maryland in 1972 but between 1976 and 1982 there were numerous reports of resistant populations occurring throughout Virginia, New York, Delaware, Pennsylvania, Massachusetts, West Virginia and Illinois (Le Baron & Gressel, 1982; Le Baron pers. comm., 1987).

Although the majority of species to evolve triazine-resistance in the 1970s were dicotyledons, in the 1980s there has been an increase in *Poa annua* in California and in six different European countries whilst in Israel no less than seven grass species evolved resistance in roadside habitats which have been sprayed with triazines for up to 15 years (Gressel, 1983).

Some species appear to have a propensity to evolve resistance to triazines possibly because genetic variation for resistance is widely dispersed through many populations. It was unlikely to have been chance that caused *P. annua* to have evolved resistance to triazines in six European countries and the USA and also to have evolved resistance to paraquat in the UK. Similarly triazine-resistant biotypes of *Chenopodium album* have occurred in 10 different countries and in 17 different States within the USA. It is conceivable that in species such as these, there is an inherently higher mutation frequency of the chloroplast *psbA* gene coding for the 32 kD thylakoid membrane protein involved in herbicide binding.

In Britain, triazine-resistant biotypes of weed species were first detected in 1981 in plant nurseries, bush fruit plantations and in orchards (Putwain, 1982) up to 15 years after triazine herbicides were first used in these locations. Since 1981 the following triazine-resistant biotypes of weeds have been found in Britain (Table 1) with the number of confirmed locations in brackets; *Chamomilla suaveolens* (1), *Epilobium ciliatum* (1), *Erigeron canadensis* (1), *Poa annua* (5), and *Senecio vulgaris* (20). The full extent of the distribution of triazine-resistant weed species in Britain is not known but resistant biotypes of *Senecio* and *Poa* are certainly more widespread than the number of confirmed locations indicates.

Resistance to herbicides other than triazines

Worldwide, the majority of cases of herbicide-resistance have been to the s-triazine herbicides. Cases of resistance to herbicides other than triazines have been scarce and in scattered locations until recently (Table 2). Resistance to paraquat has been most commonly detected. In Japan paraquat-resistant biotypes of *Erigeron philadelphicus* evolved resistance to paraquat in mulberry plantations (Watanabe *et*

Fig. 1. The number of new cases of herbicide-resistant biotypes occurring each year, species/countries not previously recorded (LeBaron, personal communication).

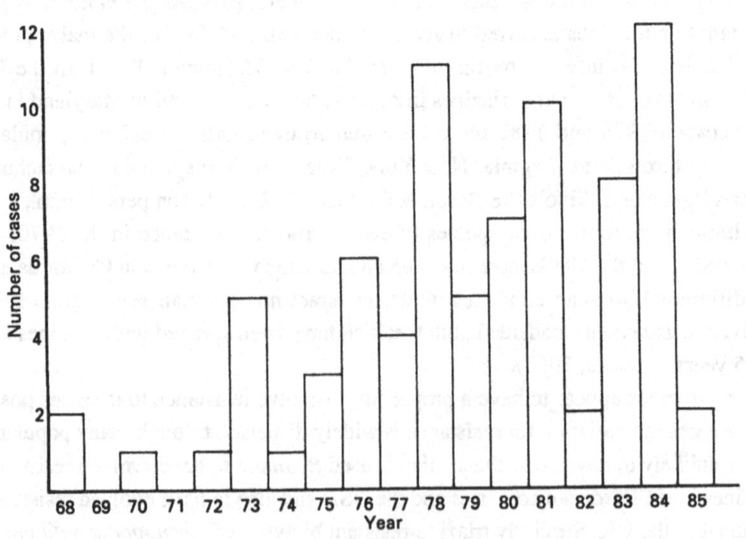

Table 1. *Occurrence of herbicide-resistant weed biotypes in the United Kingdom*

Herbicide	Species	Year Found	Number of Locations	Crop
Simazine	*Senecio vulgaris*	1981	20	Plant nurseries Bush fruits
Simazine	*Poa annua*	1982	5*	Orchards
Simazine	*Chamomilla suaveolens*	1983	1	Horticultural Research Station
Simazine	*Erigeron canadensis*	1984	1*	
Simazine	*Epilobium ciliatum*	1984	1*	Bush fruits
Chlortoluron	*Alopecurus myosuroides*	1984	1	Winter wheat
Mecoprop	*Stellaria media*	1985	1	Winter wheat
Paraquat	*Poa annua*	1978	1	Horticultural nursery

*Probably in many locations
Source: P.D. Putwain & D.Clay, pers. comm., 1987.

Table 2. *Occurrence and distribution of weed species resistant to herbicides other than triazines*

Herbicide	Species	Year Found	Location	Crop
Chlortoluron	*Alopecurus myosuroides*	1984	UK	Winter wheat
Diclofop methyl	*Lolium rigidum*	1982	Australia (many sites)	Pasture/cereal rotation
Mecoprop	*Stellaria media*	1985	UK	Winter wheat
Paraquat	*Conyza bonariensis*	1979	Egypt	Citrus
Paraquat	*Poa annua*	1978	UK	Horticultural nursery
Paraquat	*Erigeron philadelphicus*	1981	Japan	Mulberry
Pyrazon	*Chenopodium album*	1984	East Germany Hungary	Sugar beet
Trifluralin	*Eleusine indica*	1970	USA (Many sites)	Cotton

al., 1982) and the closely related *Conyza bonariensis* evolved paraquat-resistance in citrus plantations in Egypt. In Australia, a population of *Hordeum glaucum* evolved resistance to paraquat after 15 years of repeated use of the herbicide (Powles, 1986). A paraquat-resistant population of *Poa annua* was detected in a horticultural nursery in Britain (Le Baron & Gressel, 1982). In each of these cases there were many repeated applications of paraquat which produced a relatively intense selection pressure.

Other occurrences of herbicide-resistance include populations of *Eleusine indica* which are resistant to trifluralin. These have appeared in several locations in South Carolina after about 10 years use of the herbicide (Mudge, Gossett & Murphy, 1984). In Australia, *Lolium rigidum* resistant to diclofop-methyl has been found in 15 widely scattered locations (Heap & Knight, 1982; Knight pers. comm. 1986).

In Britain, Moss & Cussans (1985) detected substantially enhanced tolerance, verging on resistance to the substituted urea herbicide chlortoluron in *Alopecurus myosuroides* collected from winter wheat where substituted urea herbicides were used intensively for over ten years. Recently, resistance to mecoprop and dichlorprop has appeared in a population of *Stellaria media* (Lutman & Lovegrove, 1985). This is the first known occurrence of substantial field resistance to phenoxy-herbicides, despite their extensive use worldwide over many years.

To conclude it is clear that the only feature common to all recorded cases of herbicide-resistance is that there has been recurrent (usually annual) use of a herbicide over many years in crop monoculture or perennial plantation crops.

Definition of resistance, tolerance and susceptibility

In order to adequately discuss the inheritance and mechanisms of resistance, it is necessary to precisely define what is meant by the term. Le Baron & Gressel (1982) recognised the FAO definitions of resistance (Anon., 1965) which is a decreased response of a population of animal or plant species to a pesticide or control agént as a result of their application. Resistant populations are unharmed by a concentration of herbicide that completely kills all unselected populations. This concentration of herbicides is one that in field practice produces an effective kill.

At the other end of the scale, susceptibility encompasses natural variability in sensitivity to herbicides within unselected populations. This is the natural 'background' variability. It is possible for LD_{50} values of different populations to vary by a factor of 2× or greater (Sykes, 1980). Tolerance is reduced susceptibility (or conversely enhanced resistance) which may sometimes result from selection by herbicides. It is a response rather less than 100% survival by zero reduction in growth at herbicide concentrations which normally give 100% kill of susceptible biotypes. However, at lower herbicide concentrations, tolerant biotypes have higher survival and growth greater than susceptible biotypes.

There is no clear division between 'highly tolerant' and 'resistant' biotypes and low levels of tolerance merge into susceptibility. Relative resistance is therefore a continuum of response and all terms and divisions are somewhat arbitrary.

Mechanism of herbicide resistance

It is important to understand the mechanism of resistance to herbicides before applying intact plant and cell selection methods to increase herbicide-resistance in crop plants, and essential if genetic engineering methods are to be attempted (Gressel, 1985, 1987). The mechanism of resistance of the three herbicides, s-triazine, glyphosate and paraquat is discussed below.

Triazines

The mode of action of the s-triazine herbicides is by inhibition of photosynthesis. The site of inhibition is a 32 kD polypeptide which forms part of the Photosynthesis II complex located within the thylakoid membrane of the chloroplasts (Gardner, 1981; Van Rensen, 1982). This polypeptide, referred to as Q_B, is the secondary electron acceptor of Photosystem II and regulates electron transport between Photosystem II and the plastoquinone pool (Kyle, 1985) (Fig. 2). The effect of the triazine herbicides is to bind to the 32 kD polypeptide and block the flow of electrons.

Earlier studies have shown that resistance to s-triazines in weed species may be due to a reduced inhibitory activity of triazines on the chloroplast electron transfer chain (Radosevich & Devilliers, 1976; Souza Machado *et al.*, 1977), and, specifically, a decreased affinity for the B protein. Measurement of leaf chlorophyll fluorescence in the presence and absence of the s-triazine herbicides provides a very rapid method of assessing the presence and level of resistance to these herbicides (Arntzen, Pfister & Steinbeck, 1982). In an examination of resistant and susceptible biotypes of *Sinapsis arvensis* exposed to 10^{-4} M atrazine, the leaf chlorophyll fluorescence remained unchanged in resistant plants but more than doubled in susceptible biotypes (Ali, McLaren & Souza Machado, (1986) (Table 3). Another measure of electron flow in chloroplasts is the activity of the Hill reaction. In isolated chloroplasts of resistant plants of *S. arvensis* the Hill reaction was unaffected by 10^{-5} or 10^{-4} M atrazine but was inhibited in susceptible plants (Fig. 3). An analysis of another atrazine resistant weed, *Brachypodium distachyon* showed that whole leaf and plastid fluorescence was increased in the presence of 10^{-5} M atrazine when material was obtained from susceptible biotypes but was unaffected when from resistant biotypes (Gressel *et al.*, 1983) (Fig. 4). The Hill reaction of isolated chloroplasts from resistant biotypes was also unaffected.

A more detailed analysis of the electron transport chain on the reducing side of Photosystem II was made by Jansen *et al.*, (1986) using chloroplasts from atrazine resistant and susceptible *Chenopodium album*. In the absence of triazine herbicides the rate of electron transport between water and plastoquinone was actually lower in the chloroplasts isolated from the resistant plants. There was, however, no difference in the overall rate and quantum yield of the whole electron transport chain. This was explained by the fact that the rate limiting step in the electron chain was at the stage of plastoquinone oxidation, i.e. after the stage normally inhibited by triazines. Plants from both resistant and susceptible biotypes showed the same rate of photosynthesis in response to increasing light intensities (Fig. 5). Clearly the changes in the electron transport chain in the resistant plants had no effect on the growth of the plants under herbicide free conditions. This situation is not always true for all weed species since atrazine-resistant biotypes of *Senecio vulgaris* and *Amaranthus retroflexus* (Conard & Radosevich, 1979), *Amaranthus hybridus* (Ahrens & Stoller, 1983) and *Chenopodium album* (Warwick & Black, 1981) showed a reduction in relative fitness.

Mechanisms which confer resistance to the atrazine resistant biotypes will include a change in the binding site of the herbicide or a reduction in the concentration of the

Fig. 2. Electron transport pathway from Photosystem II showing the position of the secondary electron acceptor of Photosystem II, Q_B.

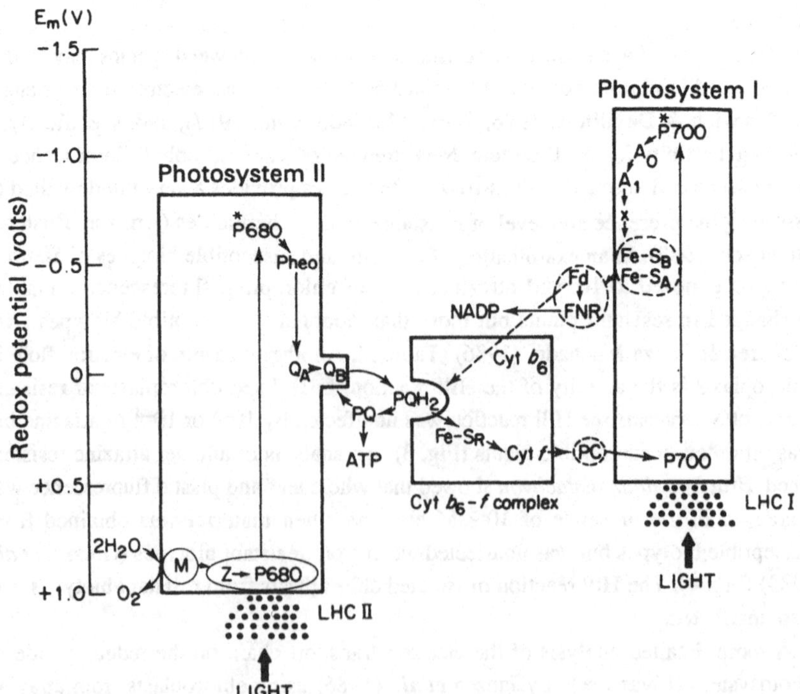

Table 3. *Leaf chlorophyll fluorescence (LCF) in leaf sections of two triazine susceptible and one triazine resistant biotype of* Sinapis arvensis *exposed to 10^{-4} M atrazine (Ali* et al., *1986).*

Plants	Leaf chlorophyll fluorescence (arbitary units)		
	Untreated	10^{-4} M atrazine	
	LCF	LCF	Increased % Control
Susceptible			
S. arvensis	0.30 ± 0.03	0.68 ± 0.06	131
S. arvensis	0.28 ± 0.02	0.60 ± 0.04	119
Resistant			
S. arvensis	0.37 ± 0.01	0.40 ± 0.01	7

herbicide at the binding site. The change in the binding site implies that the structure and composition of the binding site of the herbicide, the kD polypeptide, has been altered in resistant biotypes. The gene which encodes this polypeptide, designated *psbA*, has been studied in triazine resistant and susceptible biotypes of *Amaranthus hybridus* (Hirschberg & McIntosh, 1983). These authors found from the nucleotide sequence of the chloroplast DNA that the 32 kD protein in a triazine-resistant biotype of *A. hybridus* differs from the polypeptide of a herbicide susceptible biotype by only one amino acid. The amino acid serine is substituted by glycine in the resistant biotype which must cause the change in affinity of the binding site for the herbicide. Analysis of atrazine resistant mutants of *Chlamydomonas reinhardi* has shown that there are additional amino acid substitutions in the 32 kD polypeptide beside that of serine (Erickson, Rahire & Rochaix, 1985). Since the binding side of the herbicide on the polypeptide is not known, it is difficult to see how such subtle changes can alter the binding affinity for the herbicide. It is possible that the change from serine to glycine, or the other amino acid substitution, occurs within the binding site itself and thus alters binding affinity. A further suggestion is that the tertiary structure of the polypeptide is modified causing a change to the entire binding site (Hirschberg & McIntosh, 1983), and possibly to binding sites used by other photosynthetic inhibitors, such as the herbicide diuron. This would explain why the affinity of more than one herbicide may be influenced by the development of resistance to one herbicide alone (Gressel, 1985).

An alternative mechanism for explaining the difference in response to triazine herbicides of resistant and susceptible plants is that the concentration of herbicide at the binding site is effectively reduced by detoxification. For example, maize plants

Fig. 3. Photochemical activity as measured by the Hill reaction of isolated chloroplasts from triazine susceptible and resistant biotypes of *S. arvensis* exposed to OM (○---○, □---□); 10^{-5} M (◑---◑, ◧---◧); and 10^{-4} M (●---●, ■---■) atrazine (Ali *et al.*, 1986).

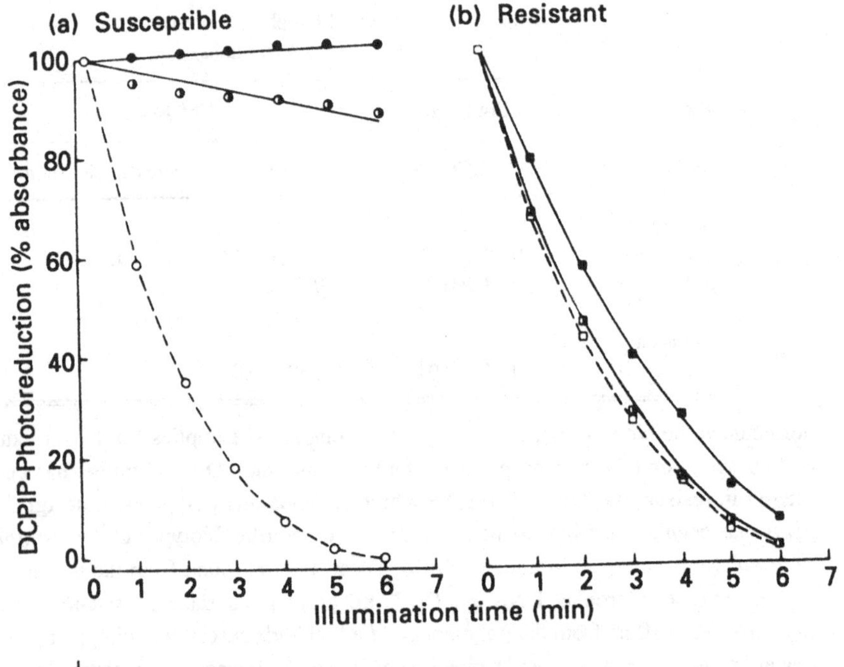

Fig. 4. Fluorescence of chloroplasts isolated from leaves of resistant and susceptible biotypes of *Brachypodium distachyon* in the presence of 10^{-5} M atrazine (Gressel *et al.*, 1983). A. Fluorescence scan of dark-adapted plastids from susceptible plants (control). B. Plastids from resistant plants treated as in A, dark adapted and 10^{-5} M divron was added.

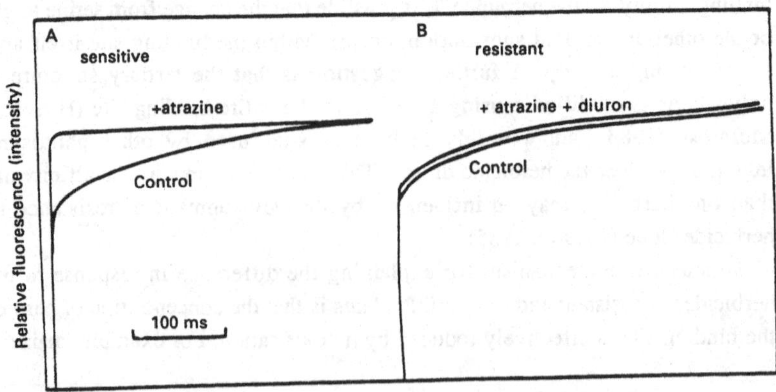

Table 4. *Atrazine-^{14}C metabolism in leaf discs of resistant and susceptible corn lines showing formation of glutathione-atrazine conjugation (G-S-A) (Shimabukuro* et al., *1971).*

Time intervals (hr)	Distribution of ^{14}C activity					
	Resistant			Susceptible		
	Atrazine	G-S-A (dpm)	%G-S-A	Atrazine	G-S-A (dpm)	%G-S-A
0	8100	1690	17.3	9225	30	0.3
0.5	3155	2390	43.1	5755	25	0.4
1.0	3205	3849	54.6	4770	10	0.2
2.5	1800	3895	68.4	4025	25	0.6

resistant to atrazine possess an enzymic and non enzymic detoxification mechanism (Shimabukuro *et al.*, 1971). The enzymic mechanism involves the synthesis of a conjugate between glutathione and atrazine, catalysed by the enzyme glutathione s-transferase. The difference between resistant and susceptible maize lines is shown both by the increased conjugation of atrazine to form glutathione s-atrazine (Table 4) and the higher activity of the glutathione s-transferase in the resistant lines (Table 5). The non-enzymic pathway in which atrazine is hyroxylated to hydroxy atrazine was considered not to be of importance in the resistance mechanism. The s-triazine resistant weed *B. distachyon* also showed a detoxification mechanism for dealing with the herbicide (Gressel *et al.*, 1983), in addition to its plastid-based mechanism of resistance. The existence of a detoxification mechanism in this species was shown by the increased proportion of ^{14}C radioactivity in water soluble and chloroform soluble atrazine metabolites in resistant plants compared with susceptible plants (Fig. 6). A further degradation pathway was examined in atrazine susceptible and resistant *Senecio vulgaris* but did not appear to be related to tolerance (Gressel, Shimabukuro & Dugsen, 1983). The presence of three mechanisms of triazine degradation may be more common in triazine resistant weeds than was first supposed.

Glyphosate

Although the mechanism of action of glyphosate on plants in the field at the concentrations which are normally used is not clear (Gressel, 1985), there are indications from work with bacteria, fungi and cell cultures that the toxicity is a result of a specific inhibition of one enzyme step in the aromatic pathway in all these organisms. Most of the work on mechanisms of action has been carried out using simple cultured systems. In plant cell cultures this step has been identified as the conversion of shikimic-3-phosphate to 5-enol pyruvylshikimic-3-phosphate (EPSP) which is catalysed by the enzyme EPSP synthase (Amrhein *et al.*, 1983; Steinrücken *et al.*,1986). The mechanism of inhibition by glyphosate in all these organisms is by competitive inhibition. As a result of exposing plant cells to glyphosate,

Table 5. *Glutathione s-transferase activity in corn leaves (Shimabukuro* et al., *1971).*

Corn line	Response to atrazine	Specific activity (nmoles G-S-atrazine per mg protein^{-1} hr^{-1})
GT112RfRf	Resistant	1.63
GT112	Susceptible	0.03
NDKE47101	Resistant	2.73
B49	"	3.28
CI.31A	"	2.31
B52	"	1.46
Oh43	"	3.62
WF9	"	2.32

Fig. 5. Rate of photosynthesis in intact leaves of triazine-resistant and susceptible biotypes of *Chenopodium album* biotypes at various light intensities. (-●-) resistant plants (-O-) susceptible plants (Jansen *et al.*, 1986).

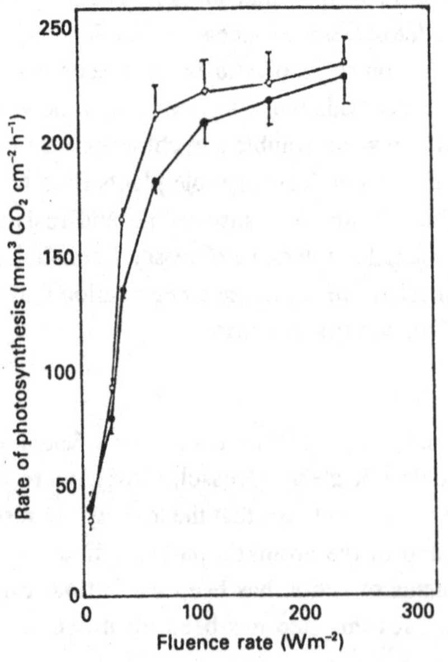

shikimic acid and shikimic acid-3-phosphate accumulate, particularly shikimic acid since plants possess the ability to dephosphorylate shikimic acid-3-phosphate (Amrhein *et al.*, 1980; Hollander-Czytko & Amrhein, 1983).

The inhibition can be reversed in algae, bacteria and some plant cell cultures by the addition of the three aromatic acids, phenylalanine, tyrosine and tryptophan (Lee, 1980), but not when these are applied or injected into intact plants (Rubin, Gaines & Jensen, 1982). Care must be taken therefore when extrapolating from plant cell cultures to intact plants.

Fig. 6. Metabolism of ^{14}C-(ring)-atrazine by sensitive and resistant biotypes of *Brachypodium distachyon* showing radioactivity in parent atrazine and metabolites (A) and metabolies only (B). Each symbol shape represents a separate experiment (Gressel *et al.*, 1983).

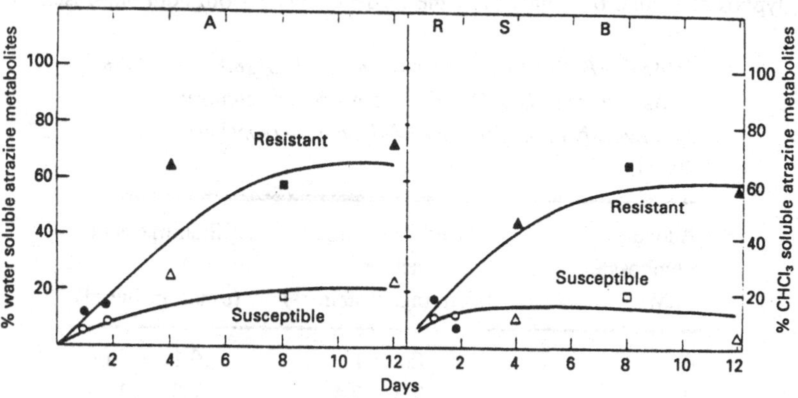

Fig. 7. Growth of glyphosate sensitive cells (-○-) in the absence of glyphosate and glyphosate tolerant cells of *Corydalis sempervirens* tissue cultures in the presence of (-●-) 5 mM glyphosate (Amrhein *et al.*, 1985).

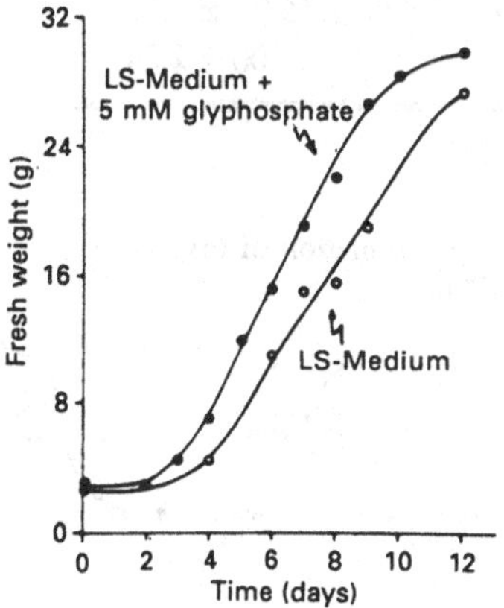

Although no glyphosate resistance has been detected in weeds, glyphosate resistance has been selected in bacterial cultures (Duncan, Lewendon & Coggins, 1983; Rogers *et al.*, 1983; Comai *et al.*, 1983 and Schulz, Sost & Amrhein, 1984) and in plant cell cultures of *Corydalis sempervirens* (Amrhein *et al.*, 1983), *Daucus carota* (Mafziger *et al.*, 1984) and *Petunia hybridia* (Steinrücken *et al.*, 1986). In bacterial cultures selected for resistance, the mechanism of resistance may be due to over-production of enzyme EPSP synthase, or structural changes in the EPSP synthase; whereas in plant cell cultures resistance appears to be due only to over-production of the enzyme. Thus repeated selection of cell cultures of *C. sempervirens* in the presence of glyphosate produced a clone tolerant to 5.0 mM glyphosate (Amrhein *et al.*, 1983; Amrhein, Johanning & Smart, 1985) (Fig. 7). The resistant cells showed an enhanced level of EPSP synthase activity with each increase in glyphosate concentration up to 5.0 mM glyphosate (Table 6). Analysis of the EPSP synthase from both non-resistant and

Table 6. *EPSP-synthase activity and levels of shikimic acid in tissue culture cells of* Corydalis sempervirens *cultured continuously in the presence of glyphosate (Amrhein* et al., *1983).*

Added glyphosate (mM)	EPSP-synthase activity (nkat. mg. protein $^{-1}$)	Shikimic acid (µmol. g. fwt.$^{-1}$)
0	0.5 ± 1	0.18 ± 0.04
0.5	7.3 ± 2.5	7.88 ± 1.44
1	7.0 ± 3.7	8.84 ± 1.84
2	11.7 ± 4.0	10.03 ± 1.99
3	12.2 ± 3.8	10.06 ± 2.26
5	12.8 ± 2.7	15.74 ± 2.69
10	16.6 ± 3.0	18.00 ± 2.84

Fig. 8. Site of action of paraquat showing formation of toxic superoxide and hydroxyl radicals (Shaaltiel & Gressel, 1986).

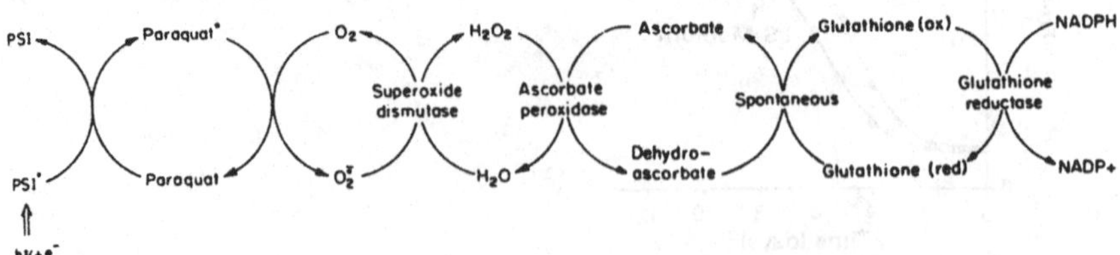

resistant cells by physical and chemical methods, showed no difference between the enzyme from the two sources. When the enzymes were separated by SDP electrophoresis, then visualised by immunoblot techniques, over-production of the EPSP synthase was established in the resistant biotype (Amrhein *et al.*, 1983, 1985). Cell cultures of *P. hybrida* selected for resistance to glyphosate showed the same pattern (Steinrücken *et al.*, 1986). The specific activity of the EPSP synthase from cells resistant to 10.0 mM glyphosate showed a 10–20 fold increase in enzyme activity. This increase in enzyme activity was also due to enhanced production of the enzyme. Since resistance to glyphosate has appeared as a result of selection in a range of organisms, it may be just a question of time before resistance appears in weed populations under natural conditions. Already there has been one report of enhanced tolerance in *Elymus repens* (Sampson, personal communication 1987).

Paraquat

Paraquat causes rapid chloroplast degradation and desiccation of plants. The site of action is in the chloroplasts at Photosystem I where it competes for electrons from the primary electron acceptor. The superoxide radical anions are broken down to give oxygen and hydrogen peroxide. The hydroxyl radicals released from the hydrogen peroxide cause membrane damage in the chloroplasts and finally rapid death of the plants. Normally hydrogen peroxide is removed from the chloroplasts by an ascorbate glutathione cycle (Fig. 8). The important enzymes in this cycle of paraquat detoxification are superoxide dismutase, ascorbate peroxidase and glutathione reductase.

Fig. 9. Effect of paraquat on ^{14}C-O_2 assimilation by paraquat tolerant (-●-) and susceptible (--●--) biotypes of *Conyza bonariensis* (Shaaltiel & Gressel, 1986).

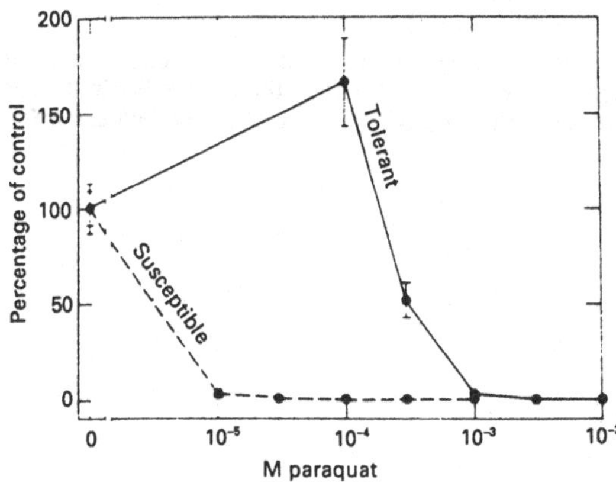

Analysis of resistant and susceptible *Conyza* plants showed higher levels of superoxide dismutase (Youngman & Dodge, 1981), but these increased levels could not be confirmed by Vaughn & Fuerst (1985). However, higher superoxide dismutase levels were found in paraquat resistant *Lolium* (Harvey & Harper, 1982). Shaaltiel & Gressel (1986) showed that in *Conyza bonariensis* biotypes, resistant plants showed no reduction in photosynthesis up to 10^{-4} M paraquat (Fig. 9). Measurements of the activities of the three enzymes from intact chloroplasts of resistant and susceptible biotypes indicated that the activities of the enzymes were enhanced in the resistant plants compared to the susceptible plants (Table 7). It was suggested that the increased

Table 7. *Enzyme activities of superoxide dismutase, ascorbate peroxidase and glutathione reductase in extracts of intact chloroplasts of paraquat susceptible and resistant biotypes of* Conyza bonariensis *(Shaaltiel & Gressel, 1986).*

Enzyme	Biotype	Activity (U mg. protein^{-1})	%Control (susceptible biotype)
Superoxide dismutase	Susceptible	1.04	100
	Resistant	1.67	160
Ascorbate peroxidase	Susceptible	0.74	100
	Resistant	1.85	185
Glutathione reductase	Susceptible	0.64	100
	Resistant	1.87	292

Fig. 10. Regression of mean paraquat tolerance of progeny on tolerance of 39 female parents. Regression coefficient = 0.361 ± 0.08. Heritability in the narrow sense was estimated as 0.772, twice the regression coefficient (Faulkner, 1974).

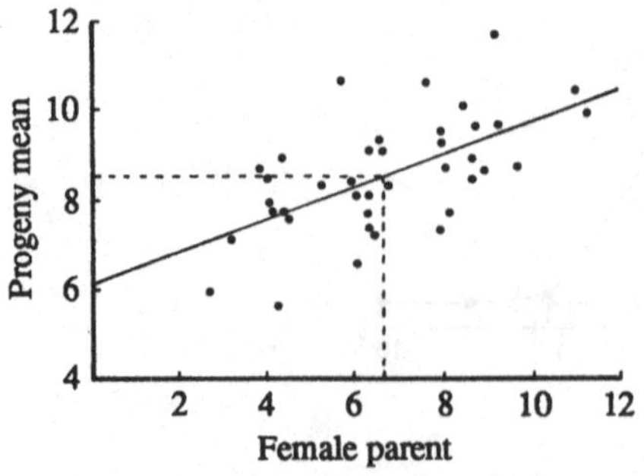

Table 8 *Examples of quantitative inheritance of herbicide resistance.*

Herbicide	Species	Heritability	Authority
Chloramben methyl ester	*Cucumis sativus*	0.36–0.87 Narrow sense	Miller *et al.* (1973)
Atrazine	*Linum usitatissimum*	0.29–0.34	Comstock & Anderson (1968)
MCPA	*Linum usitatissimum*	0.07–0.36	Stafford *et al.* (1968)
Paraquat	*Lolium perenne*	0.51–0.72 Narrow sense	Faulkner (1974)
Diclofop-methyl	*Zea mays*	0.95 Broad sense	Geadelmann & Anderson (1977)
Simazine	*Brassica oleracea* (wild population)	0.32–0.37 Narrow sense	Sykes (1980)
Barban	*Avena fatua*	0.12–0.63	Price *et al.* (1983)

activities of these enzymes enable the chloroplasts to remove the hydrogen peroxide and superoxide formed after the paraquat treatment.

However, this may not be the full story since evidence has been provided by Fuerst *et al.* (1985) that in the same species, *C. bonariensis* resistance to paraquat is due to exclusion of the paraquat from its site of action in the chloroplasts. This exclusion may be by a sequestration mechanism. Additionally, the paraquat may be excluded from the site of action by an inhibition of distribution in the resistant biotypes, such as in paraquat resistant *Erigeron philadelphicus* (Tanaka, Chisaka & Saka, 1986). In another paraquat-resistant weed, *Hordeum glaucum* Steud, the germinating seeds of the resistant biotypes were unaffected by paraquat in the dark, whereas the germination and growth of the susceptible biotype was inhibited (Powles, 1986). The resistance may be due to an inhibition of uptake and distribution as in the previous weed species.

Inheritance of resistance to herbicides

Quantitative inheritance

Quantitative patterns of inheritance in which relative resistance is a polygenically controlled character have been found in several crop and weed species. Generally differential tolerance is quantitatively inherited. The concept of heritability is applicable since it is a measure of the amount of genetic as opposed to environmental control of variation. Examples of quantitative inheritance of tolerance to herbicides are given in Table 8 together with heritability estimates where they exist.

It is rarely possible to link inheritance studies with biochemical mechanisms of tolerance involving degradation or conjugation of a herbicide. A tenuous link was established between inheritance and mechanism of herbicide-resistance in a population of *Lolium perenne*. Faulkner (1974) analysed half-sib families and correlated the paraquat tolerance of progeny with parents. The narrow sense heritability was in the range of 0.51 to 0.72 (Fig. 10). This indicates the probability of effective response of the population to natural or artificial selection. Tolerant lines of *L. perenne* showed consistently high activites of superoxide dismutase and catalase and in most instances, also high activity of peroxidase (Harvey & Harper, 1982). Activities of superoxide dismutase, catalase and peroxidase were respectively 56%, 32% and 35% higher in paraquat tolerant lines than in susceptible cultivars of *L. perenne*. These data were extended by the later work of Shaaltiel & Gressel (1986) previously discussed, on the mechanism of paraquat-resistance in *C. bonariensis*.

In Israel, the grass species *B. distachyon* has evolved both full chloroplastic resistance to triazine herbicides but also demonstrates enhanced tolerance to triazines (Gressel *et al.*, 1983). Tolerant biotypes degraded atrazine more rapidly than the susceptible ones. At present (Gressel pers. comm. 1987), it is not known whether tolerance is inherited in the nucleus in contrast to inheritance of resistance in the plastome. Tolerance to triazines has been observed in other species which also evolve chloroplastic resistance (e.g. *Senecio vulgaris*, Holliday & Putwain, 1980). Degradation of atrazine is probably a usual mechanism of triazine-tolerance in those species where it has evolved.

In other species/herbicides combinations, low levels of heritability were found. Broad sense heritability of 0.29 and 0.34 were found by Comstock & Anderson (1968) in the progeny of a cross between a selected atrazine-tolerant line and a susceptible cultivar of flax. Tolerance to MCPA in flax was polygenic (Stafford *et al.*, 1968) and heritability was low (0.01 to 0.36).

Particularly high heritability was found by Miller *et al.* (1973) for tolerance of chloramben methyl ester in cucumber. A study of progeny of crosses between tolerant and susceptible lines gave narrow sense heritability in the range of 0.36 to 0.87. An investigation by Geadelmann & Anderson (1977) of the inheritance of tolerance to diclofop-methyl in maize also gave high levels of broad sense heritability. The estimates were based on single cross hybrids of tolerant and susceptible inbred lines, and values were in the vicinity of 0.95. There are several other studies (Table 8) demonstrating quantitative inheritance of herbicide-resistance but in none of these is there any data concerning herbicide metabolism. There is, however, further evidence from investigation of the inheritance of major (nuclear) genes.

Nuclear inherited major genes

Inheritance of susceptibility to atrazine in maize (Grogan, Eastin & Palmer, 1963) is controlled by the action of a single recessive allele. In maize the mechanism

Table 9. *Segregation of paraquat resistance in* Conyza

	Sensitive	Resistant	Ratio R/S
F_1			

S x R	0	50	
R x S	0	50	
F_2			

S x R	61	170	2.8:1
R x S	52	170	3.3:1
Total	113	340	3.0:1

Data from Shaatiel *et al.*, pers. comm. 1987

of resistance involves the synthesis of a conjugate between glutathione and atrazine. Inheritance of triazine-resistance in soybean (Edwards, Barrentine & Kilen, 1976) and tolerance in tomato (Souza Machado, 1982) is controlled by a single nuclear gene. In the case of soybean, resistance was transferred to an important sensitive cultivar, Tracy (Eastin, 1981) and the resistant cultivar, Tracy M, was found to metabolise metribuzin (based on methanol extraction) more rapidly than did Tracy.

Recently, atrazine-tolerance in a serious weed species, *Abutilon theophrasti*, was shown to correlate with formation of glutathione conjugate of atrazine. The rate of formation of this conjugate was 2- to 4-times greater in the tolerant biotype (Gronwald *et al.*, 1987). Tolerance was not inherited in the chloroplast genome but as a nuclear, single partially dominant allele (Anderson & Gronwald, 1987).

There is important new evidence concerning the inheritance of resistance to paraquat in *C. bonariensis* (Shaaltiel *et al.*, unpublished) in which elevated levels of enzyme involved in conferring resistance co-segregate with resistance. Reciprocal pair crosses were made between resistant and sensitive individuals. All F_1 reciprocal plants were highly resistant (Table 9). Thus maternal inheritance was not involved. The activities of F_1 plants of superoxide dismutase, ascorbate peroxidase and glutathione reductase were as high as in the resistant parent. In the F_2 generation there was a distribution of 3 resistant to 1 sensitive. It was concluded that probably one dominant nuclear gene controls the levels of the enzymes involved in resistance via superoxide detoxification. It is possible (but improbable) that more than one gene is involved in control of the enzymes but with very tight linkage.

Other examples where variability in resistance or tolerance has been found, due to single gene differences include a) resistance to 2,4-D in *Sorghum bicolor* (Wiese & Quinby, 1969), where resistance was dominant, b) sensitivity to metoxuron in winter wheat which was determined by a single recessive gene in some cases, while in others two genes appear to be involved (Lupton & Oliver, 1976), and c) inheritance studies in F_2 populations of barley derived from crosses of parents which differed in their reaction to barban chlorosis and which showed that susceptibility was dominant to resistance and that resistance was controlled by a single recessive gene, independently inherited (Hayes, Pfeiffer & Rana, 1965).

In the weed species *Hordeum jubatum*, tolerance to siduron was determined by at least three dominant complementary major genes (Schooler, Bell & Nalewaja, 1972).

Some further evidence of control of herbicide-resistance by major nuclear genes is provided by work on tobacco plant tissue cultures (Radin & Carlson, 1978). They obtained tobacco mutants tolerant to bentazon and phenmediphan using haploid plants and an *in situ* selection method with subsequent tissue culture. The regenerated tolerant plants (diplodized with colchicine) were crossed to wild-type plants. Resulting analysis of F_1 and F_2 progeny indicated single recessive gene mutations in the majority of cases.

Picloram tolerant mutants in tobacco were isolated from cell suspensions by Chaleff & Parsons (1978). In four tolerant lines (out of seven originally derived from cell culture), tolerance was conferred by dominant single gene mutations. There was linkage between two of the four mutations (Chaleff, 1980).

Maternal inheritance of triazine-resistance

The majority of triazine-resistant biotypes of weed species are protected from interference with photosystem II of photosynthesis since triazines do not bind to the plastids of resistant biotypes. The chloroplast *psbA* gene codes for a thylakoid membrane protein, which is involved in the prevention of herbicide binding. There is evidence from several species with triazine-resistant biotypes that this trait is maternally inherited, presumably in the plastid genome.

Maternal inheritance of triazine-resistance has been demonstrated in *Brassica campestris* (Souza Machado, 1982), *Senecio vulgaris* (Scott & Putwain, 1981), *Amaranthus retroflexus* (Solymosi, 1981) and *Chenopodium album* (Warwick & Black, 1980).

Maternal inheritance of triazine-resistance has also been demonstrated in *Solanum nigrum* (Gasquez, Darmency & Compoint, 1981) and in *Poa annua* (Darmency & Gasquez, 1981). However, in these two species, a few hybrids inherited the paternal chloroplast characteristics. In *Poa annua*, for example, there was 0.1% transmission of triazine-resistance via pollen. Thus paternal transmission, although possible, is not important as a means of dispersing resistant genes into susceptible populations.

A few species have been found which possess biotypes with intermediate characteristics of triazine-resistance. The intermediate biotypes are characterised by intermediate chlorophyll fluorescence induction curves in which return to a higher steady-state fluorescence lies between the curves for resistant and susceptible biotypes.

Such characteristics have been found in *C. album* (Gasquez, AlMovemar & Darmency, 1985), *P. annua* (Gasquez & Darmency, 1983) and in *C. polyspermum* and *A. bouchonii* (Solymosi, Kostyal & Lehoczki, 1986). This latter work is of considerable interest because in both species, inheritance studies have shown that the intermediate characteristic and resistance were both maternally inherited. There was no segregation for resistance or the intermediate characteristic in F_2 progeny. It appears likely that there is a chloroplast gene coding for intermediate characteristics.

From the available evidence it is clear that there is a wide diversity in the mode of inheritance of herbicide-resistance. This occurs even with a single chemical such as atrazine where resistance is controlled by different genetic systems in different species. There is a strong contrast with resistance to insecticides where the mode of inheritance is similar for a wide range of species (e.g. monofactorial inheritance of resistance to DDT and dieldrin). The apparent flexibility of plant genetic systems provides a warning for the future. Over-dependence on a few chemicals may lead to ever-increasing occurrences of evolution of herbicide-resistance in weeds.

Cross-resistance

Evolution of cross-resistance in weeds is becoming more common. This is a worrying development since, in some cases, the options for using alternative herbicides appear very limited. For example, in Australia, *Lolium rigidum* has evolved resistance to diclofop-methyl (Heap & Knight, 1986) in many locations but also appears to be cross-resistant to the herbicides, fluazifop-butyl and chlorsulfuron and also to two experimental herbicides CGA 82725 and DPX-T6376. *L. rigidum* was not previously field-treated with these chemicals. The sulfonylurea herbicides chlorsulfuron and DPX-T6376 have a completely different site and mode of action from diclofop-methyl. There appears to be no suitable alternative herbicides available to control diclofop-methyl-resistant *L. rigidum*.

Similar inexplicable cross-tolerances have been reported for *A. myosuroides*. This species is highly tolerant to chlortoluron but also possesses tolerance to chemically unrelated chlorsulfuron and diclofop-methyl (Moss & Cussans, 1987).

Co-resistance of triazine-resistant biotypes is not unexpected with herbicides such as bromacil and pyrazon where inhibition of photosynthetic electron transport is presumably the primary site of action of the latter two herbicides (Fuerst *et al.*, 1986). However, an unexpected occurrence of cross-resistance was reported by Rubin, Yaacoby & Schenfeld (1985). In Israel the triazine-resistant weeds *A. myosuroides* and *Phalaris paradoxa* are also cross-resistant to diclofop-methyl. This could cause problems if the triazine-resistant biotype of *A. myosuroides* spread into cereal crops. Clearly, there is a great deal more to be learnt concerning the fundamental biochemistry and physiology of herbicide cross-resistance. The possible appearance in weeds of widespread general cross-resistance is a concern for the future.

Conclusions

The evolution in weed species of resistance to an increasingly wide range of herbicides has created opportunities for the transfer of resistance to crop cultivars. Creation of herbicide-resistant crops can be a considerably more cost-effective approach to the solution of problems of weed infestations than the development of new chemical herbicides. Herbicide-resistant crop cultivars may eventually be developed by either traditional plant breeding methods (e.g. Faulkner, 1982; Darmency & Pernes, 1985; Maehode *et al.*, 1983), by cell culture selection (e.g. Anderson, Georgeson and Hibberd, 1984), by protoplast fusion (e.g. Gressel, Cohen & Binding, 1983) and by genetic engineering (e.g. Shah *et al.*, 1986).

Unfortunately, if the usage of herbicides is concentrated on fewer compounds that are more widely used in a greater range of crops and which are applied year after year, this will only serve to increase the rate of evolution of new herbicide-resistant strains of weeds This is not a speculation but a confident prediction based on our current knowledge. For example, the prediction (Putwain, 1982) that *Alopecurus myosuroides* would evolve resistance to the substituted urea herbicide, chlortoluron, was fulfilled within 18 months.

Therefore, allied to the development of new herbicide-resistant crop varieties, there must be a parallel development of strategies of weed population management which will minimise the probability of evolution of resistance (e.g. Gressel & Segal, 1982) and which will allow containment of resistance if it were to occur. It would be irresponsible to deliberately encourage the evolution of herbicide-resistant weeds if the only alternative herbicides were so costly that they were uneconomic. For example, in bush fruit crops in the UK the alternative herbicides (napropamide, oxyfluorfen, propyzamide) which might replace simazine when an infestation of a resistant weed species had occurred, are approximately ten times more expensive. The alternatives often would be uneconomic in blackcurrants where profit margins are slim whereas in high value nursery stock the alternative herbicides are not precluded on economic grounds.

Evolved cross-resistance is a very recent problem which appears to be getting more common. Although some cross-resistances are explicable since the compounds are all inhibitors of photosystem II (e.g. Solymosi, Lehoczki & Lasky, 1986), Gressel (1986) has suggested that such resistances to photosystem II inhibiting herbicides are predictable due to the existence of plastome mutator genes. Other cross-resistances (e.g. Heap & Knight, 1986) are not as easily explained. Cross-resistances are a potential new concern for the crop-protection industry and for growers.

Potential cross-resistances are not the only concern for growers and herbicide manufacturers. The steady occurrence of new cases of evolved herbicide-resistance will create new challenges in the long term. Many weed species possess persistent buried dormant seed populations. Thus, once a species has evolved resistance to a herbicide at a particular location, it is unlikely that the herbicide could be used again until after an elapsed period of 10 or even 20 years. A small residual seed population of a resistant biotype would rapidly increase again to infestation levels.

Acknowledgements

We are most grateful for useful discussion and critical comment by Professor J. Gressel and for information provided by Mr D. Clay.

References

Ahrens, W.H. & Stoller, E.W. (1983). Competition, growth rate and CO_2 fixation in triazine susceptible and resistant smooth pigweed (*Amaranthus hybridus*). *Weed Science*, **31**, 438–444.

Ali, A., McLaren, R.D. & Souza Machado, V. (1986). Chloroplastic resistance to triazine herbicides in *Sinapis arvensis* L. (wild mustard). *Weed Research*, **26**, 39–44.

Amrhein, N., Deus, B., Gehrke, P. & Steinrücken, H.C. (1980). The site of inhibition of the shikimate pathway by glyphosate. II Interference of glyphosate with chorismic acid formation in *in vivo* and *in vitro*. *Plant Physiology*, **66**, 830–834.

Amrhein, N., Johanning, D., Schab, J. & Schulz, A. (1983). Biochemical basis for glyphosate-tolerance in a bacterium and a plant tissue culture. *FEBS Letters*, **157**, 191–196.

Amrhein, N., Johanning, D. & Smart, C.C. (1985). A glyphosate-tolerant plant tissue culture. In *Primary and Secondary Metabolism of Plant Cell Cultures*. eds. K.H. Newman, W. Barz, & E. Reinhard. Proceedings in Life Sciences, pp. 356–361. Springer-Verlag, Berlin.

Anderson, P.C., Georgeson, M.A. & Hibberd, K.A. (1984). Cell culture selection of herbicide resistant corn. *Crop Science Society of America, Meeting Abstracts*, p56. Quoted by Gressel, J. (1987). Genetic manipulation for herbicide-resistant crops. In: *Biological and Chemical Approaches to Combating Resistance to Xenobiotics*. eds. M. Ford, D. Hollomon, B. Khambay, R. Sawicki., London, Society of Chemistry and Industry.

Anderson, R.N. & Gronwald, J.W. (1987). Inheritance of atrazine-tolerance in velvetleaf (*Abutilon theophrasti* Medik.). *Meeting Weed Science Society of America, Abstracts*, No.125.

Anonymous (1965). Report of the first session of the FAO Working Party of Experts on Resistance of Pests to Pesticides, Rome, October 4th–9th.

Arntzen, C.J., Pfister, K. & Steinbeck, K.E. (1982). The mechanism of chloroplast triazine resistance. Alterations in the site of herbicide action. In: *Herbicide Resistance in Plants*. eds. H. Le Baron and J. Gressel, pp. 185–214. John Wiley & Sons, New York.

Chaleff, R.S. (1980). Further characterization of picloram-tolerant mutants of *Nicotiana tabacum*. *Theoretical and Applied Genetics*, **58**, 91–95.

Chaleff, R.S. & Parsons, M.F. (1978). Direct selection *in vitro* for herbicide-resistant mutants of *Nicotiana tabacum*. *Proceedings of the National Academy of Science, USA*, **75**, 5104–5107.

Comai, L., Sen, L.C. & Stalker, D.M. (1983). An altered aroA gene product confers resistance to the herbicide glyphosate. *Science*, **221**, 370–371.

Comstock, V.E. & Andersen, R.N. (1968). An inheritance study of tolerance to atrazine in a cross of flax (*Linum usitatissimum* L.). *Crop Science*, **8**, 508–509.

Conard, S.G. & Radosevich, S.R. (1979). Ecological fitness of *Senecio vulgaris* and *Amaranthus retroflexus* biotypes susceptible or resistant to triazine. *Journal of Applied Ecology*, **16**, 171–177.

Darmency, H. & Gasquez, J. (1981). Inheritance of triazine resistance in *Poa annua*: consequences for population dynamics. *New Phytologist*, **89**, 487–493.

Darmency, H. & Pernes, J. (1985). Use of wild *Setoria viridis* to improve triazine resistance in cultivated *S. italica* by hybridization. *Weed Research*, **25**, 175–179.

Duncan, K., Lewendon, A. & Coggins, J.R. (1983). The purification of 5-enol pyruvyl shikimate-3-phosphate synthase from an over producing strain of *Escherichia coli*. *FEBS Letters*, **165**, 121–127.

Eastin, E.F. (1981). Movement and fate of metribuzin in Tracy and Tracy M soybeans. *Proceedings Southern Weed Science Society*, **34**, 263–267.

Edwards, Jr. C.J., Barrentine, W.L. & Kilen, T.C. (1976). Inheritance of sensitivity to metribuzin in soybeans. *Crop Science*, **16**, 119–120.

Erickson, J.M., Rahire, M. & Rochaix, J.D. (1985). Herbicide resistance and cross resistance: changes at three distinct sites in the herbicide binding-protein. *Science*, **228**, 204–207.

Faulkner, J.S. (1974). Heritability of paraquat tolerance in *Lolium perenne* L. *Euphytica*, **23**, 281–288.

Faulkner, J.S. (1982). Breeding herbicide-tolerant crop cultivars by conventional methods. In: *Herbicide Resistance in Plants*, eds., H.M. Le Baron & J. Gressel. pp. 235–256. New York: John Wiley & Sons.

Fuerst, E.P., Nakatani, H.Y., Dodge, A.D., Penner, D. & Arntzen, C.J. (1985). Paraquat resistance in *Conyza*, *Plant Physiology*, **77**, 984–989.

Fuerst, E.P., Arntzen, C.J., Pfister, K. & Penner, D. (1986). Herbicide cross-resistance in triazine-resistant biotypes of four species. *Weed Science*, **34**, 344–353.

Gardner, G. (1981). Azidoatrazine: photo affinity label for the site of triazine herbicide action in chloroplasts. *Science*, **211**, 937–940.

Gasquez, J. & Darmency, H. (1983). Variation of chloroplast properties between two triazine resistant biotypes of *Poa annua* L. *Plant Science Letters*, **30**, 99–106.

Gasquez, J., Darmency, H. & Compoint, J.P. (1981). Etude de la transmission de la résistance chloroplastique aux triazines chez *Solanum nigrum* L. *Compte rendus de l'Academie des. Sciences. Paris*, (D), **292**, 847–849.

Gasquez, J., AlMovemar, A. & Darmency, H. (1985). *Pesticide Science*, **16**, 392.

Gressel, J. (1983). Spread and action of herbicide tolerances as used in crop breeding. *Proceedings 10th International Congress of Plant Protection*, 608–615.

Gressel, J. (1985). Biotechnologically conferring herbicide resistance in crops: the present realities. In: *Molecular Form and Function of the Plant Genome*. ed. L. van Vloten-Doting, G.S.P. Groot & T.C. Hall, NATO ASI Series, Vol. 83, pp. 489–504, Plenum Press, New York, London.

Gressel, J. (1986). Modes and genetics of herbicide resistances in plants. In: *Pesticide Resistance: Strategies and Tactics for Management*. Washington, National Academy Press, pp. 54–73.

Gressel, J. (1987). Genetic manipulation for herbicide-resistant crops. In *Biological and Chemical Approaches to Combating Resistance to Xenobiotics*, eds. M. Ford, D. Hollman, B. Khamborg, and R. Sawicki, pp. 266–280, London, Society of Chemistry and Industry.

Gressel, J. & Segel, L.A. (1982). Interrelating factors controlling the rate of appearance of resistance: the outlook for the future. In: *Herbicide Resistance in Plants*, eds. H.M. LeBaron and J. Gressel pp. 325–347. New York: John Wiley & Sons.

Gressel, J., Cohen, N. & Binding, H. (1984). Somatic hybridization of an atrazine resistant biotypes of *Solanum nigrum* with *Solanum tuberosum*. II. Segregation of plastomes. *Theoretical and Applied Genetics*, **67**, 131–134.

Gressel, J., Regev, Y., Malkin, S. & Kleifeld, Y. (1983). Characterization of a s-triazine resistant biotype of *Brachypodium distachyon*. *Weed Science*, **31**, 450–456.

Gressel, J., Shimabukuro, R.H. & Duysen, M.E. (1983). N-dealkylation of atrazine and simazine in *Senecio vulgaris* biotypes: a major degradation pathway. *Pesticide Biochemistry and Physiology*, **19**, 361–370.

Grogan, C.O., Eastin, E.F. & Palmer, R.D. (1963). Inheritance of susceptibility of a line of maize to simazine and atrazine. *Crop Science*, **3**, 451.

Gronwald, J.W., Anderson, R.N. & Yee, C. (1987). Enhanced atrazine metabolism in an atrazine-tolerant velvetleaf (*Abutilon theophrasti* Medik.) accession. *Weed Science Society of America, Meeting Abstracts*, No.154.

Hartmann, F. (1979). The atrazine resistance of *Amaranthus retroflexus* L. and the expansion of resistant biotypes in Hungary. *Novenyvedelem*, **15**, 491–495.

Hartwig, E.G., Barrentine, W.L. & Edwards, C.J. (1980). Registration of Tracy-M soybeans. *Crop Science*, 20, 825.

Harvey, B.M.R. & Harper, D.B. (1982). Tolerance to bipyridylium herbicides. In: *Herbicide Resistance in Plants*, eds. H.M. LeBaron & J. Gressel, pp. 215–233, J. Wiley, New York.

Hayes, J.D., Pfeiffer, R.K. & Rana, M.S. (1965). The genetic response of barley to DDT and barban and its significance in crop protection. *Weed Research*, **5**, 191–206.

Heap, J. & Knight, R. (1982). A population of ryegrass tolerant to the herbicide diclofop methyl. *Journal Australian Institute of Agricultural Sciences*, **48**, 156–158.

Heap, I. & Knight, R. (1986). The occurrence of herbicide cross-resistance in a population of annual ryegrass, *Lolium rigidum*, resistant to diclofop-methyl. *Australian Journal of Agricultural Research*, **37**, 149–156.

Hirschberg, J. & McIntosh, L. (1983). Molecular basis of herbicide resistance in *Amaranthus hybridus*. *Science*, **222**, 1346–1348.

Hollander-Czytko, H. & Amrhein, N. (1983). Sub-cellular compartmentation of shikimic acid and phenylalanine in buckwheat cell suspension cultures grown in the presence of shikimate pathway inhibitors. *Plant Science Letters*, **29**, 89–96.

Holliday, R.J. & Putwain, P.D. (1980). Evolution of herbicide resistance in *Senecio vulgaris*: variation in susceptibility to simazine between and within populations. *Journal of Applied Ecology*, **17**, 779–791.

Jansen, M.A.K., Hobe, J.H., Wesselius, J.C. & van Rensen, J.J.S. (1986). Comparisons of photosynthetic activity and growth performance in triazine-resistant and susceptible biotypes of *Chenopodium album*. *Physiologie Vegetale*, **24**, 475–484.

Kyle, D.J. (1985). The 32,000 dalton Q_B protein of photosystem II. *Photochemistry and Photobiology*, **41**, 107–116.

Le Baron, H.M. & Gressel, J. (1982). *Herbicide Resistance in Plants*. John Wiley & Sons, New York.

Lee, T.T. (1980). Characteristics of glyphosate inhibition of growth in soyabean and tobacco callus cultures. *Weed Research*, **20**, 365–369.

Lupton, F.G.H. & Oliver, R.H. (1976). The inheritance of metoxuron susceptibility in winter wheat. *Proceedings 13th British Weed Control Conference*, 473–478.

Lutman, P.J.W. & Lovegrove, A.W. (1985). Variations in the tolerance of *Galium aparine* (Cleavers) and *Stellaria media* (Chickweed) to mecoprop. *Proceedings of the 1985 British Crop Protection Conference-Weeds*, 411–418.

Moss, S.R. & Cussans, G.W. (1985). Variability in the susceptibility of *Alopercurus myosuroides* (black-grass) to chlortoluron and isoproturon. In: The biology and control of weeds in cereals. *Conference of the Association of Applied Biologists, Aspects of Applied Biology*, **9**, 91–98.

Moss, S.R. & Cussans, G.W. (1987). Detection and practical significance of herbicide resistance with particular reference to the weed *Alopecurus myosuroides* (Black-grass). In *Biological and Chemical Approaches to Combating Resistances to Xenobiotics*. eds. M. Ford, D. Holloman, B. Khambay, and R. Sawicki, pp. 200–213, London: Society for Chemical Industry.

Mudge, L.C., Gossett, B.J. & Murphy, T.R (1984). Resistance to goosegrass (*Eleusine indica*) to dinitroaniline herbicides. *Weed Science*, **32**, 591–594.

Nafziger, E.D., Widholm, J.M. Steinrücken, H.C. & Kilmer, J.L. (1984). Selection of characterization of a carrot cell line tolerant to glyphosate. *Plant Physiology*, **76**, 571–574.

Powles, S.B. (1986). Appearance of a biotype of the weed, *Hordeum glaucum* Steud., resistant to the herbicide paraquat, *Weed Research*, **26**, 167–172.

Price, S.C., Hill, J.E. & Allard, R.W. (1983). Genetic variability for herbicide reaction in plant populations. *Weed Science*, **31**, 652–657.

Putwain, P.D. (1982). Herbicide resistance in weeds – an inevitable consequence of herbicide use? *Proceedings of the 1982 British Crop Protection Conference – Weeds*, 719–728.

Radin, D.N. & Carlson, P.S. (1978). Herbicide-tolerant tobacco mutants selected *in situ* recovered via regeneration from cell culture. *Genetic Research, Cambridge*, **32**, 85–89.

Radosevich, S.R. & Devilliers, O.T. (1976). Studies on the mechanism of s-triazine resistance in common groundsel. *Weed Science*, **24**, 229–232.

Rogers, S.G., Brand, L.A., Holder, S.B., Sharp, E.S. & Brackin, M.J. (1983). Amplication of the aroA gene from *E. coli* results in tolerance to the herbicide glyphosate. *Applied Environmental Microbiology*, **46**, 37–43.

Rubin, J.L., Gaines, C.G. & Jensen, R.A. (1982). Enzymological basis for herbicidal action. *Plant Physiology*, **70**, 833–839.

Rubin, B., Yaacoby, Y. & Schonfeld, M. (1985). Triazine resistant grass weeds: cross resistance with wheat herbicide, a possible threat to cereal crops. *1986. British Crop Protection Conference – Weeds*,. 1171–1178.

Scott, K.R. & Putwain, P.D. (1981). Maternal inheritance of simazine resistance in a population of *Senecio vulgaris*. *Weed Research*, **21**, 137–140.

Schooler, A.B., Bell, A.R. & Nalewaja, J.D. (1972). Inheritance of siduron tolerance in foxtail barley. *Weed Science*, **20**, 167–169.

Shah, D.M., Horsch, R.B., Klee, H.J., Kishore, G.M., Winter, J.A., Turner, N.E., Hironaka, C.M., Sanders, P.R., Gasser, C.S., Aykent, S., Siegel, N.R., Rogers, S.G., Fraley, R.T. (1986). Engineering herbicide tolerance in transgenic plants. *Science*, **233**, 478–481.

Schultz, A., Sost, D. & Amrhein, N. (1984). Insensitivity of EPSP-synthase to glyphosate confers resistance to this herbicide in a strain of *Aerobacter aerogenes*. *Archives of Microbiology*, **137**, 121–213.

Shaaltiel, Y. & Gressel, J. (1986). Multienzyme oxygen radical detoxifying system correlated with paraquat resistance in *Conyza bonariensis*. *Pesticide Biochemistry and Physiology*, **26**, 22–28.

Shimabukuro, R.H., Frear, D.S., Swanson, H.R. & Walsh, W.C. (1971). Glutathione conjugation – an enzymic basis for atrazine resistance in corn. *Plant Physiology*, **47**, 10–14.

Solymosi, P. (1981). Inheritance of herbicide resistance in *Amaranthus retroflexus*. *Novenyterneies*, **30**, 57–60.

Solymosi, P., Lehoczki, E. & Laskay, G. (1986). Difference in herbicide resistance to various taxonomic populations of common lambs quarters (*Chenopodium album*) and late-flowering goosefoot (*Chenopodium strictum*) in Hungary. *Weed Science*, **34**, 175–180.

Solymosi, P., Kostyal, Z. & Lehoczki, E. (1986). Characterization of intermediate biotypes in atrazine-susceptible populations of *Chenopodium polyspermum* L. and *Amaranthus bouchonii*. Thell. in Hungary. *Plant Science*, **47**, 173–179.

Souza Machado, V. (1982). Inheritance and breeding potential of triazine tolerance and resistance in plants. In: *Herbicide Resistance in Plants*, eds. H.M. Le Baron & J. Gressel, pp. 251–273, John Wiley, New York.

Souza Machado, V., Banden, J.D., Stephenson, G.R. & Jensen, K.I.N. (1977). Differential atrazine interference with the Hill reaction of isolated chloroplasts from *Chenopodium album* L. *Weed Research*, **17**, 407–413.

Souza Machado, V., Ali, A. & Shupe, J. (1983). Breeding chlorotriazine herbicide resistance into ruta baga (*Brassica napus* L.) and Chinese cabbage (*Brassica campestris* L.) *Cruciferae Newsletter*, No.8, 21–22.

Stafford, R.E., Comstock, V.E. & Ford, J.H. (1968). Inheritance of tolerance in flax (*Linum usitatissimum* L.) treated with MCPA. *Crop Science*, **8**, 423–426.

Steinrücken, H.C., Schulz, A., Amrhein, N., Porter, C.A. & Fraley, R.T. (1986). Overproduction of 5-enolpyruvyl shikimate-3-phosphate synthase in a glyphosate tolerant *Petunia hybrida* cell lines. *Archives of Biochemistry and Biophysics*, **244**, 159–178.

Sykes, S.R. (1980). Selection for herbicide resistance in Cruciferous crop plants. Ph.D. thesis, University of Liverpool.

Tanaka, Y., Chisaka, H. & Saka, H. (1986). Movement of paraquat in resistant and susceptible biotypes of *Erigeron philadelphicus* and *E. canadensis*. *Physiologia Plantarum*, **66**, 605–608.

Van Rensen, J.J.S. (1982). Molecular mechanisms of herbicides action near Photosystem II. *Physiological Plantarum*, **54**, 440–442.

Vaughn, K.C. & Fuerst, E.P. (1985). Structural and physiological studies of paraquat-resistant *Conyza*. *Pesticide Biochemistry and Physiology*, **24**, 86–94.

Warwick, S.I. & Black, L. (1980). Uniparental inheritance of atrazine resistance in *Chenopodium album*. *Canadian Journal of Plant Sciences*, **60**, 751–753.

Warwick, S.I & Black, L. (1981). The relative competitiveness of atrazine susceptible and atrazine resistant populations of *Chenopodium album* and *C. strictum*. *Canadian Journal of Botany*, **59**, 689–693.

Watanabe, Y., Hanma, T., Ito, K. & Miyahara, M. (1982). Paraquat resistance in *Erigon philadelphicus*. *Weed Research* (Japan), **7**, 49–54.

Whitehead, C.W. & Switzer, C.M. (1963). The differential response of strains of wild carrot to 2,4-D and related herbicides. *Canadian Journal Plant Sciences*, **43**, 580–94.

Wiese, A.F. & Quinby, J.R. (1969). Inheritance of 2,4-D and propazine resistance in grain sorghum. *Abstracts of Meeting of the Weed Science Society of America*, 1969, 29.

Youngman, R.J. & Dodge, A.D. (1981). On the mechanism of paraquat resistance in *Conyza* sp. In: *Proceedings 5th International Congress of Photosynthesis*, ed. G. Akoyunoglou, pp. 537–544. Balaban, Philadelphia.

J. GRESSEL

Conferring herbicide resistance on susceptible crops

Introduction

The chemical industry is continually finding it harder and far more expensive to develop new selective herbicides. Conversely, there are already many herbicides in existence with excellent spectra of weed control which would be useful in a given crop, except for a minor problem; they also kill the crop. The chemical industries biorational approach to this problem is to study the herbicide degrading enzymatic pathways specific to that crop. The herbicidal moiety could then be modified by adding a chemical group that will be recognised by those enzymes, thus aiding in the herbicide degradation by the crop. Another biorational approach is to make current herbicides more effective by adding synergists that either prevent their breakdown, or prevent detoxification of toxic products in some weeds (cf. Gressel & Shaaltiel, 1988).

The approach taken by the biotechnologists in industry and the public sector is diametrically opposed to that of the chemist: to make crops genetically and biochemically amenable to herbicides by conferring resistance. There are vast differences in the levels of sophistication of the biotechnologies involved. The biotechnological approaches have the advantage of often dealing with time-tested herbicides with well elucidated weed control properties and toxicology. Registration requirements can be rather simple if it can be shown that the metabolites and residues of the herbicide match those found in crops where the herbicide is already used. Where this is not the case, more extensive research provisions might be required. If a major crop is biotechnologically transformed, and the herbicide in question is one that has hitherto been minor, a new problem can ensue; ground water residues and other environmental contaminants may increase to the levels of detectability, as usage increases.

Choice of crop and herbicides. The farmer needs the best possible cost-effective weed control. This would mean to obtain resistance to herbicides that are off-patent and thus less expensive, and in those major crops that still have less than desirable herbicides. The chemical industry wants to make more crops resistant to herbicides that are still on patent. This may not materialise; the herbicide may go off-patent before resistant crops are ready for marketing. Commercial biotechnology companies that are not associated with the chemical industry may be expected to confer resistance to herbicides that are off-patent, unless they have agreements with herbicide producers to gain part of the profit from added herbicide sales. They will wish to cover their R&D costs by seed sales. As the biotechnology products are man-made they will have relatively broad patent protection in the USA compared to standard varieties (cf. Lesser, 1986). Other countries are expected to evolve towards the same levels of intellectual protection. Farmers save seed of many crop species, for the next year. These crops will be less desirable for the biotechnology companies, even with patent protection, as such varieties are usually purchased for about a fourth of the areas seeded. This may preclude deep commercial interest in small-grains, soybeans and other crops which are not sold as F_1 hybrids that self destruct in the F_2. Only maize, sunflowers, rape, and high cost (but low volume) vegetable seeds remain as objects of major interest. Novel small-grain and soybeans are left for the non-profit university and government biotechnologists. There is a need for resistance in the vegetable crops, where it often is not economically feasible to develop specific herbicides. Another area with great need for herbicide resistance is the forestry industry (Haissig, Nelson & Kidd, 1987). The economic aspects of selecting for resistance are well summarised by Benbrook & Moses (1986).

Where are the results? The number of reviews on potential is far greater than that of 'hard-data' papers with concrete results. Biotechnology oversold itself by suggesting that with genetic engineering, you need only take the gene for resistance, introduce it into the crop and there you have it. It was not quickly enough discovered that except for kanamycin resistance, the genes were not available, and when found, were not as amenable as had been thought. Prokaryotic genes do not always work well in eukaryotes. Initially, the need for gene-constructs giving the correct levels of expression as well as tissue and organ specific targeting were not realised, and only recently have areas such as environmental controls of expression come under study (cf. Fluhr *et al.*, 1986). It was overlooked that there were no methods to transform most major crops. Now, as many of the problems have been pinpointed and solved, there is some loss of interest on the part of investors. The desire for immediate results often precluded performing the necessary agronomic, physiological, biochemical and molecular biological groundwork needed before embarking on genetic engineering. Too many projects started out as massive efforts and then dwindled. Clearly the lack of needed basic background information should have engendered small-scale probes to obtain missing data and only then escalate efforts. Finally there are some real results, with seemingly good genes to report from the major efforts invested.

Resistance genes of choice

At least three different sometimes interrelated routes can lead to resistance: (a) a modification in the structure or the quantity of the target enzyme; (b) enhanced degradation or dissipation of the herbicide; (c) enhanced rates of detoxifying phytotoxic entities whose synthesis is caused by the herbicide.

Different weeds have been known to evolve one or more different mechanisms for resistance to a given herbicide. Crop resistance to a herbicide is often different from mechanisms that evolve conferring resistance to weeds. Additionally, it is not always known how a crop is resistant to a given herbicide. Microorganisms may yet have different methods of resistance to plants. For these reasons, the biotechnologist has a choice of genes. It is well known in evolution that selection for a given new trait most often has a cost that is seen as a loss in competitive-fitness as compared to the wild type. In our case, it means that most herbicide resistances in crops may well have a cost in yield. This yield penalty is best measured by comparing the resistant variety with the susceptible parent variety in hand-weeded situations. This cost can be quantified in many ways; the best way is to measure yield in comparison with alternative herbicides under weedy situations. Hand-weeding is clearly costly in itself. Because there may always be competing approaches for a given crop, the biotechnologist conferring good resistance, with a herbicide having a good weed control spectrum, at lowest crop yield loss due to resistance, should succeed. These parameters can be estimated well in advance. Consideration of the alternative resistance pathways can assist in choosing crops and herbicides.

Resistance through the target enzyme

Target site modification. Many herbicides act as anti-metabolites, blocking the activity of an enzyme by binding to it. Single gene mutations modifying a single amino acid can change the structure of the protein, thus precluding herbicide binding. This has been shown to be the case in the *psb A* gene product giving rise to atrazine resistance (Gressel, 1985), enol-pyruvate-shikimate phosphate (EPSP) synthase giving rise to glyphosate resistance (Comai, Sen & Stalker, 1983; Fillatti *et al.*, 1987), and acetolactate synthase (ALS) giving rise to resistance to imidazolinone and sulfonyl urea herbicides (Falco, Dumas & McDevitt, 1985). Single site mutations are clearly the easiest to obtain, both in plants and micro-organisms. For this reason, this has been the first route of choice with biotechnologists, as outlined in the later sections. The protein modification precluding herbicide binding may have an effect on the binding of the natural substrate or other properties of the modified enzyme. Evolution of optimal enzyme properties has continued for millenia; few or no mutations are neutral, i.e. without effects. Otherwise they would be the wild type. Thus, it is hard to expect that target site mutations should be neutral. The case with atrazine resistance is clearly the most extreme. Maize is resistant to atrazine because it degrades it.

Only *Abutilon* of the more than 50 species that evolved resistance to atrazine has been clearly shown to possess degradation as a mode of resistance (Andersen & Gronwald, 1987). Another weed, *Brachypodium* seems to have partially co-evolved this mechanism (Gressel *et al.*, 1983). Most other weeds have evolved resistance through a mutation in the chloroplast-inherited *psb*A gene. Its product is probably one of the most highly conserved proteins known, and all of the triazine-resistant mutation sequences show an amino acid transversion near the purported active site (Chapter 2). This mutation results in at least a ten-fold slower flow of electrons through this protein, which is part of photosystem II of photosynthesis. There is debate if this slower flow can decrease photosynthesis but there is a 20 to 30% loss of yield and a considerably lower competitive fitness in all cases of rapidly growing species that were closely analysed (Conard & Radosevich, 1981; Gressel & Ben-Sinai, 1985). The triazine resistant biotype of one of the grasses *Phalaris paradoxa* may not be unfit (Schonfeld *et al.*, 1987*a*). This greater fitness has been suggested to be the result of a difference in another amino acid, which is also peculiar to the wild type of this species (Schonfeld *et al.*, 1987*b*). These and other results have led the molecular biologists to try to truly engineer multi-amino acid changes by casette mutagenesis, where whole sequences of changed genes for amino acids are inserted as a block in an effort to decrease the unfitness. This technique allows inserting changes that cannot be caused by mutagens that modify single nucleotides. The results of these efforts have not yet been reported.

The lack of fitness of mutant EPSP-synthase the target enzyme for glyphosate, and the seeming fitness of a mutant ALS, the target of sulfonylurea and imidazolinone herbicides are described in later sections.

Resistance through elevated levels of target enzyme. There are many cases of resistance to drugs (Schimke, 1982) and insecticides (Moches *et al.*, 1986) which have been shown to be due to replications of the target site enzyme. Plant cell cultures have evolved resistance to glyphosate by much higher production of EPSP-synthase (cf. Smith *et al.*, 1986) and to glufosinate by replication of the gene for glutamine synthase (Donn *et al.*, 1984). No one has been able to regenerate plants from any of these cultures. Such increases could theoretically be due to modifications in genes controlling the rates of enzyme synthesis. Instead, the resistance has been due to a specific replication of the gene coding for the enzymes binding the herbicide. If the herbicide binding is an infinitesimal proportion of the total cellular protein, and the herbicide is used at very low rates, even a 50 fold increase in the enzyme could have little cost in crop yield. This may not be the case with EPSP-synthase as it is estimated as being a few percent of the soluble chloroplast proteins in resistant cells (Amrhein pers. comm.). Such plant cells would require at least a few times more EPSP-synthase to be resistant to the highest used field levels, without the margin of error required of selective herbicides.

Toxicological considerations with target site resistance. Most naturally resistant crops degrade their selective herbicides and no residues appear in the edible portions or in the trash. This information is routinely required for registration. The herbicide may remain unchanged in the plants bearing some target site modifications. The producers will then have to show that residue in the edible portion has no human toxicity, and that the residue in the trash degraded in the soil. Both the consumer and regulatory climates at present are such that there should be no pesticide residues in foodstuffs, even if they are non-toxic. These are important considerations in choosing biotechnological systems. With many herbicides there is little problem, e.g. most crops slowly degrade s-triazines, just not rapidly enough to remain alive under field conditions. Conversely, there is no information on the fate of glyphosate in resistant plants, and no one has found metabolic products of glyphosate when it was applied at sub-lethal doses.

Herbicide degrading enzymes

Many different enzyme systems are known to degrade herbicides, both in crops, weeds and micro-organisms, especially *Pseudomonas* (see review by Comai & Stalker, 1986). Many microorganisms are being selected or engineered to degrade herbicide residues and spills. Many degrading enzymes are only amenable to the genetic engineering approaches, and not the simpler selection approaches of conferring resistance (described in later sections). If a plant has a low level of a herbicide degrading enzyme system, it is conceivable that this level can be increased by selection techniques. If there is no degradation at all, selection is nigh impossible. It is a basic tenet of genetics that it takes a very long time, even in evolutionary terms, to evolve a new gene, except by small modifications of a pre-existing gene. Thus, if there is no gene for degradation, then it will not evolve *de novo*, and must be transferred from another species. Degradases were not the first choices of the biotechnologists because initial enzymological studies were required with 'alien' organisms (i.e. not *E. coli* or tobacco). The route to resistance with degradases can be more involved than with target site mutations. Conversely, if the degradase is efficient, the cost in yield should be much lower.

Degradation systems that have simple substrates and high herbicide specificity, e.g. sugar-transferases that glycosylate herbicides, acetyl transferases that add ubiquitous acetyl CoA are cases in point. Enzymes using substrates that are often less available can be problematic, e.g. glutathione in herbicide degradation, or glycosyl transferases using rare sugars. Many microbial degradation systems are part of 'scavenger' systems, providing minerals and metabolites to the organism. Such systems may be incompatible with higher plants, as the higher plant cannot 'excrete' unwanted metabolic products into the medium.

Many herbicides are degraded by plants and micro-organisms using cytochrome-P_{450} mixed function oxygenases (MFOs). These monooxygenases are membrane bound and very hard to work with, and microbial oxygenases could be deleterious in a plant host. MFOs have other properties that must be fully studied before being genetically engineered in a crop; they often confer multiple cross-resistances to many chemically unrelated pesticides. This has been well documented with insecticide resistances and has been implicated in some of the recently evolved multi-herbicide resistances (cf. Gressel, 1987). Herbicides are often used to control crops, i.e. the 'volunteers' that appear due to crop seed shed the previous year. Maize appearing in a soybean field is clearly a weed. If selected or engineered MFOs were to confer too many co-resistances, there would be no way to selective control such volunteer ex-crops.

Detoxification of herbicide generated toxic products

Many herbicides, especially those interacting with photosynthesis or otherwise photodynamically generating active oxygen species, lipoxidise plant membranes causing death. Elevation of the levels of the enzymatic pathway that detoxified naturally produced oxygen radicals has been correlated with resistance to many such herbicides. At least three enzymes are elevated (Shaaltiel & Gressel, 1986) and they co-segregate with resistance through the F_2 generation in *Conyza* species that evolved paraquat resistance (Shaaltiel *et al.*, 1988). It would be hard, with our present knowledge, to conceive how to engineer crops with such a pleiotropic, multienzyme system probably under a gene that controls transcription or translation levels. More information on how such control genes work will clearly be needed.

Approaches and achievements

Whole plant breeding and selection

Field selection for resistance should be very hard to perform because of the numbers of plants involved, the low selection pressure that can be applied, and the season-long generation time (Fig. 1). Large expense and long times can be involved. Even with the highest selection pressures (kill rates) there are many more susceptible escapees (plants that are somehow missed), than plants that are genetically resistant (i.e. the natural mutation frequency). If resistance is dominantly inherited on an allele of a single gene, one plant in 10^5 to 10^7 should be resistant; if it is recessively inherited on a single gene, the frequency is calculated at being 10^{-10} to 10^{-14}. The recessive frequency may be much higher than this estimate due to somatic recombination (cf. Williams, 1976). Very high seeding rates would usually be required, and there would be many progeny in the first year, and many years will be required to select in the field. There is always the possibility that there will not be a single gene for resistance in the genome and many years of selection will be wasted.

Interestingly, one group has been successful (at the first level) in obtaining resistance using field screening for resistance in alfalfa (lucerne). They used chlorsulfuron at highly toxic levels, and chlorsulfuron is a herbicide with one of the highest residual activities. They were thus able to apply a very high selection pressure. Twenty million alfalfa seeds were sown and 15 putatively resistant plants appeared after a preplant incorporation of 16 g ha^{-1} (Stannard & Fay, 1987).Four of these retained the resistance after post-emergence and laboratory retreatments, and have been cloned up for breeding purposes. If the seeds had been planted at the normal density of 1500 seeds m^{-2}, then 1.3 ha would have been used. Obviously they were more densely planted or it would have been hard to find those twenty plants. None of the plants were resistant at the level of ALS, the target enzyme.

Fig. 1. Rate of increase in truly resistant individuals using field selection. The rates were calculated using the equations in Gressel & Segel (1978), assuming that germination is uniform and immediate, and that no fitness differential is allowed to be expressed between resistant and wild-type strains. The number of repeated treatments represents the number of generations treated. In the field, this equals seasons, unless second post-emergence treatments can be used. In cell cultures, this equals doublings in the presence of the herbicide. The estimated increase needed to attain 100% resistant individuals with inheritance at different allelic situations is shown on the right axis. The point for a single recessive gene may be much closer to that of a single dominant gene if the data from lower plants (Williams, 1976) can be extrapolated. The frequency of mutants in field situations which have maternally inherited PSII herbicide resistance is probably less than 10^{-20}. Note that the frequency of truly resistant individuals will still be undiscernable (less than 1% resistance) until one or two treatments before 100% resistance is achieved.

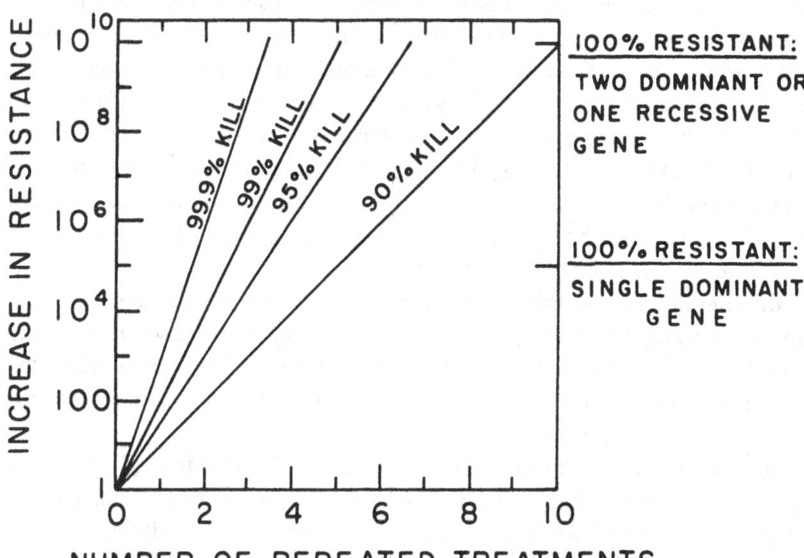

Laboratory selection at the whole plant level. Mutagenised seeds of *Arabadopsis thaliana* (not a crop) have been used by various groups, often as a prelude to isolating resistant genes for genetic engineering. The tiny seeds were evenly treated with pre-emergence type herbicides. *Arabidopsis* is less amenable to post-emergence herbicides; a group that is gaining in preference as they can be sparingly used only when needed. Another system gaining in interest for plant selection is the fern *Ceratopteris*. Selection pressure is applied on the photoautotrophic haploid gametophytes allowing easier selection of recessive mutants for resistance. Mutants resistant to paraquat, acifluorfen and other xenobiotics have been isolated. Paraquat resistance was monogenically inherited (Hickok, Warne & Slocum, 1987). These mutants can be used after analysis of fitness for isolation of the resistance genes for genetic engineering. If the gene for resistance is a pleiotropic control gene, there may be problems in recognising it for isolation. Using similar methods, large numbers of mutagen treated soybean seeds were selected for chlorsulfuron tolerance (Sebastian & Chaleff, 1987). None of the mutants had target site resistance at the level of ALS.

Attempts have been made to use isolated embryos from seeds to find enhancement of asulam resistance in barley (Giffard, Collin & Putwain, 1986). While the selected embryos had a greater level of tolerance, the plants derived from them did not have a normal growth rate in the presence of the herbicide. This could be due to the small (1500) initial population used for selection.

Plant breeding. There have been cases where herbicide resistance has been found in the gene pool of a given crop species. Breeders had to transfer the resistance into the varieties of choice by crossing, back-crossing and field testing. Even using two generations per year, there are at least five years of work before a herbicide resistant variety can be field tested and released. Resistance is often achieved in non-varietal material by the selection or engineering approaches outlined in later sections and must be transferred to 'name' varieties by such a five-year process. Biotechnologists have much to learn from the breeder, especially in the initial use of large numbers of biotypes which seemingly have the desired trait.

Metribuzin resistance in soybeans is inherited as a single dominant gene (Edwards, Barrentine & Kilen, 1976), and was transferred to an important yet sensitive variety (Eastin, 1981). Surprisingly the breeding was done by the public sector, as neither the herbicide producer nor the seed companies saw a market need at that time for performing this simple transfer.

Paraquat tolerance in *Lolium perenne* evolved in no-till wheat in England and the polygenically inherited tolerant material was used as the basis of a breeding project. The resulting material allows the use of low rates of paraquat in establishing stands of perennial rye-grass in pastures (cf. Faulkner, 1982).

Maternally inherited triazine resistance has evolved in many weed species (Chapters by van Rensen and by Putwain & Collin). Two of the weeds were sufficiently related to crops to allow crossing the trait into related crop varieties. The crosses from *Setaria viridis* (green foxtail) into the crop *Setaria italica* (foxtail millet) were relatively straightforward (Darmency & Pernes, 1985), but the transfer from *Brassica campestris* into *Brassica napus* (rape) was complicated by differences in chromosome numbers between the species (Souza-Machado, 1982). The yields of the resulting crop varieties in both cases are considerably lower than the original variety (cf. Ricroch *et al.*, 1987; Grant & Beversdorf, 1985; Gressel & Ben-Sinai, 1985, Vaughn, 1986). This loss of yield can be somewhat alleviated by further hybridisation (Grant & Beversdorf, 1985) but the resulting hybrids always yielded less than the reciprocal hybrids whose offspring were triazine sensitive (Hume & Beversdorf, 1986; Beversdorf, Hume & Donnelly-Vanderloo, 1987). Cytoplasmic male sterility, a trait allowing inexpensive hybrid seed production has been introduced by protoplast fusion into the triazine resistance rape lines (Yarrow *et al.*, 1986). This will allow some yield increase at lower cost. This loss of yield is probably due to the low fitness that is usually (cf. Gressel, 1985) but possibly not always (Schonfeld *et al.*, 1987*a*, *b*), inherent in the chloroplast genome mutation conferring triazine resistance. The triazines are excellent candidates for crop resistance as they are among the least expensive herbicides. Conversely, there is the danger of evolution of more triazine-resistant weeds if triazines are used without rotation with other non-photosystem II inhibiting herbicides.

Tissue and cell culture selection techniques

Cell cultures have short 'generation times' (doubling in 3–5 days), 10^5–10^6 cells per ml, with each cell theoretically representing a whole plant. This allows the possibility of very uniform treatment with herbicides at high selection pressures. Thus, cell cultures have many advantages over the field for selection of resistance. Unfortunately, many species cannot yet be regenerated from tissue cultures and often only some varieties of some species can be regenerated. Reports continually appear of previously recalcitrant crop species being regenerated from cell culture, including soy, wheat and maize.

Selection with non-photosynthetic cells. Tissue culture selection techniques were pioneered with tobacco; a dangerous weed! Using cell culture techniques there is now tobacco resistant to picloram (Chaleff & Parsons, 1978), sulfonylureas (Chaleff & Ray, 1984), paraquat (Furasawa *et al.*, 1984; Hughes, 1986), and other herbicides using the techniques outlined in Fig. 2.

In vitro selection has finally been achieved with a modicum of success with a 'real' crop, maize. It was first necessary to develop methods for reproducible regeneration.

Selection was then applied for resistance to the imidazolinone group of herbicides and partially dominant resistance was achieved. Better resistance was achieved when both resistant alleles were present (Anderson, 1986). The regenerable maize was not a commonly used inbred for generating commercial varieties. The resistant maize strains have been transferred to a seed producer for introducing the gene into commercial inbreds, testing for effectivity, and eventual release after fulfilling the concomitant requirements for 'new use' registration. The resistance is due to a site modification in acetolactate synthase (Chapter 7), and the imidazolinone resistant maize has cross resistance to their competitors' sulfonylurea herbicides, which act by inhibiting the same enzyme (Anderson, 1986). This may be a commercially daunting problem with other herbicides as well.

Microspore selection, and selection at the level of protoplasts derived from microspores have been used to select for *Brassica napus* mutants resistant to chlorsulfuron. Regenerated resistant plants were fertile after colchicine doubling. Some of the regenerants were resistant at the level of ALS, the target enzyme, and some had some other mechanism of resistance (Swanson *et al.*, 1988).

Selection with green tissues. None of the herbicides described above affect photosynthesis, a system lacking in most cell cultures. Methods had to be worked out to treat photosynthetically competent mutagenised leaves with herbicides and then

Fig. 2. Typical methods used to select herbicide resistance to herbicides that do not specifically inhibit photosynthesis. Initial selection can be in suspension cultures, large callus pieces on micro-calli derived from plating suspension cultures on medium and then overlaying the herbicide. *Source*: From Meredith & Carlson (1982), by permission.

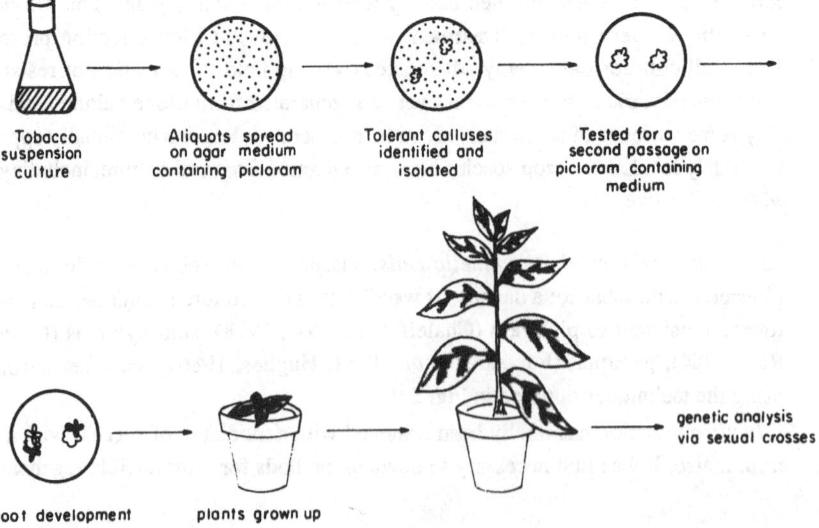

isolate small, resistant green islands of cells (Fig.3). These were excised to tissue culture medium for propagation. Thus, there is now tobacco with nuclearly inherited resistance to the photosynthesis inhibitors bentazon and phenmedipham (Radin & Carlson, 1978). More recently, tobacco having chloroplast inherited streptomycin resistance was obtained after using a specific chloroplast DNA mutagen (Fig. 3) (Fluhr *et al.*, 1985). This technique using seedlings and isolating 'green islands' may be amenable to finding resistance to photosystem II inhibiting herbicides that interact with the *psb*A gene product. This will depend on problems resulting from possible recessiveness.

Protoplast fusion transfer of resistance

There are cases where resistance appears in one species, and there is a desire to transfer it to a related but not cross-breeding species. Interspecific protoplast fusions have been achieved with a transfer of genetic material, but the resultant plants are usually infertile (Melchers *et al.*, 1978; Binding *et al.*, 1982). The procedure has some use only when the genes for resistance have unknown products (and thus are hard to isolate by molecular biological procedures for genetic engineering), or there is

Fig. 3. Methods of isolating mutants to herbicides specifically inhibiting photosynthesis. An example of a tobacco cotyledon from mutagenised seed showing green islands resistant to streptomycin treatment is shown. *Source*: Modified, consolidated and redrawn from Meredith & Carlson (1982) and Fluhr *et al.* (1985).

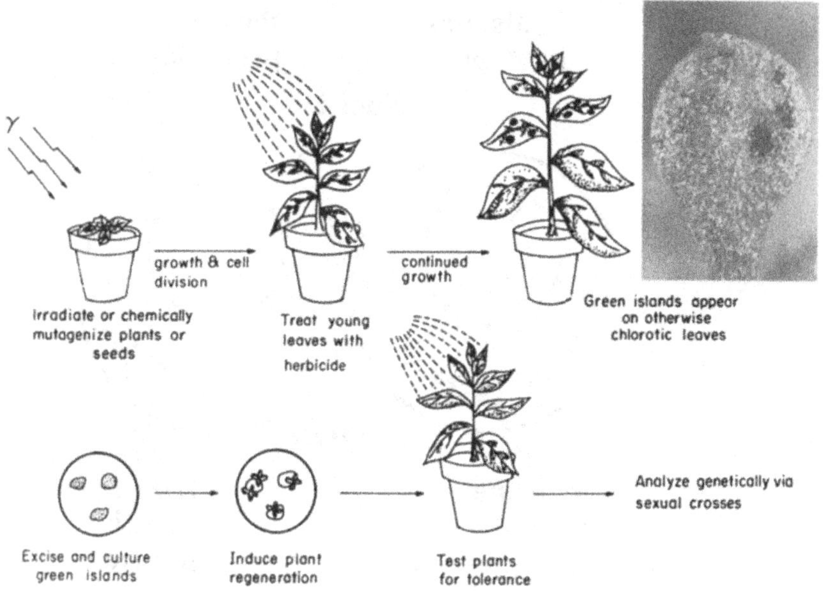

no easy way to introduce the known gene or genes. The latter has been the case for chloroplast inherited genes; there is no simple way yet to transform chloroplasts by genetic engineering with the necessary levels of gene expression.

Protoplast-fusion transfer of triazine resistance was suggested as a possible method (Gressel, Zilkah & Ezra, 1978), just after the first indication appeared that resistance was maternally inherited. The protocol suggested was to modify a technique used to transfer cytoplasmic male sterility (Zelcer, Aviv & Galun, 1978). The triazine resistant donor was first X-irradiated to inactivate its nucleus as the nuclear genome is undesirable in the product (Fig. 4). Most protoplasts from a name variety will regenerate back to that variety after fusion. Some will have the introduced maternal traits, if donor-nucleus inactivating protocols are used. If the technologies for regenerating plants from rape-seed protoplasts that are now available had been available eight years ago, it might have been simpler to separately transfer resistance into each variety by this means, instead of the multiple back-crosses now required. Triazine resistance has been introduced into tobacco (Menczel *et al.*, 1986) and into

Fig. 4. The transfer of chloroplast-inherited resistance by protoplast fusion (hybridisation). The protoplasts of the resistant *Solanum nigrum* were irradiated to cause functional enucleation. They were then fused with potato protoplasts. For reasons as yet unclear, there is a segregation to one or the other plastid type during subsequent cell divisions giving calli and then plants with one or other plastid types. This type of protocol has been successful in transferring triazine resistance to varieties Mirka and Desiree (Gressel *et al.*, 1985). *Source*: From Weller *et al.* (1987), by permission.

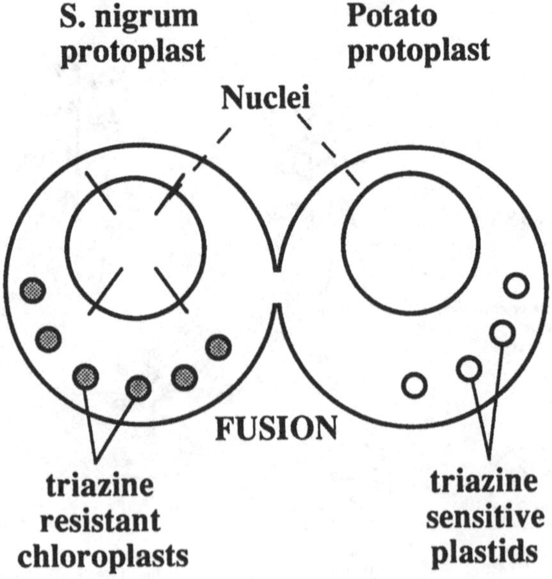

potatoes (Gressel, Aviv & Perl, 1985) by this procedure. Although there seems to be a need for inexpensive, good, broad-spectrum, season-long weed-control in potatoes (Weller, Masiunas & Gressel, 1987), it may not be easy to commercialise the triazine resistant potato varieties, even if they prove to have good yields. Present weed-control costs are a small proportion of the total production costs in potatoes. Potato seed companies feel that they could not gain much of a premium for triazine-resistant potato varieties. They have shown little interest in testing and marketing the resistant strains of varieties Desiree and Mirka that have been generated.

Genetic engineering of resistance

Much effort has been made to learn about the processes involved in successfully genetically engineering plants; and engineering in crops for herbicide resistance has become a major economic target. The process has been over-simplified in popular exposes: simply isolate the gene for resistance, insert it into a plasmid, then into a vector system to introduce it into plant cultures, and regenerate the cultures as a newly resistant crop (Fig. 5). Kanamycin resistance is regularly engineered into tobacco protoplasts and regenerated to resistant plants in undergraduate university courses. The problems entailed in actually doing this with real herbicides and real crops are discussed below. Much of the effort has been towards engineering glyphosate resistance into crops, with commercial laboratories competing and encountering many obstacles that were not even envisaged when the research began.

One of the advantages of the genetic engineering approach is that once the gene is successfully engineered into a plasmid vector system, with all the necessary control

Fig. 5. General scheme for genetically engineering plants for herbicide resistance. This scheme does not include the use of selectable markers commonly used to enrich for and test for transformants.

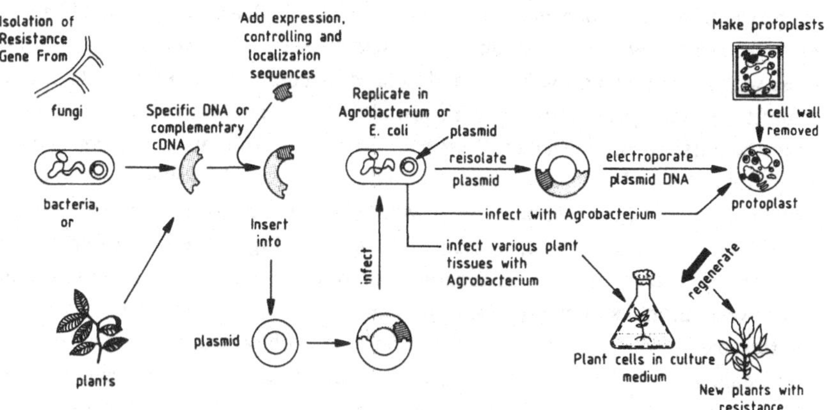

sequences, it can be used to transform many different plant species. Besides the solanaceous species which have been used as models, there are reports of transforming soybean, cotton, sunflower and maize using one of the simple *Agrobacterium* infection systems, albeit with model marker genes. Many species have been transformed by incubating protoplasts with plasmid DNA and pushing it through the cell membrane with polyethylene glycol, or electroporating it through the membrane with a high voltage pulse or even pelting cells with DNA on microprojectiles (Klein *et al.*, 1987). The major remaining obstacles were the cereals. This may have been overcome by de la Pena, Lorz & Schell (1987) who injected DNA into floral tillers of rye, where some was presumably taken up into the pollen and ovules. Injected plants were cross pollinated and the dominant marker gene was expressed in a low 0.07% of the resulting seeds.

Engineering glyphosate resistance

It has been elegantly shown that glyphosate is a strong inhibitor of the enzyme EPSP-synthase (Chapter 6). EPSP-synthase is clearly the sole target site of glyphosate in bacteria, as mutants with modified enzyme are totally resistant to glyphosate (Comai *et al.*, 1983; Sost, Schulz & Amrhein, 1984). EPSP-synthase was presumed to be the sole site of glyphosate action in plants as the effect of glyphosate in cell cultures could be reversed by supplying aromatic amino acids. Two approaches were initially taken to introduce glyphosate resistance; mutated target site, and overproduction of EPSP-synthase.

Modified target site enzyme. Glyphosate resistant *Salmonella* and *Aerobacter* mutants with target-site resistance were isolated (Comai *et al.*, 1983; Schulz, Sost & Amrhein, 1984, respectively). Sost *et al.* (1984) carefully characterised highly purified enzyme from *Klebsiella* and found a 16-fold lower affinity for one of the natural substrates in the resistant bacteria. Such an enzyme would have too high a cost in a plant. Comai *et al.* (1985) carefully selected among glyphosate resistant *Salmonella* mutants to find those with near normal growth rates and isolated mutants with better natural substrate affinities (Comai, pers. comm.), and showed that the mutation was due to a single amino acid transversion (Stalker, Hiatt & Comai, 1985). The natural promoters of the T_i plasmid of *Agrobacterium* were fused to the resistant *Salmonella* genes to obtain tomato plants that were tolerant to otherwise lethal rates of glyphosate (Fillatti *et al.*, 1987). The non-specific expression of a gene at high levels is often costly, unnecessary, and can lead to the accumulation of toxic products. In these cases, the plants were still not sufficiently resistant to field levels of the glyphosate. One reason for this may have been that EPSP-synthase is a chloroplast localised enzyme (Mousdale & Coggins, 1985), and the engineered resistant enzyme was not getting into the plastids. The nuclear coded gene has now been fused to a gene sequence for the transit peptide (Della-Cioppa *et al.*, 1987). Using such a construct with a mutant

EPSP-synthase from bacteria, they have transgenic tobacco plants which appear to be resistant to 3.6 kg/ha glyphosate (Kishore *et al.*, 1988) close to the maximum used of 4.5 kg/ha for controlling perennial weeds. The plants still appear to be visibly inhibited and chlorotic at their growing points and yield data have yet to be presented. The mutant EPSP, synthase used has a 40-fold higher K_m for the natural substrate of the enzyme than the wild type. This is partially offset by using a promoter that causes the plant to synthesise more enzyme. The resistant enzyme is produced in addition to the sensitive enzyme, and is probably only active while glyphosate is present.

Amplified sensitive EPSP-synthase. Glyphosate resistant *E. coli* that grossly overproduces EPSP-synthase has been isolated. Rogers *et al.* (1983) have engineered multiple copies of the gene for normal EPSP-synthase into plasmids and then back into the host bacteria. These bacteria then make nearly 100-fold more enzyme. The K_m and other properties of the enzyme remained unchanged, and such mutants were used in their early genetic engineering efforts. Numerous laboratories have found plant cell lines with high levels of resistance, but none would regenerate to plants. The mode of resistance was traced to an overproduction of EPSP-synthase wherever checked (cf. Steinrücken *et al.*, 1986), presumably due to gene duplication. Unlike bacteria, no target-site mutants were found in any of the many cell culture systems subjected to selection pressure. One of these replicated genes was spliced to control sequences from a viral plant pathogen that confers high expression of viral coat protein, obtaining *Petunia* tolerant to 0.8 kg ha^{-1} glyphosate (Shah *et al.*, 1986*b*).

Is EPSP-synthase the sole site of glyphosate action in all plants? This question, critical for genetic engineering of glyphosate resistance in crops should clearly have been answered in a preliminary series of physiological and biochemical studies. Only bits of circumstantial evidence are available. The effort of glyphosate on many cell cultures and whole duckweed plants can be reversed by aromatic amino acids. The aromatic amino acids do not reverse the glyphosate action in some other cell cultures and have never been shown to reverse glyphosate action on a whole terrestrial plant. Indeed, other disparate compounds have been found to reverse glyphosate action in some cell cultures including simple organic acids (Killmer, Widholm & Slife, 1981) and auxins (Lee, 1980). 2,6-Dihydroxyacetophenone has been found to protect higher plants from glyphosate action (Lee & Starratt, 1986). It is rather clear that EPSP-synthase is the primary site of action of glyphosate, insofar as its activity is suppressed at the lowest glyphosate concentration. This information is not enough to understand the complete herbicidal activity of glyphosate. Keen, Holliday & Yoshikawa (1982) found that a spray treatment of 10 μM glyphosate suppressed EPSP-synthase activity in leaves, but without any visible signs of lethality. Glyphosate is sprayed in the field at rates equivalent to 1,800 to 71,000 μM (Worthing, 1987). Many secondary sites of glyphosate action may well be hidden in this huge span from the concentration needed to suppress EPSP-synthase, to that required for herbicidal activity.

Little is known about the natural degradation of glyphosate in plants, as dead plants cannot degrade glyphosate. It may be wise to consider using 'glyphosatases' for conferring resistance, if such genes would not be deleterious to plants. Glyphosate is biodegraded in the soil and various organisms can utilise glyphosate as a sole phosphorus source (cf. Comai & Stalker, 1986). Such genes might provide the next round of glyphosate-resistant engineered plants or may be added to the mutant EPSP-synthase or replicated natural EPSP-synthase type resistant transgenic plants.

Engineering progress for other resistances

Herbicide resistant ALS. The imidazolinone and sulfonylurea herbicides inhibit acetolactate synthase (Chapter 7). Tobacco (Chaleff & Ray, 1984), maize (Anderson, 1986), yeast and bacterial mutants have been isolated with a resistant enzyme. The ALS genes from bacteria and yeast have been isolated, cloned, and sequenced. A single amino acid difference between the resistant and susceptible ALS was found in bacteria (Falco *et al.*, 1985). There seem to have been problems transforming plants using the bacterial genes, possibly because of promoters. The work with the microorganisms was not wasted. It has not been possible to sufficiently purify the target-site modified ALS that appeared in resistant tissue cultures of tobacco, to prepare antibodies against it, or micro sequence it. This hampered isolation of the resistance gene from tobacco, that should have the ideal control sequences. A hybridisation probe made from the yeast ALS gene was used to 'fish' the plant gene with a native control sequence from an enzymatically chopped plant DNA genome, and this will be engineered for transformation. The target site ALS mutants may well be the rare case with no effective fitness loss, as none had been seen in the microbes or plants with the modified gene.

Atrazine degradation. Maize degrades atrazine with a glutathione-S-transferase and mutants lacking this enzyme are sensitive (Shimabukuro *et al.*, 1971). A gene for the atrazine specific glutathione-S-transferase has been cloned and sequenced (Shah *et al.*, 1986*a*). G. Helmer and colleagues engineered transgenic tobacco plants containing a triazine detoxifying glutathione-S-transferase (Ludlow, 1986). One-fourth of the F_2 plants were tolerant to an otherwise lethal atrazine dose but not to rates anywhere near those used under field conditions. Atrazine is lethal to many species at nearly two orders of magnitude below the field dose. The transgenic plants may be useful only in rotation after maize, when there is some residual atrazine in the soil that inhibits normal tobacco. Such low-level resistance can be the result of insufficient expression of the gene in most other species (cf. Gressel *et al.*, 1978). Watanabe *et al.* (1986) attained high glutathione levels by engineering multiple copies of the two genes responsible for the biosynthesis of this tripeptide. This approach may have to be added to the engineering scheme to obtain atrazine resistance. Even this may not be enough if the amino acids required for glutathione synthesis are limiting. This is especially true of cysteine, which is not recycled during atrazine degradation.

Bromoxynil resistance. The gene coding for a nitrilase that degrades bromoxynil has been isolated from *Klebsiella ozaenae*. It was then cloned into *E. coli* where it was expressed (Stalker & McBride, 1987). The gene was being tailored for expression in crops.

Glufosinate resistance – the model for success. Glufosinate resistance was found in alfalfa cell cultures by stepwise increase in herbicide concentration (Donn *et al.*, 1984), using techniques similar to those in Fig. 2. Resistance was found to be due to elevated levels of sensitive glutamine synthase, the specific target of this herbicide (Chapter 8). Although much has been learnt about the molecular biology of the gene and its amplification (Das Sarma, Tischer & Goodman, 1986), resistant crops have not emanated from this research approach.

A different direction was taken by Thompson *et al.* (1987). They reasoned that *Streptomyces hygroscopicus* the organism producing bialaphos (glufosinyl-alanyl-alanine) should be somehow resistant to its own secretion product. They found that the resistance was due to an acetyl-CoA-transferase which also acetylated bialaphos and glufosinate, inactivating their toxicity. They were unable to purify the enzyme and produce a polyclonal antibody against it. Resistant transformants of a sensitive *Streptomyces* sp. were isolated using a genomic bank of the bialaphos producing actinomycete and the *bar* gene was characterised (Murakami *et al.*, 1986; Thompson *et al.*, 1987). They were able to subclone a fragment into *E. coli* and get high expression of this '*bar*' gene. The *bar* gene fused to a viral promoter was introduced into a T_i plasmid having kanamycin resistance as a second selectable marker, and then it was co-integrated into *Agrobacterium* (De Block *et al.*, 1987). Tobacco protoplasts as well as tomato and potato leaf discs were infected with the *Agrobacterium* and transgenic plants were regenerated and tested (Fig. 6). Some of the plants of all three species were resistant to commercial preparations of glufosinate (max. tested 4 kg ha^{-1}) (De Block *et al.*, 1987). Two tobacco transformant lines with a 100-fold difference in *bar* gene expression were both equally resistant to this high herbicide level (Leemans *et al.*, 1987). These herbicide rates are well over the field levels normally used, providing a sufficient level for farmer error. Biochemical testing showed that resistance was due to the acetylation, and that the actinomycete gene and its product were present. The transformants were fully resistant to 1 kg/ha in a field trial, and there was no significant effect on yield, as measured by leaf length (Leemans *et al.*, 1987). The acetyl-transferase is present in such low quantities that it should not present a significant genetic load. It should be relatively easy to use the same

Agrobacterium or its plasmid DNA to transform many crop species, conferring resistance to this post-emergence herbicide.

Concluding remarks

The rapid progress in the conferring of herbicide resistance on crops is fascinating to follow. Many mistakes have been made and it is clear that the morals have been learnt. There is already the first case of conferring resistance which was quick, easy and relatively inexpensive. Not all cases will be as easy as with glufosinate: a milestone success which others will be measured against. Projects can now be better evaluated in advance as to likelihoods of success and costs, based on accumulated experience. It requires multipronged approaches, initial selections of large numbers of putative resistant mutants and slow evaluation at many levels to find the best. Herbicide resistant varieties have been released and more and better ones are sure to come. Conferring herbicide resistance will not replace finding new and better herbicides, it will only complement doing so.

Fig. 6. Glufosinate resistant tobacco and potato plants. The plants were transformed with the *Streptomyces bar* gene for glufosinate-acetyl-CoA transferase, showing glufosinate resistance. *Source*: Results of De Block *et al.* (1987). Photo of plants kindly supplied by Plant Genetics Systems, Ghent, Belgium.

Note: The International Standards Organisation common name for this herbicide is glufosinate (cf. Worthing, 1987), although it often appears in the scientific literature as phosphinothricin.

Acknowledgements

The author thanks his collaborators for their research efforts quoted here and various colleagues who have provided unpublished results and manuscripts. The author's research on transfer of triazine resistance was supported by the Schmidt Fund for Applied Research. The author holds the Gilbert de Botton Chair of Plant Sciences.

References

Anderson, R.M., & Gronwald, J. (1987). Non-cytoplasmic inheritance of atrazine-tolerance in velvet leaf (*Abutilon theophrasti*) *Weed Science*, **35**, 496–8.

Anderson, P.C. (1986). Cell culture selection of herbicide tolerant corn and its ramifications. In *Proceedings of the Forty-First Annual Corn and Sorghum Industry Research Conferences*, Publication No. 41, Washington DC: American Seed Trade Association.

Benbrook, C.M., & Moses P.B. (1986). Herbicide resistances; environmental and economic issues. *Proceedings of the BioExpo86 Conference*, pp. 27–54, Boston: Butterworths.

Beversdorf, W.D., Hume, D.J. & Donnelly-Vanderloo, M.J. (1988). Agronomic performance of triazine-resistant and susceptible reciprocal spring Canola hybrids. *Crop Science*, **28**, 932–4.

Binding, H., & Jain, S.M., Finger, J., Mordhorst, G., Nehls, R. & Gressel, J. (1982). Somatic hybridization of an atrazine resistant biotype of *Solanum nigrum* with *S. tuberosum*: I. Clonal variation in morphology and in atrazine sensitivity. *Theoretical and Applied Genetics*, **63**, 272–7.

Chaleff, R.S. & Parson, M.F. (1978). Direct selection *in vitro* for herbicide resistant mutants of *Nicotiana tabacum*. *Proceedings of the National Academy of Science, USA*, **75**, 5104–7.

Chaleff, R.S. & Ray, T.B. (1984). Herbicide-resistant mutants from tobacco cell cultures. *Science*, **223**, 1148–51.

Comai, L., Sen, L.C. & Stalker, D.M. (1983). An altered *aro A* gene product confers resistance to the herbicide glyphosate. *Science*, **221**, 370–1.

Comai, L., Faccioti, D., Hiatt, W.R., Thompson, G., Rose, R.E. & Stalker, D.M. (1985). Expression in plants of a mutant *aro A* gene from *Salmonella typhimurium* confers tolerance to glyphosate. *Nature*, **317**, 741–4.

Comai, L. & Stalker, D. (1986). Mechanism of action of herbicides and their molecular manipulation. *Oxford Surveys of Plant Molecular & Cell Biology*, **3**, 167–95.

Conard, S.G. & Radosevich, S.R. (1979). Ecological fitness of *Senecio vulgaris* and *Amaranthus retroflexus* biotypes susceptible or resistant to triazine. *Journal of Applied Ecology*, **16**, 171–7.

Darmency, H.M. & Pernes, J. (1985). Use of wild *Setaria viride* (L) Beauv. to improve triazine resistance in cultivated *S. italica* (L) by hybridization. *Weed Research*, **25**, 175–9.

Das Sarma, S., Tischer, E. & Goodman, H.M. (1986). Plant glutamine synthase complements a *glu A* mutation in *E. coli*. *Science*, **232**, 1242–4.

De Block, M., Botterman, J., Vandewiele, M., Dockz, J., Thoen, C., Gossele, V., Movva, N.R., Thompson, C., Van Montagu, M. & Leemans, J. (1987). Engineering herbicide resistance in plants by expression of a detoxifying enzyme. *EMBO Journal*, **6**, 2513–18.

de la Pena, A., Lorz, H. & Schell, J.(1987). Transgenic rye plants obtained by injecting DNA into young floral tillers. *Nature*, **325**, 274–6.

Della-Cioppa, G., Bauer, S.C., Taylor, M.L., Rochester, D.E., Klein, B.K., Shah, D.M., Fraley, R.T. & Kishore, G.M. (1987). Targetting a herbicide-resistant enzyme from *Escherichia coli* to chloroplasts of higher plants. *Bio/Technology*, **5**, 579–84.

Donn, G., Tischer, E., Smith, J.A. & Goodman, H.M. (1984). Herbicide resistant alfalfa cells; an example of gene amplification in plants. *Journal of Molecular and Applied Genetics*, **2**, 621–635.

Eastin, E.F. (1981). Movement and fate of metribuzin in Tracy and Tracy M soybeans. *Proceedings of the Southern Weed Science Society*, **34**, 263–7.

Edwards Jr., C.J., Barrentine, W.L. & Kilen, T.C. (2976). Inheritance of sensitivity to metribuzin in soybeans. *Crop Science*, **16**, 119–20.

Falco, S.C., Dumas, K.S. & McDevitt, R.E. (1985). Molecular genetic analyis of sulfonylurea herbicide action resistance in yeast. In *Molecular Form and Function of the Plant Genome*, eds. L Van Vloten-Doting *et al.*, pp. 467–78, New York: Plenum.

Faulkner, J. (1982). Breeding herbicide-tolerant crop cultivars by conventional methods. In *Herbicide Resistance in Plants*, pp. 235–56. eds. H.M. LeBaron & J. Gressel, New York: Wiley.

Fillatti, J.J., Kiser, J., Rose, R., & Comai, L. (1987). Efficient transfer of a glyphosate tolerance gene into tomato using a binary *Agrobacterium tumefaciens* vector. *Bio/Technology*, **5**, 726–30.

Fluhr, R., Aviv, D., Galun, E. & Edelman, M. (1985). Efficient induction and selection of chloroplast-coded antibiotic-resistant mutants in *Nicotiana*. *Proceedings of the National Academy of Science (USA)*, **82**, 1485–9.

Fluhr, R., Kuhlemeier, C., Nagy, F. & Chua, N.-H. (1986). Organ specific and light-induced expression of plant genes. *Science*, **232**, 1106–12.

Furusawa, I., Tanaka, K., Thanutong, P., Mizuguchi, A., Yazaki, M. & Asada, K. (1984). Paraquat resistant tobacco calluses with enhanced superoxide dismutase activity. *Plant & Cell Physiology*, **25**, 1247–54.

Giffard, S.C., Collin, H.A. & Putwain, P.D. (1986). Enhancement of asulam resistance in barley. *Weed Research*, **26**, 389–96.

Grant, I. & Beversdorf, W.D. (1985). Heterosis and combining ability estimates in spring-planted oilseed rape (*Brassica napus*). L. *Canadian Journal of Genetics and Cytology*, **27**, 472–8.

Gressel, J. (1985). Herbicide tolerance and resistance alteration of site of activity. In *Weed Physiology* vol. 2, ed. S.O. Duke, pp. 159–89, Boca Raton: CRC Press.

Gressel, J. (1987). Appearance of single and multi-group herbicide resistances, and strategies for their prevention. *British Crop Protection Conference – Weeds*. Vol. 2. pp. 479–88.

Gressel, J., Aviv, D. & Perl, A. (1985). Methods of producing herbicide resistant plant varieties and plants produced thereby. *US Patent Application* **707**, 416.

Gressel, J. & Ben-Sinai, G. (1985). Low intra-specific competitive fitness in a triazine resistant, nearly nuclear-isogenic line of *Brassica napus*. *Plant Science*, **38**, 29–32.

Gressel, J., Regev, Y., Malkin, S. & Kleifeld, Y. (1983). Characterization of an s-triazine resistant biotype of *Brachypodium distachyon*. *Weed Science*, **31**, 450–6.

Gressel, J. & Segel, L.A. (1978). The paucity of genetic adaptive resistance of plants to herbicides: possible biological reasons and implications. *Journal of Theoretical Biology*, **75**, 349–71.

Gressel, J., & Shaaltiel, Y. (1988). Biorational synergists for oxidant generating herbicides. In *Biotechnology in Crop Protection*, pp. 4–24, eds. J.J. Menn, R.M. Hollingsworth & P.A. Hedin. Washington DC: American Chemical Society Symposium Series No. 39.

Gressel, J., Zilkah, S. & Ezra, G. (1978). Herbicide action, resistance and screening in cultures vs. plants. In *Frontiers in Plant Tissue Cultures*, – 1978, ed. T.A. Thorpe, pp. 427–36, Calgary: University Calgary Press.

Haissig, B.E., Nelson, N.D. & Kidd, G.H. (1987). Trends in the use of tissue culture in forest improvement. *Bio/Technology*, **5**, 52–9.

Hickok, L.G., Warne, T.R. & Slocum, M.K. (1987). *Ceratopteris richardii*. Application for experimental plant biology. *American Journal of Botany*, **74**, 1304–16.

Hughes, K.W. (1986). Tissue culture derived crossing barriers. *American Journal of Botany*, **73**, 323–9.

Hume, D.J. & Beversdorf, W.D. (1986). Triazine resistance and heterosis in rapeseed F_1 hybrids. American Society of Agronomy, New Orleans, Abstracts p.114.

Keen, N.T., Holliday, M.J. & Yoshikawa, M. (1982). Effects of glyphosate on glyceollin production and the expression of resistance to *Phytophthora megasperma f. sp. glycinea* in soybean. *Phytopathology*, **72**, 1471–4.

Killmer, J., Widholm, J. & Slife, F. (1981). Reversal of glyphosate inhibition of carrot cell culture growth by glycolytic intermediates and organic and amino acids. *Plant Physiology*, **68**, 1299–1302.

Kishore, G.M. (1988). EPSP-synthase – from biochemistry to engineering of glyphosate tolerance. In *Biotechnology in Crop Protection*, pp. 37–48, eds. J.J. Menn, R. M. Hollingsworth & P.A. Hedin, Washington DC: American Chemical Society Symposia No. 379.

Klein, T.M., Wolf, E.D., Wu, R. & Sanford, J.L. (1987). High-velocity microprojectiles for delivering nucleic acids into living cells. *Nature*, **327**, 70–3.

Lee, T.T. (1980). Characteristics of glyphosate inhibition of growth in soybean and tobacco callus cultures. *Weed Research*, **20**, 365–9.

Lee, T.T. & Starratt, A.N. (1986). Phenol–glyphosate interactions: Effects on IAA metabolism and growth in plants. In *Proceedings of the First International Symposium on Adjuvants for Agrochemicals*. Brandon, Canada, 5–7 Aug. 1986.

Leemans, J., De Block, M., D'Halluin, K., Botterman, J. & De Greef, W. (1987). The use of glufosinate as a selective herbicide on genetically engineered resistant tobacco plants. *British Crop Protection Conference – Weeds* Vol. 3. pp. 867–70.

Lesser, W. (1986). Seed-patent forecast. *Bio/Technology* **4**, 783–5.

Ludlow, K.A. (1986). Agricultural biotechnology research institute initiates first field test. *Broadcast*, (Research Triangle NC: Ciba-Geigy Corp. Agricultural Division) Aug. p. 2.

Melchers, G., Sacristan, M.D. & Holder, A.A. (1978). Somatic hybrid plants of potato and tomato regenerated from fused protoplasts. *Carlsberg Research Communications*, **43**, 203–18.

Menczel, L., Polsby, L.S., Steinback, K.E. & Maliga, P. (1986). Fusion mediated transfer of triazine resistant plastids. *Abstracts VI International Congress Plant Tissue Cell Culture*. Minneapolis MN p.81.

Meredith, C.P. & Carlson, P.S. (1982). Herbicide resistance in plant cell cultures. In *Herbicide Resistance in Plants*, eds. H.M. LeBaron & J. Gressel, pp. 275–92, New York: Wiley.

Mouches, C., Pasteur, N., Berge, J.B., Hyrien, O., Raymond, M., de Saint Vincent, B.R., de Silvestri, M. & Georghiou, G.P. (1986). Amplification of an esterase gene is responsible for insecticide resistance in a California *Culex* mosquito. *Science*, **233**, 778–80.

Mousdale, D.M. & Coggins, J.R. (1985). Subcellular localization of the common shikimate-pathway enzymes in *Pisum sativum* L. *Planta*, **163**, 241–9.

Murakami, T., Anzai, H., Imai, S., Satoh, A., Nagaoka, K. & Thompson, C.J. (1986). The bialaphos biosynthetic genes of *Streptomyces hygroscopicus*: Molecular cloning and characterization of the gene cluster. *Molecular and General Genetics*, **205**, 42–50.

Radin, D.N. & Carlson, P.S. (1978). Herbicide tolerant tobacco mutant selected *in situ* and recovered via regeneration from cell culture. *Genetic Research*, **32**, 85–9.

Ricroch, A., Mousseau, M., Darmency, H. & Pernes, J. (1987). Comparison of triazine-resistant and -susceptible cultivated *Setaria italica* growth and photosynthetic capacity. *Plant Physiol. Biochem.* **25**, 29–34.

Rogers, S.G., Brand, L.A., Holder, S.B., Sharp, E.S. & Brackin, M.J. (1983). Amplification of the *aro A* gene from *E. coli* results in tolerance to the herbicide glyphosate. *Applied and Environmental Microbiology*, **46**, 37–43.

Schimke, R.T. ed. (1982). *Gene Amplification*. Cold Spring Harbor Laboratory, New York.

Schonfield, M., Yaacoby, T., Michael, O. & Rubin, B. (1987*a*). Triazine resistant with reduced vigor in *Phalaris paradoxa*. *Plant Physiology*, **83**, 329–33.

Schonfield, M., Yaacoby, T., Ben-Yehuda, A., Rubin, B. & Hirschberg, J. (1987*b*). Triazine resistance in *Phalaris paradoxa*: Physiological and molecular analyses. *Zeitschrift fur Naturforschung*, **42c**, 779–82.

Schulz, A., Sost, D. & Amrhein, N. (1984). Insensitivity of EPSP-synthase to glyphosate confers resistance to this herbicide in a strain of *Aerobacter aerogenes*. *Archives of Microbiology*, **137**, 121–4.

Sebastian & S.A., Chaleff, R.S. (1987). Soybean mutants with increased tolerance for sulfonylurea herbicides. *Crop Science*, **27**, 948–52.

Shaaltiel, Y, Chua, N.-H., Gepstein, S. & Gressel, J. (1988). Dominant pleiotropy controls enzymes cosegregating with paraquat resistance in *Conyza bonariensis*. *Theoretical and Applied Genetics*, **75**, 850–6.

Shaaltiel, Y. & Gressel, J. (1986). Multienzyme oxygen radical detoxifying system correlated with paraquat resistance in *Conyza bonariensis*. *Pesticide Biochemistry and Physiology*, **26**, 22–8.

Shah, D.M., Hironaka, C.M., Wiegand, R.C., Harding, E.I., Krivi, G.G. & Tiemeier, D.C. (1986*a*). Structural analysis of a maize gene coding for glutathione-S-transferase involved in herbicide detoxification. *Plant Molecular Biology*, **6**, 203–11.

Shah, D.M., Horsch, R.B., Klee, H.J., Kishore, G.M., Winter, J.A., Turner, N.E., Hironaka, C.M., Sanders, P.R., Gasser, C.S., Aykent, S., Siegel, N.R., Rogers, S.G. & Fraley, R.T. (1986*b*). Engineering herbicide tolerance in transgenic plants. *Science*, **233**, 478–81.

Shimabukuro, R.H., Frear, D.S., Swanson, H.R. & Walsh, W.C. (1971). Glutathione conjugation – an enzymatic basis for atrazine resistance in corn. *Plant Physiology*, **47**, 10–14.

Smith, C.M., Pratt, D. & Thompson, G.A. (1986). Increased 5-enolpyruvyl-shikimic acid 3-phosphate synthase activity in a glyphosate tolerant variant strain of tomato cells. *Plant Cell Reports*, **5**, 298–301.

Sost, D., Schulz, A. & Amrhein, N. (1984). Characterization of a glyphosate-insensitive 5-enolpyruvyl-shikimic acid-3-phosphate synthase. *FEBS Letters*, **173**, 238–42.

Souza-Machado, V. (1982). Inheritance and breeding potential of triazine tolerance and resistance in plants. In *Herbicide Resistance in Plants*, eds. H.M. LeBaron & J. Gressel, pp. 257–74, New York: Wiley.

Stalker, D.M., Hiatt, W.R. & Comai, L. (1985). A single amino acid substitution in the enzyme EPSP synthase confers resistance to the herbicide glyphosate. *Journal of Biological Chemistry*, **260**, 4724–8.

Stalker, D.M.& McBride, K.E. (1987). Cloning and expression in *Escherichia coli* of a *Klebsiella ozaenae* plasmid borne gene encoding a nitrilase specific for the herbicide bromoxynil. *Journal of Bacteriology*, **169**, 955–60.

Stannard, M.E. & Fay, P.K. (1987). Selection of alfalfa seedlings for tolerance to chlorsulfuron. *Weed Science Society of America Abstracts*, St. Louis. p.61.

Steinrücken, H.C., Schulz, A., Amrhein, N., Porter, C.A. & Fraley, R.T. (1986). Overproduction of 5-enolpyruvylshikimate-3-phosphate synthase in a glyphosate-tolerant *Petunia hybrida* cell line. *Archives of Biochemistry and Biophysics*, **244**, 169–78.

Swanson, E.B., Coumans, M.P., Brown, G.L., Patel, J.P. & Beversdorf, W.D. (1988). The characterization of herbicide tolerant plants in *Brassica napus* after *in vitro* selection of microspores and protoplasts. *Plant Cell Reports*, **7**, 83–7.

Thompson, C., Movva, N.R., Tizard, R., Cromeri, R., Davies, J.E., Lauwereys, M. & Botterman, J. (1987). Characterization of the herbicide resistance gene *bar* from *Streptomyces hygroscopicus*. *EMBO Journal*, **6**, 2519–23.

Vaughn, K.C. (1986). Characterization of triazine-resistant and susceptible isolines of canola (*Brassica napus*. L.). *Plant Physiology*, **82**, 859–63.

Watanabe, K., Yamano, Y., Murata, K. & Kimura, A. (1986). Glutathione production by *Escherichia coli* cells with hybrid plasmid containing tandemly polymerized genes for glutathione synthase. *Applied Microbiology and Biotechnology*, **24**, 375–8.

Weller, S.C., Masiunas, J.B. & Gressel, J., (1987). Biotechnologies of obtaining herbicide tolerance in potato. In *Biotechnology of Plant Improvement* Vol.3, ed. Y.P.S. Bajaj, pp. 281–97, Berlin: Springer.

Williams, K. (1976). Mutation frequency at a recessive locus in haploid and diploid strains of a slime mould. *Nature*, **260**, 785–7.

Worthing, C.R. ed. (1987). *The Pesticide Manual – A World Compendium* – 8th edition. Croydon: British Crop Protection Council.

Yarrow, S.A., Wu, S.C., Barsby, T.L., Kemble, R.J. & Shepard, J.F. (1986). The introduction of CMS mitochondria to triazine tolerant *Brassica napus*, L., var. Regent, by micromanipulation of individual heterokaryons. *Plant Cell Reports*, **5**, 415–8.

Zelcer, A., Aviv, D. & Galun, E. (1978). Interspecific transfer of cytoplasmic male sterility by fusion between protoplasts of normal *Nicotiana sylvestris* and X-irradiated protoplasts of male sterile *N. tabacum*. *Zeitschrift fur Pflanzenphysiologie*, **90**, 397–407.

HERBICIDE GLOSSARY

(Includes all structures not in the text)

AC 222164

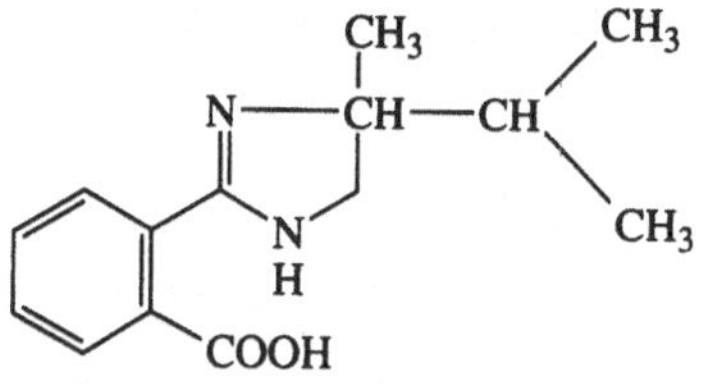

Acifluorfen

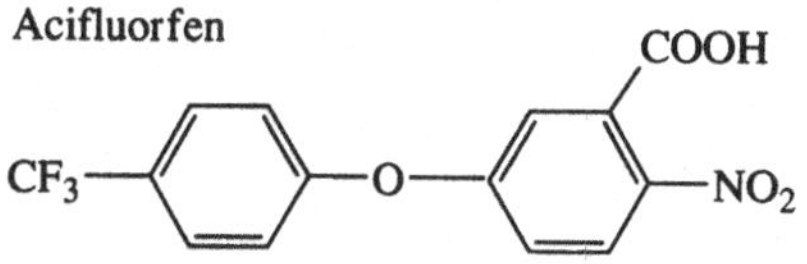

Alloxydim

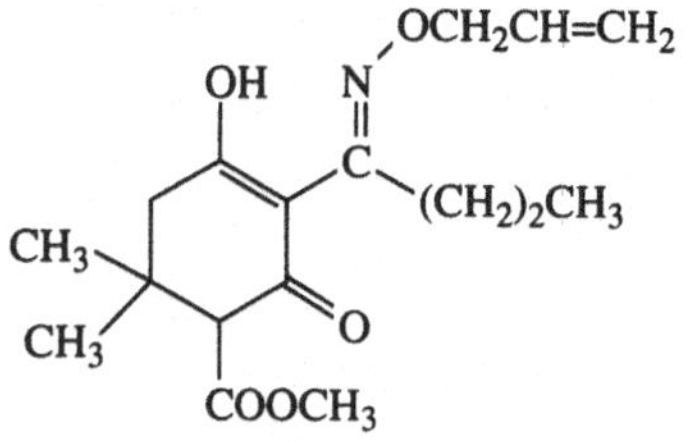

Aminobenzotriazole

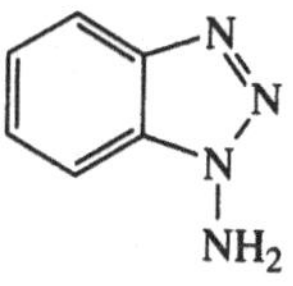

Aminotriazole

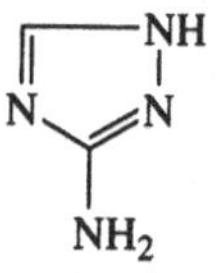

Barban

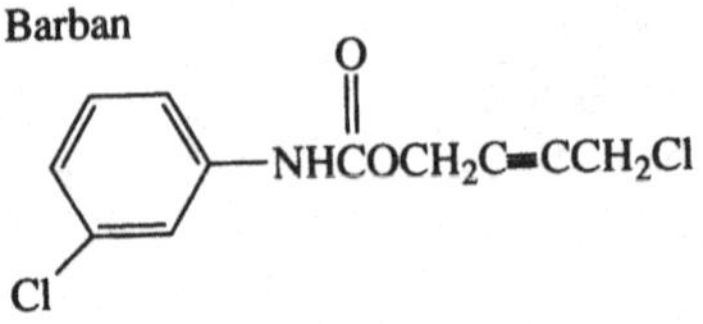

O

—NHCOCH$_2$C≡CCH$_2$Cl

Cl

Bromoxynil

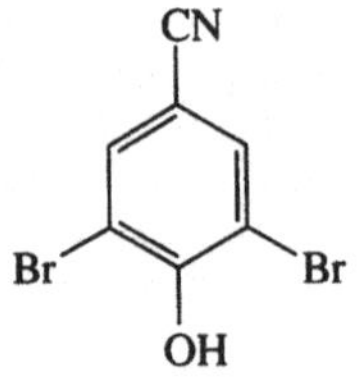

CN

Br Br

OH

Butachlor

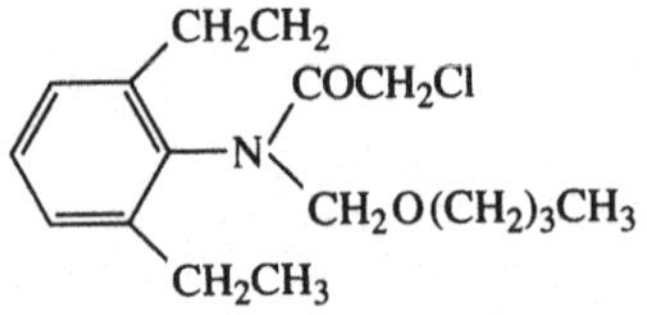

CH$_2$CH$_2$

COCH$_2$Cl

N

CH$_2$O(CH$_2$)$_3$CH$_3$

CH$_2$CH$_3$

Buthidazole

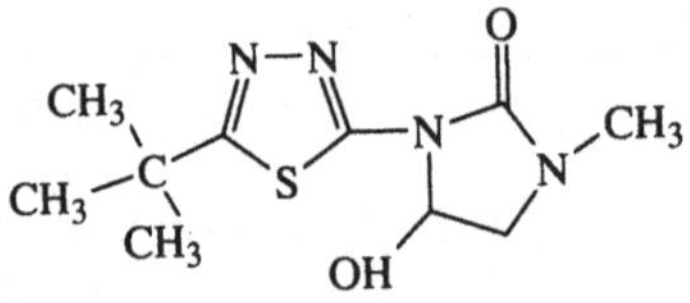

O

N—N

CH$_3$ CH$_3$

CH$_3$—C N N

S

CH$_3$

OH

Butylate

O

[(CH$_3$)$_2$CH CH$_2$]$_2$NCSCH$_2$CH$_3$

Chloramben

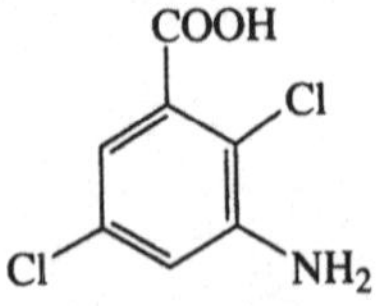

COOH

Cl

Cl NH$_2$

Clofop–butyl

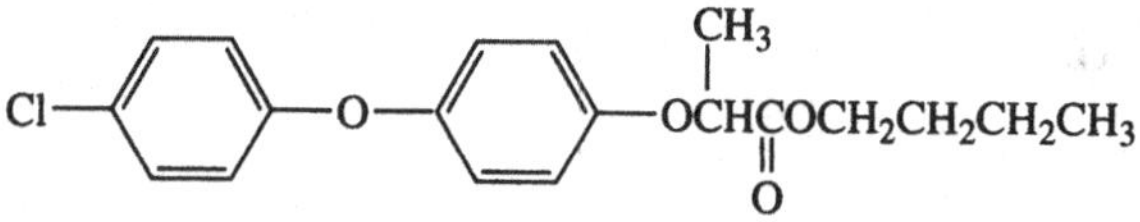

$Cl-C_6H_4-O-C_6H_4-OCH(CH_3)C(=O)OCH_2CH_2CH_2CH_3$

Cyanazine

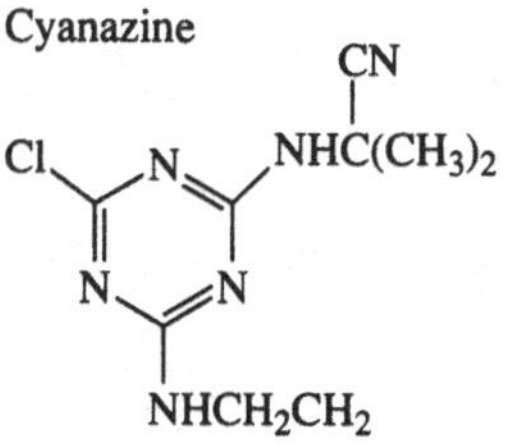

Cl, N, NHC(CH_3)_2 with CN; N, N; $NHCH_2CH_2$

Dichlormate

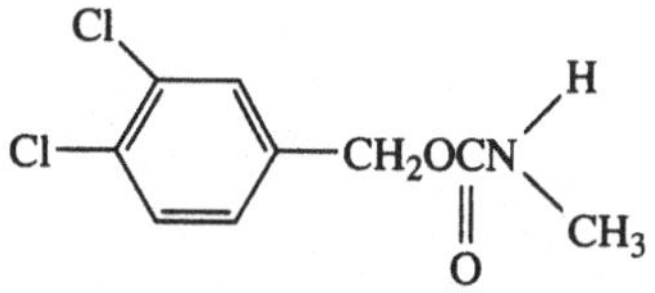

Cl, Cl, $CH_2OC(=O)N(H)CH_3$

Dicloran

NH_2, Cl, Cl, NO_2

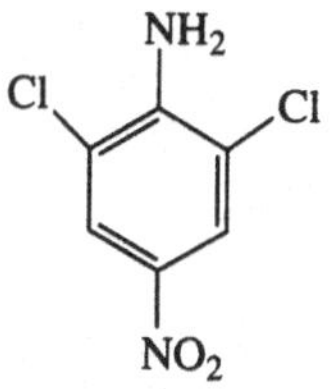

Difunone

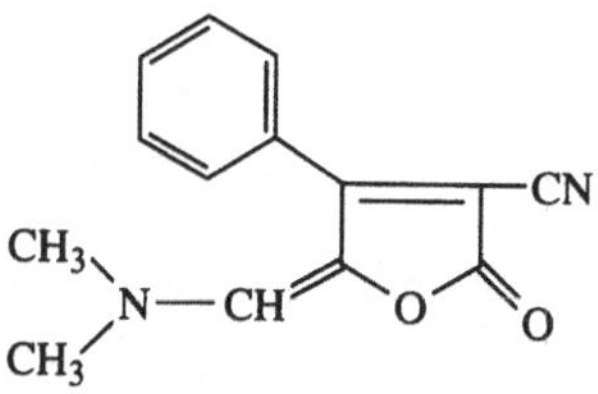

CN, $(CH_3)_2N-CH=$, O, O

Dinoseb

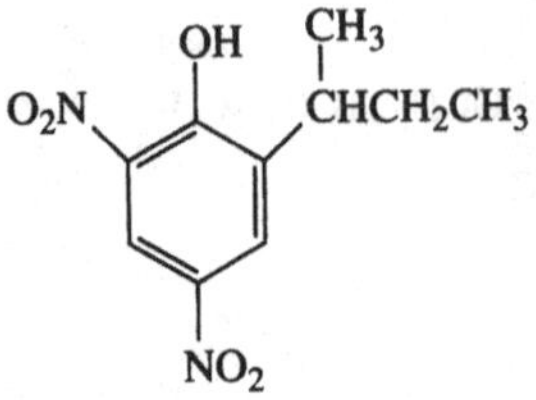

Diquat

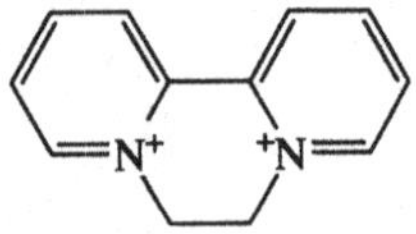

DNOC

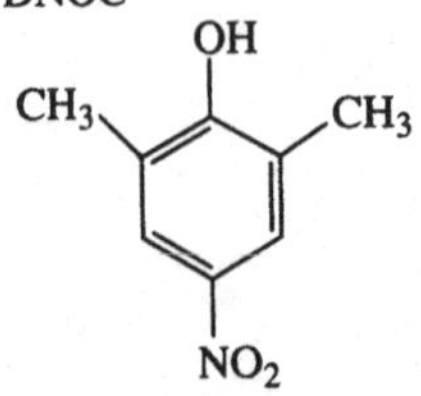

Fenoxaprop-ethyl

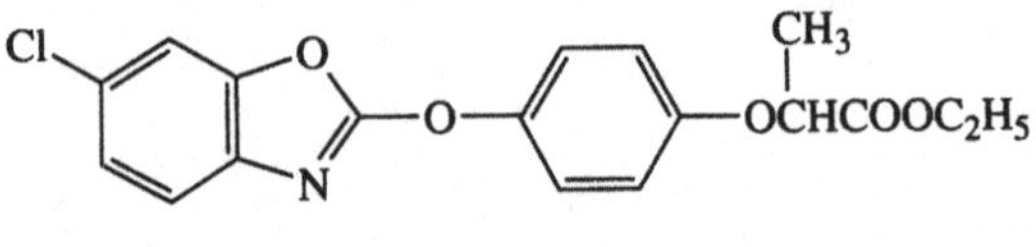

Fenthiaprop-ethyl

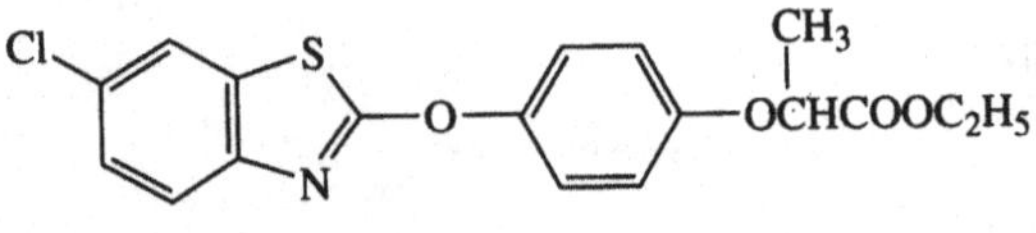

Flamprop–methyl

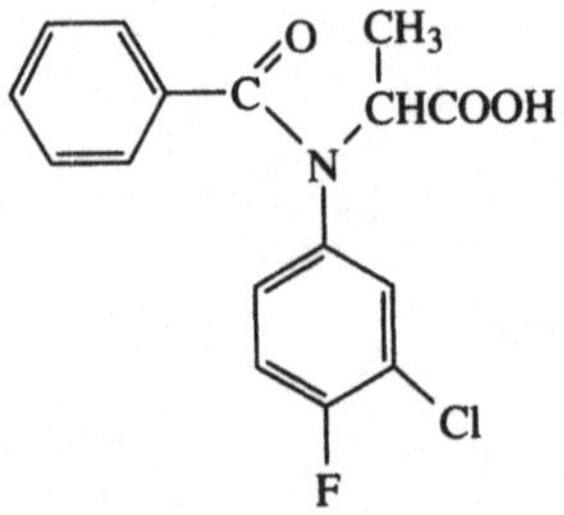

Fluridone

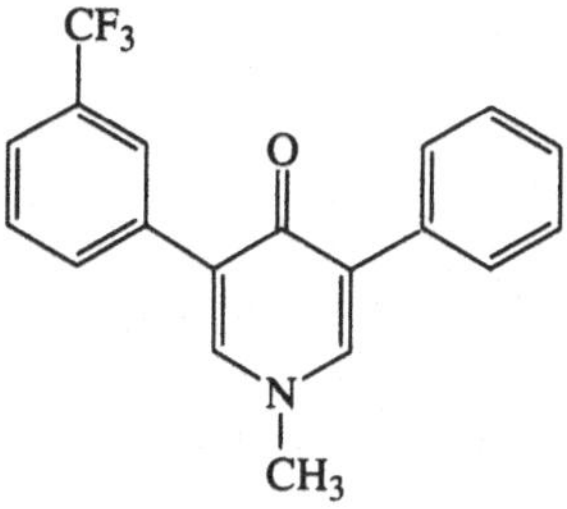

Gabaculin

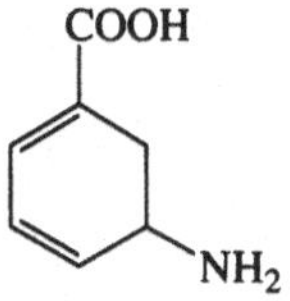

Imazapyr

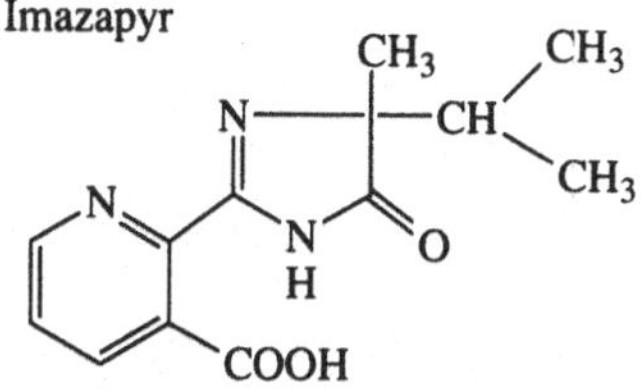

Imazaquin

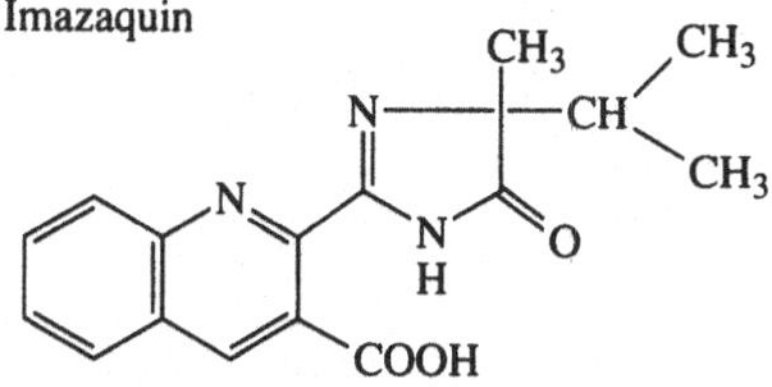

J334

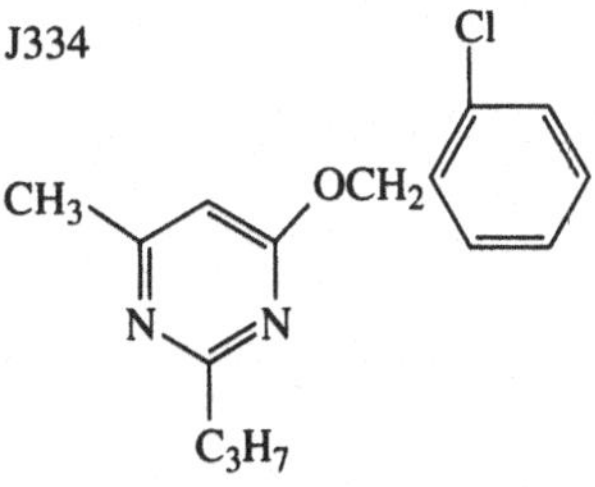

J 739

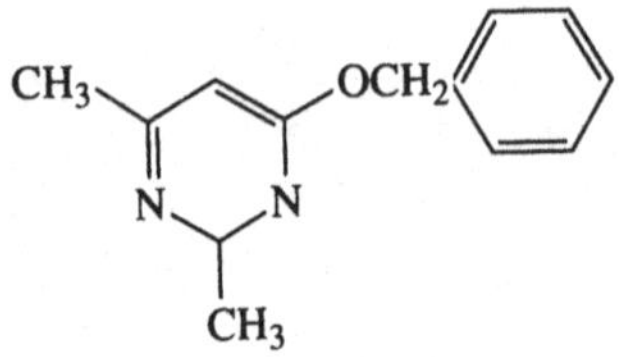

CH_3 OCH_2 N N CH_3

J 852

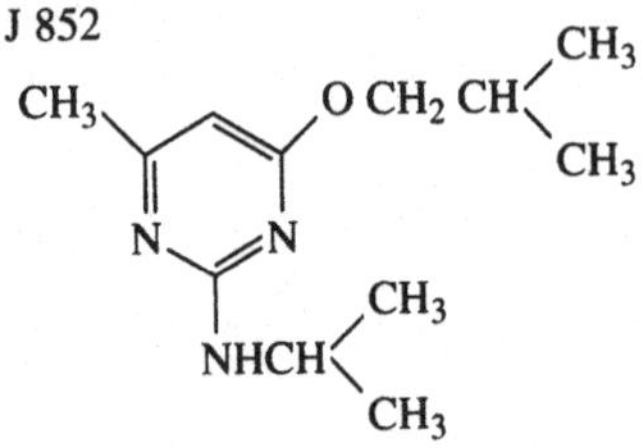

CH_3 $O\ CH_2\ CH$ CH_3 CH_3 N N $NHCH$ CH_3 CH_3

MCPA

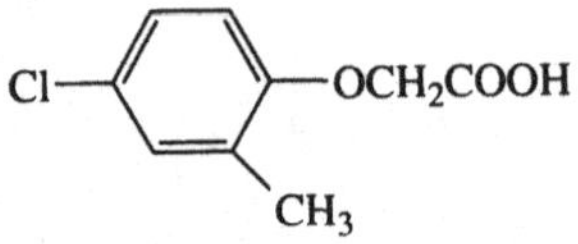

Cl OCH_2COOH CH_3

Mecoprop

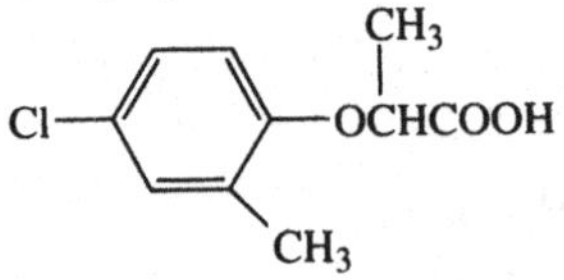

CH_3 Cl $OCHCOOH$ CH_3

Metamitron

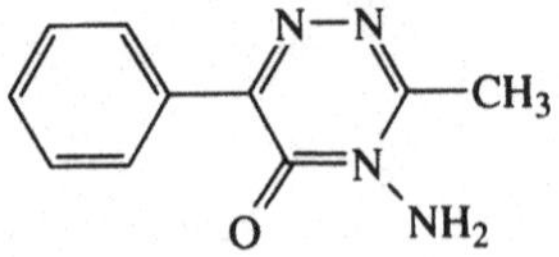

N–N CH_3 N NH_2 O

Metoxuron

CH_3 $NHCON(CH_3)_2$ Cl

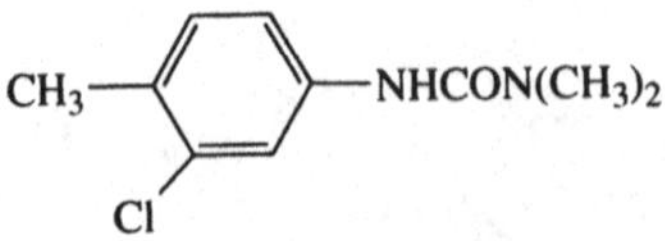

Naproanilide

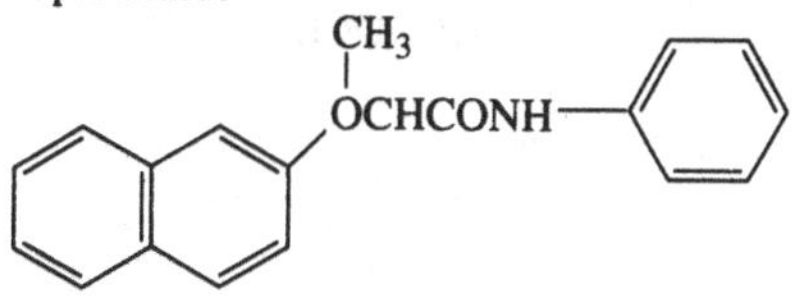

CH_3
$OCHCONH$

Nitrofluorfen

Cl
O_2N O CF_3

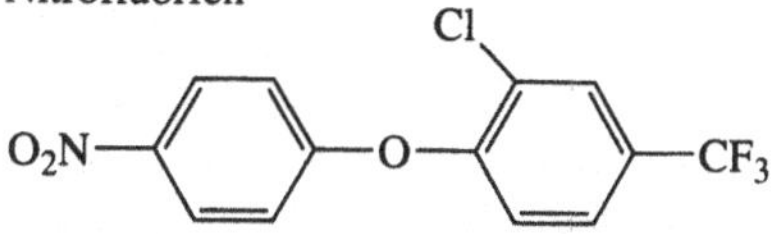

2-nitro-6-methyl sulphonanilide

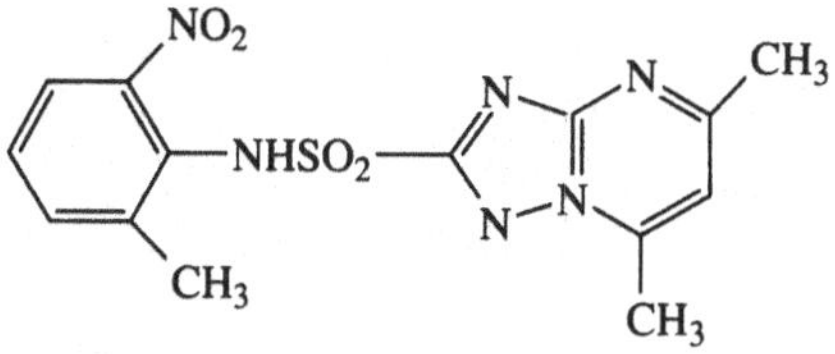

NO_2
N N CH_3
$NHSO_2$
N N
CH_3
CH_3

Oxyfluorfen

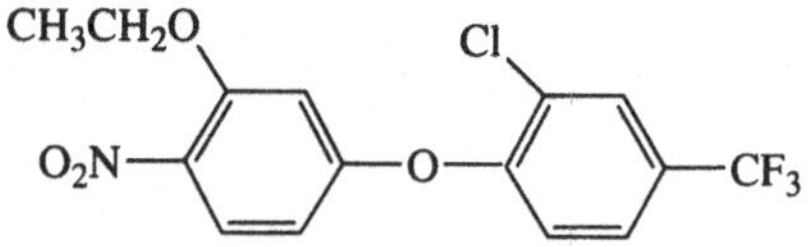

CH_3CH_2O Cl
O_2N O CF_3

Paraquat

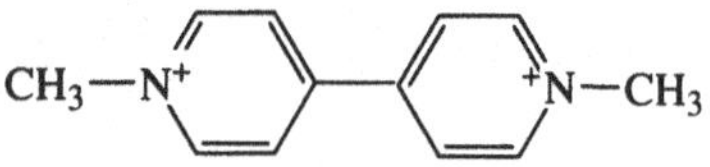

CH_3-N^+ $^+N-CH_3$

Pebulate

O
$CH_3(CH_2)_3-NCSCH_2CH_2CH_3$
CH_2CH_3

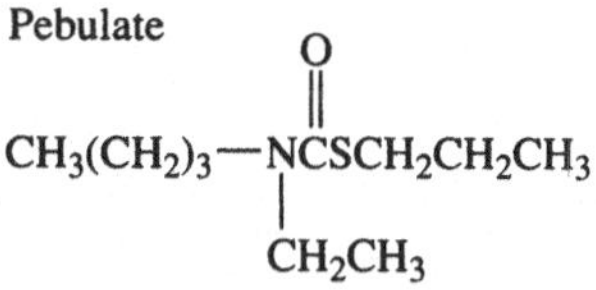

Pentachlorophenol

CH_3
CH_3 $NHCOCH(CH_2)_2CH_3$
Cl

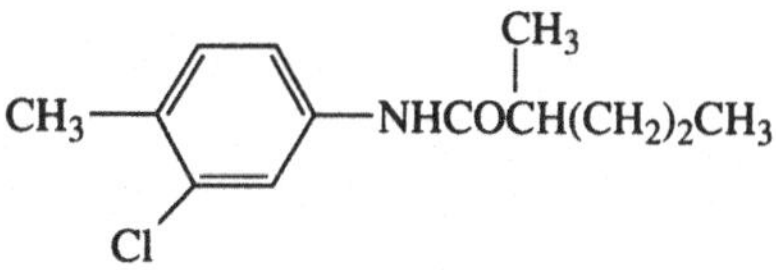

Phenmedipham

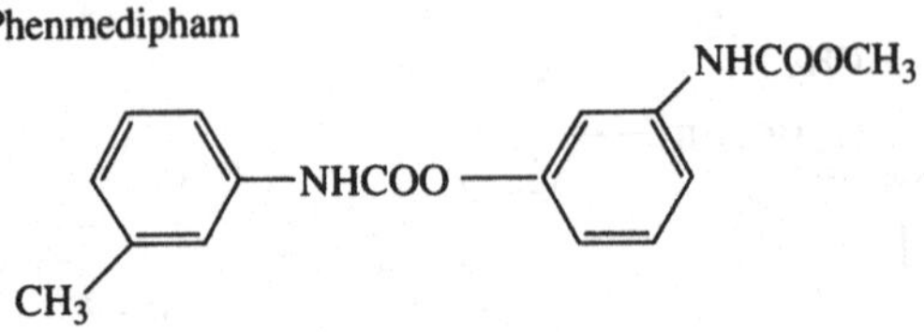

$NHCOOCH_3$

NHCOO

CH_3

N-phthalyl-L-valine anilide

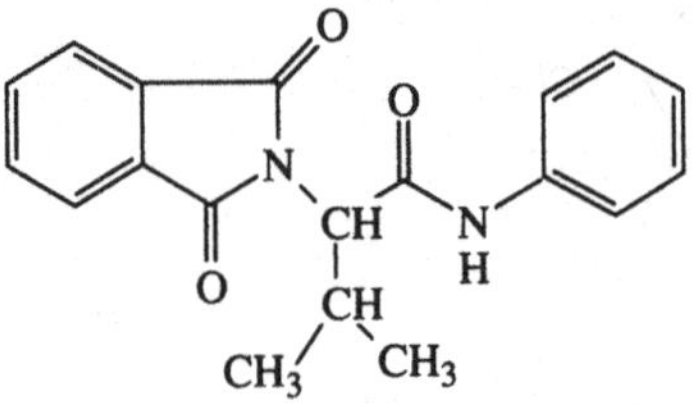

O O N CH N H O CH CH_3 CH_3

Picloram

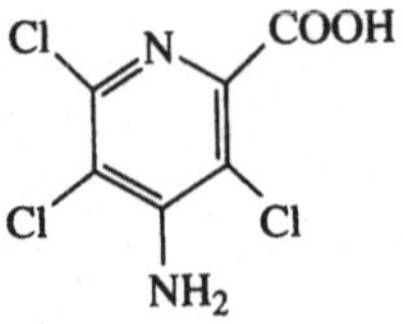

Cl N COOH Cl Cl NH_2

Propachlor

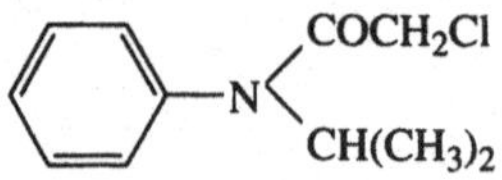

$COCH_2Cl$ N $CH(CH_3)_2$

Propanil

Cl $NHCOCH_2CH_3$ Cl

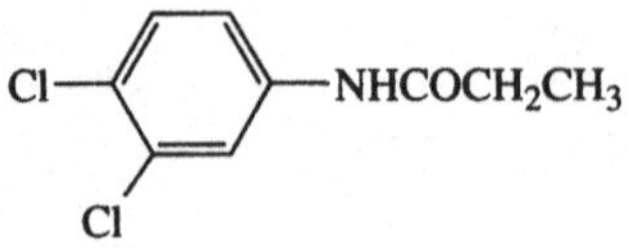

Pyrazon (chloridazone)

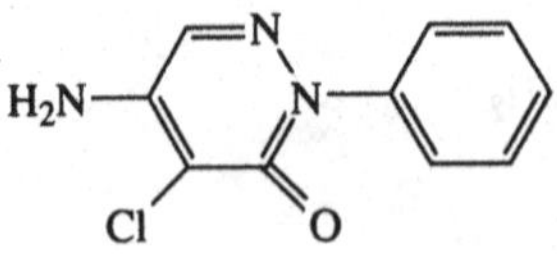

N H_2N N Cl O

Siduron

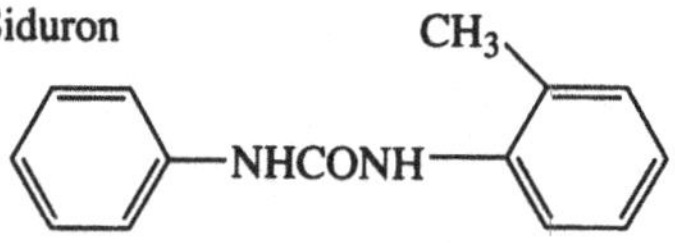

CH_3 — NHCONH —

Simazine

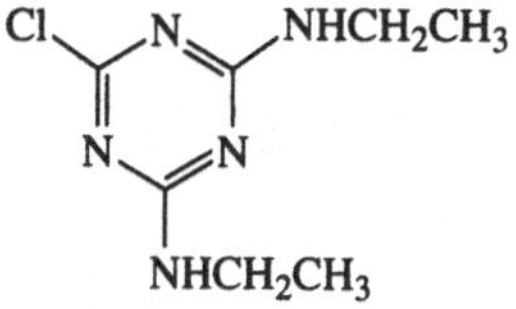

Cl, N, $NHCH_2CH_3$, N, N, $NHCH_2CH_3$

Simeton

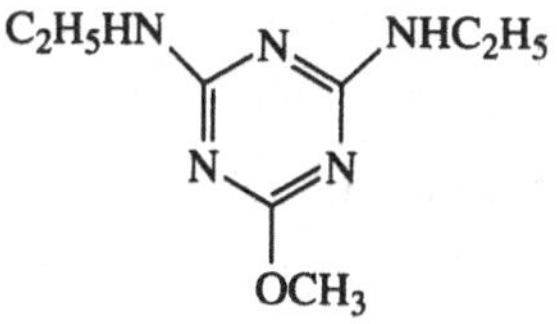

C_2H_5HN, N, NHC_2H_5, N, N, OCH_3

Sulfometuron methyl

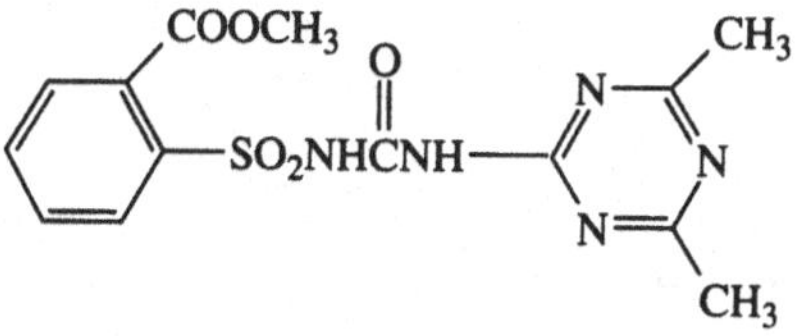

$COOCH_3$, O, SO_2NHCNH, N, N, N, CH_3, CH_3

2, 4, 5-T

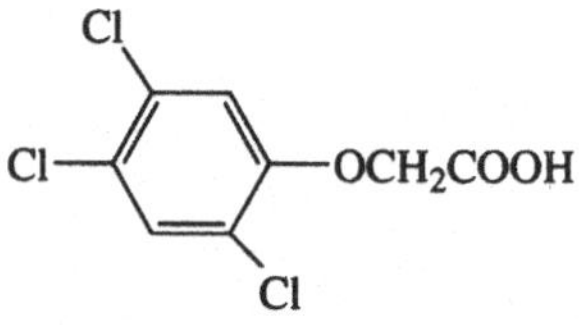

Cl, Cl, OCH_2COOH, Cl

Terbutryn

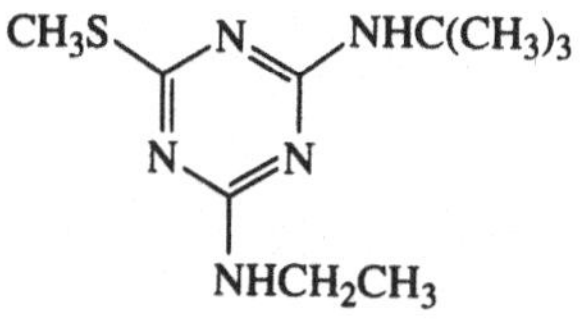

CH_3S, N, $NHC(CH_3)_3$, N, N, $NHCH_2CH_3$

Tetcyclasis

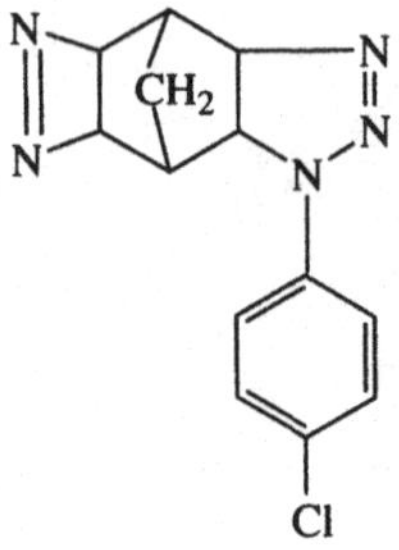

Tralkoxydim

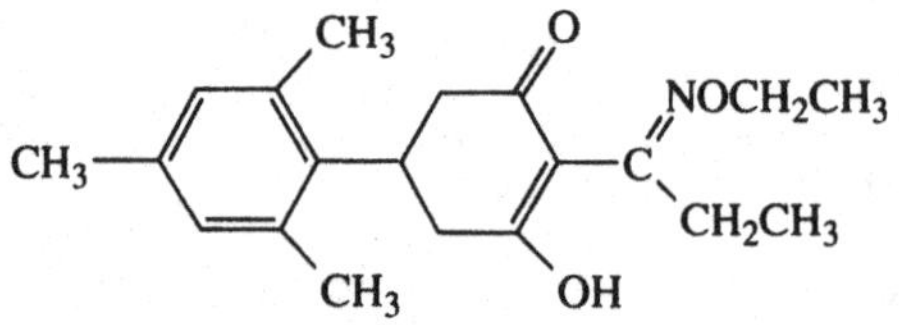

Triclopyr

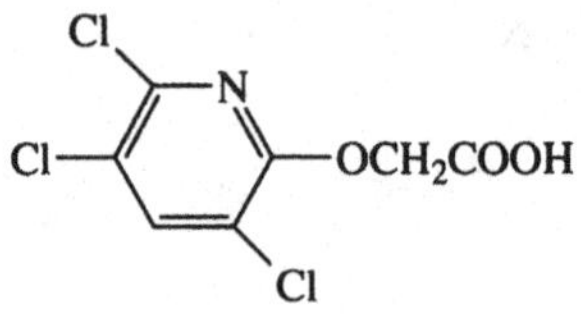

Trietazine

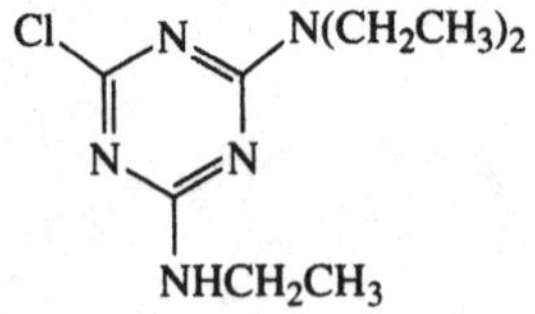

Vernolate

$$(CH_3CH_2CH_2)_2N\overset{\overset{\displaystyle O}{\|}}{C}SCH_2CH_2CH_3$$

HERBICIDE INDEX

GENERAL INDEX